0349284-

KV-350-216

WHITTLE, FRANK, SIR
GAS TURBINE AERO-THERMODYNAMIC
000349284

621.438

W62

OTHER PERGAMON TITLES OF INTEREST

BABISTER	Aircraft Dynamic Stability and Response
CHERINGTON	Airline Price Policy
HEARN	Mechanics of Materials
KÜCHEMANN	The Aerodynamic Design of Aircraft
MANN & MILLIGAN	Aircraft Fatigue Design, Operation and Economic Aspects
MARSCHALL & MARINGER	Dimensional Instability
NAPOLITANO	Space Activity, Impact on Science and Technology
	Space Stations, Present and Future
	Space and Energy
	A New Era in Space Transportation
	Using Space - Today and Tomorrow Volume 1 - Space Based Industry Symposium Volume 2 - Communications Satellite Symposium
	Astronautics for Peace and Human Progress
OPPENHEIM	Gasdynamics of Explosive and Reactive Systems
REAY	The History of Man Powered Flight

RELATED JOURNALS PUBLISHED BY PERGAMON

Acta Astronautica

Advances in Earth Oriented Applications of Space Technology

Planetary and Space Science

Progress in Aerospace Sciences

Vertica

SIR FRANK WHITTLE, K.B.E., C.B.

GAS TURBINE AERO-THERMODYNAMICS

With Special Reference to Aircraft Propulsion

SIR FRANK WHITTLE, K.B.E., C.B.,
Air Commodore R.A.F. (ret.), M.A. Sc.D. (h.c.),
F.R.S., F. Eng. Hon. F.I. Mech.E., Hon. F.R.Ac.S., Hon. F.A.I.A.A.

PERGAMON PRESS

OXFORD · NEW YORK · TORONTO · SYDNEY · PARIS · FRANKFURT

U.K.	Pergamon Press Ltd., Headington Hill Hall, Oxford OX3 0BW, England
U.S.A.	Pergamon Press Inc., Maxwell House, Fairview Park, Elmsford, New York 10523, U.S.A.
CANADA	Pergamon Press Canada Ltd., Suite 104, 150 Consumers Road, Willowdale, Ontario M2J 1P9, Canada
AUSTRALIA	Pergamon Press (Aust.) Pty. Ltd., P.O. Box 544, Potts Point, N.S.W. 2011, Australia
FRANCE	Pergamon Press SARL, 24 rue des Ecoles, 75240 Paris, Cedex 05, France
FEDERAL REPUBLIC OF GERMANY	Pergamon Press GmbH, 6242 Kronberg-Taunus, Hammerweg 6, Federal Republic of Germany

Copyright © 1981 F. Whittle

All Rights Reserved. No part of this publication may be reproduced, stored in a retrieval system or transmitted in any form or by any means: electronic, electrostatic, magnetic tape, mechanical, photocopying, recording or otherwise, without permission in writing from the publishers.

First Edition 1981

British Library Cataloguing in Publication Data
Whittle, *Sir* Frank
 Gas turbine aero-thermodynamics. - (Pergamon international library).
 1. Gas-turbines
 I. Title
 629.134′353   TJ778   80-41372

ISBN 0-08-026719-X Hardcover
ISBN 0-8-026718-1 Flexicover

Printed in Great Britain by A. Wheaton & Co. Ltd., Exeter

CONTENTS

	Preface	vii
Section 1	Fundamentals	1
Section 2	Flux Density in Isentropic Compressible Flow	7
Section 3	Radial Equilibrium in Whirling Isentropic Compressible Flow	13
Section 4	A Method of Dealing with Shock Waves in Air	21
Section 5	Isentropic Flow Through Nozzles	43
Section 6	Representation of Thermal Cycles for Perfect Gases	63
Section 7	Gas Turbine Cycle Calculations Using Approximate Methods	67
Section 8	Turbo-Gas Generators	75
Section 9	Mass Flow Rate in Gas Turbines	83
Section 10	More on Part Load Performance of Gas Generators	89
Section 11	Gas Turbines with Recuperators	95
Section 12	Turbo-Compressors and Turbines	101
Section 13	Aircraft Propulsion General	145
Section 14	'Ram' Compression and Intake Design	153
Section 15	Combustion	161
Section 16	The 'Straight' Turbo Jet	169
Section 17	Effect on Cycle Calculations of More Accurate Methods	183
Section 18	Calculation of Maximum Efficiency	201
Section 19	'Two Spool' Engines	205
Section 20	Thrust Boosting	209
Section 21	Turbo Fans	217
Section 22	Effect of Height and Speed on Performance	247
Section 23	A Super-Thrust Engine	259

PREFACE

In the early 1930s, when I was deeply involved in the design of Britain's first jet engine, I was regarded as a 'crazy optimist' by the many who found it difficult to believe that a young officer of the Royal Air Force could be successful in the field of gas turbines when this field was littered by a history of failure. Yet, in retrospect, I was a pessimist. I certainly did not foresee that the day would come when (on Sept 4th 1976) I would be a passenger in the Concorde and fly across the Atlantic from London (Heathrow) to Washington DC (Dulles) in exactly 3½ hours. Neither did I foresee that, in less than four decades, engines would be in service having about twenty times the power of the W2/700 (2000 - 2500 lb static thrust at sea level)—the last of the series of engines, designed by the team of Power Jets Ltd. led by me, to be built and flown. Nor did I foresee that time between overhauls would increase to thousands of hours as against the 500 hours or so that I predicted. Also, though my 1936 notebooks contain jet engine performance for speeds of the order of 1500 mph and clearly showed that such supersonic speeds were best suited for high efficiency with this type of engine, it did not seem likely to me that the aerodynamics of supersonic aircraft could be improved to the point where long range SSTs and bombers would become possible. In those days the very few—including myself—who refused to accept that the sound barrier could never be broken, (despite ballistic proof to the contrary), were not optimistic about achieving lift/drag ratios much above about 4:1. It seems that I was a pessimist in all respects except one—the time scale for development. In 1945, since we had found that it was possible to design, build, and test a jet engine in 6 months, I was predicting the advent of jet powered civil aircraft within five years and that the turbofan would take over in quite a short time thereafter. In the event I proved to be overoptimistic on both counts. I still believe that these things could have been done had not there still existed the barrier of skepticism which obstructed them.

Perhaps the thing I least expected was that I would ever attempt to write a textbook on aero-thermodynamic theory of gas turbine design. Especially, as a consequence of the formation of the National Gas Turbine Establishment in 1946, my team and I were denied the right to continue to design and build engines. As a result, my interest in the work I had pioneered began to wane and virtually disappeared to the extent that, after 1952, when I resigned my post (honorary) as Adviser to British Overseas Airways, I lost touch with the course of development altogether for many years, my interests having turned to other fields, the chief of which was oil well drilling technology. I did not, however, stop thinking about gas turbines in general in a somewhat superficial way, but remained aloof from becoming involved with them in any practical sense, having ceased to believe that anything I could hope to do could possibly compare with my earlier achievements. I did, however, accept a number of lecture engagements over the years, mainly on the subject of the early history.

My involvement with oil well drilling technology from 1953 onwards (as "Mechanical Engineering Specialist" to Bataafsche Petroleum Maatschappij—the main operating company of the Shell Group) included the design and "paper development" of an oil well drilling motor—the Whittle turbo-drill. It was, however, outside the policy of the Shell Group to enter into the drilling equipment business beyond the point of backing innovation to the stage where it could be handed over to specialist firms. So, under the terms of my contract with Shell, all the turbo drill patents were eventually assigned to me and this gave me the opportunity to seek a sponsor for its practical development. Strangely enough this started a chain of events which led to a revival of my interest in aircraft engines.

This came about as follows: In 1959 I succeeded in arousing the interest of Bristol Siddeley Engines—then one of the two largest British aircraft engine companies—in the turbo drill project, largely by virtue of their confidence in me of Sir Arnold Hall, the Chairman and Managing Director, and Sir Stanley Hooker, the Technical Director (Aero). Both these brilliant men were friends of long standing. Arnold Hall had been a fellow undergraduate at Cambridge University taking the Mechanical Sciences Tripos and had assisted me in the design of the compressor impellor of the first experimental jet engine before we both graduated in 1936. Stanley Hooker and I first met in 1940. He was then with Rolls Royce. He became an immediate convert to the jet engine and was successful in enlisting the interest of his Chief Executive, E. W. Hives (later Lord Hives). Thereafter, Power Jets received much assistance from Rolls Royce and vice versa. This resulted in increasing technical collaboration and a ripening of the friendship between Hooker and myself. Eventually under Hooker's leadership Rolls Royce became responsible for the production of jet engines based on Power Jets designs, and began their own vigorous and highly successful development of increasingly powerful jet engines.

After I ceased to be connected with jet engine design and development in 1946 and Stanley Hooker left Rolls Royce to become Chief Engineer of Bristol Siddeley, we rather lost touch with each other for about 13 years until, in 1959, he invited me to visit Bristol "to see what they were up to." On that interesting occasion I received a warm welcome and succeeded in arousing Hooker's interest in the turbo drill. The outcome was that B.S.E. agreed to back the project and, by 1961, practical development began. Thereafter I made frequent visits to B.S.E.'s Patchway factory near Bristol for several years. Hooker and I saw much of each other and the inevitable and frequent contacts with B.S.E.'s other aero engine designers re-aroused my interest in aeronautical engineering, especially as several ambitious projects were "on the boil," including the Concorde and its Olympus 593 engines, the Harrier "jump jet" and its Pegasus vectored thrust turbo fan engine, etc. I played no official role in these projects. I was a 'spectator from my sidelines,' but was often drawn into discussions of technical problems and flatter myself that I made an occasional useful contribution.

At the same time (1961 onwards) I became involved in the patent infringement action Rateau v. Rolls Royce and accepted an invitation from Rolls Royce to act as expert witness and technical advisor. Little did I realise that this was to absorb an increasing amount of my time until the case was decided in favour of Rolls Royce in January, 1967. It did, however, oblige me to brush up on jet engine theory and this stirred my interest to such an extent that I began to explore the possibilities of improving on the Olympus 593 for a second generation SST and did indeed succeed in convincing myself that very considerable improvements were possible in the form of a low bypass ratio turbo fan. Rolls Royce (now merged with B.S.E.) and U.S. firms were also engaged on project work on similar lines. The turbo drill, however, remained my main activity until it went 'on the shelf' as a consequence of the takeover of B.S.E. by Rolls Royce in 1968. At that time it had reached an advanced stage of development and was in limited production and commercial use. But Rolls Royce were already in the financial difficulties which led to bankruptcy and the nationalisation of the aero engine divisions in 1971. This, plus a general lack of interest of the new management in a field of engineering so remote from their main sphere of activity, resulted in a rapid reduction of support for the turbo drill project until it dried up altogether. This unhappy course of events and the failure to find alternative sponsors caused me to lose interest in the drill and to increase my 'private venture' exploration of SST power plant possibilities. A strong desire to see a second generation SST come into service in my own lifetime became almost an obsession.

It was clear that such a venture would be far beyond the means of any single firm or small group of firms without massive support from government and probably beyond the means of any single government, so, in 1974-75 I took it upon myself to attempt to act as a "catalyst" in the promotion of a major cooperative scheme for the development of a second generation SST involving the U.S. and U.K. governments and, possibly, the French government and the major aircraft and engine firms of these nations. I visualised a jointly owned international company at the centre of the web which would be responsible for the project phase of the operation, for development, and for the assembly phase of production and to which the several aircraft and engine firms would contribute personnel and act as subcontractors. This scheme differed from the Anglo-French Concorde joint venture in that in the case of the latter no single central company was ever formed. There were, of course, coordinating committees.

It should be mentioned that the Boeing 7011 had been smothered to death ostensibly by the environmentalist lobby in the U.S.A. (though one may venture the surmise that the vast cost to the U.S. taxpayer was the real cause).

The opportunity for my attempt was furnished by a few invitations to lecture in the U.S. Before leaving the U.K. I had sounded out the views of Rolls Royce (1971) and obtained an encouraging reaction, but refrained from contacting the U.K. Government so as to be able to claim that I was speaking for no one but myself. In Washington D.C. I succeeded in obtaining the 'unofficial blessing' of H. M. Brittanic Ambassador, Sir Peter Ramsbotham, and had a series of encouraging reactions from officials of the F.A.A. and N.A.S.A. from whom I received much help in arranging my schedule of visits to McDonnell Douglas, Lockheed, Boeing, General Electric and Pratt and Whitney with all of whom I discussed my proposals. One of these companies was distinctly 'lukewarm' to the scheme, but, in general, the results were most encouraging and I reported accordingly on my return to the U.K.

I visited the U.S.A. again from April-June 1976 with the intention of further pursuing the matter, but suffered minor injuries in a fall after arrival at J.F.K. airport and so accomplished little in furthering the scheme which, in the event (so far as I know) came to nothing. This particular visit, however, marked the beginning of the chain of events which led to the writing of this work. In the course of it I received several important invitations, the chief of which, relevant to this book, was one from Professor A. A. Pouring to lecture at the U.S. Naval Academy in October, 1976.

For personal reasons I was then seriously considering the possibility of becoming a resident in the U.S.A., but it was necessary to return to the U.K. to make preparations for this.

I returned to the U.S.A. on September 4, 1976, as a passenger in the Concorde, to keep the engagements I had accepted, to become a U.S. resident, and, hopefully, to marry an ex-U.S. Navy nurse with whom I had been friendly over many years.

During my visit to the Naval Academy, Annapolis (to lecture) in October, 1976, Dr. Andrew A. Pouring, the then Chairman of the Aero Space Department, on hearing that I was contemplating becoming resident in the U.S., asked if I would consider joining the Faculty of the Academy as a Distinguished Visiting Research Professor. The appointment would be for one year. I indicated that the offer was of interest. This led to a number of visits to the Academy and a formal invitation from the Superintendent to become 'Navair Research Professor.' This I accepted and duly became a member of the Faculty on August 1, 1977.

At this point I must go back in time to the early days of my work on jet engine development. During that period I evolved my own special methods of dealing with the aero-thermodynamics of gas turbine design which were fundamentally far simpler than the then accepted ways of treating aero-thermodynamic problems. Stanley Hooker, for one, often remarked that this theoretical innovation was more important than the original concept of the jet engine itself.

During one of my talks with Dr. Pouring at the Naval Academy I explained my methods of dealing with thermal cycles and was extremely surprised to learn that, after a period of some thirty years, these had not found their way into textbooks and so were not available to students though I knew they had become familiar to aero engine designers. Dr. Pouring urged me to write up the subject in a thesis as part of my duties at the Academy. This I agreed to do. That became the starting point of this work.

My original intention was to do no more than explain my treatment of thermal cycles, but as time passed my so-called 'thesis' began to grow and grow. The main reasons for its expansion was the course of lectures I gave on aircraft propulsion to a class of senior midshipmen who had elected to take this subject as one of their majors. Since much of the content of these lectures was not included in available textbooks I decided it would be necessary to write every lecture in advance and distribute copies to my class. It then seemed logical to expand my thesis to include the subject matter of my lectures which covered much more ground than the treatment of thermal cycles. Moreover, as time passed, I found myself extending my methods into realms of theory beyond those with which I was familiar. This intellectual exploration proved very fruitful and I added quite substantially to my knowledge of the aero-thermodynamics of turbo machinery. The boldest of these ventures was a limited excursion into shock wave theory (Section 4). This was a field of aero-thermodynamics with which I was almost entirely unfamiliar and I was curious to find out whether, if at all, my methods could be applied. I was pleasantly surprised to find that they could. I decided that these (to me) novel results must be included in this work. It now became clear that the original thesis was becoming a book.

This piecemeal expansion has had its drawbacks. It has meant much re-writing and re-arrangement of its sections into a more logical sequence as additional material was included. I hope the reader will be tolerant with any remaining 'patchiness' he may find. I, myself, am far from satisfied with my efforts. I fear that some evidence of over-hastiness (e.g., some inconsistencies in notation) will be apparent because I have had to meet a financial deadline imposed by the fact that the funds which have been allotted for its initial publication will be cut off automatically very soon after these words are written. The completion of the work has been considerably delayed by preoccupation with other very interesting but time consuming projects plus the domestic complications arising out of transplanting myself from the U.K. to the U.S.A. and five moves of home in little more than eighteen months.

As the reader will find, the fundamental basis of my methods was to treat air as a perfect gas of constant specific heat and to deal with thermal cycles by using temperatures and temperature ratios. Also to treat velocities (or kinetic energy) as having a temperature equivalent and vice versa. These assumptions are very accurate for the temperature range 180-400°K where specific heat is, in fact, virtually constant, but become increasingly inaccurate with increase of temperature above 400°K owing to the increase of specific heat. Nevertheless, as will be shown, they give good comparative results, and make it possible to deal with even compound thermal cycles in a matter of minutes. When, in jet engine design, greater accuracy was necessary for detail design, I worked in pressure ratios, used $\gamma = 1.4$ for compression and $\gamma = 1.33$ for expansion and assumed specific heats for combustion and expansion corresponding to the temperature range concerned. I also allowed for the increase of mass flow in expansion due to fuel addition (in the range 1½ - 2%). The results, despite the guesswork involved in many of the assumptions, amply justified these methods to the point where I was once rash enough to declare that "jet engine design has become an exact science." (This statement was inspired by the fact that on the first test of the W2/500 engine every experimental point fell almost exactly on the predicted curves of performance.)

Much of Sections 1 and 2 will be familiar to advanced students, but they have been included for the benefit of students comparatively new to the subject and to provide the foundation for later sections.

I have assumed throughout that the reader will have a knowledge of the rudiments of thermodynamics, aerodynamics and calculus.

As the reader will find, I have made frequent use of numerals instead of such constants as $\frac{\gamma}{\gamma-1}$, $\sqrt{2gK_p}$ etc., thus formulae which would otherwise look most complicated are greatly simplified. I make no apology for this. I feel sure that most readers will welcome it. Any purist who would prefer otherwise can always substitute $\frac{\gamma}{\gamma-1}$ wherever he sees the exponent 3.5 and $\frac{1}{\gamma-1}$ wherever 2.5 appears as an exponent. If he wishes to translate $147.1\sqrt{\theta}$ into $\sqrt{2gK_pT_1\left[1-\left(\frac{P_2}{P_1}\right)^{\frac{\gamma-1}{\gamma}}\right]}$ he should have no trouble in doing so.

A Note on Units. I apologise to those who might have preferred the S.I. system of units. I grew up with the foot-pound-second system (except for chemistry and physics). In my young days the metric system was never used in mechanical engineering in the U.K., and I find it hard to break the habit of a lifetime. Moreover, if I had talked about thrust in kilograms [newtons], height in metres, speed in metres/second and so on, it would have been much harder to convince the skeptics in the U.K. than it was. Talking about thrust instead of horse power, of specific fuel consumption instead of lbs./H.P. hour caused quite a lot of misunderstanding. It was a bit like talking about shekels per cubic ell as a measure of density.

I have, however, always used degrees centigrade (or Celsius) and degrees Kelvin for temperature and pound calories for heat and deplore the continued use of the Fahrenheit scale in the U.S.A. It has caused me considerable trouble in my discussions with American engineers.

No system of units has a logical basis. The metric system has the merit of using multiples of ten, but otherwise its basic units such as metres, kilograms, etc. are no more logical than feet, pounds, etc. Fortunately, certain important quantities such as specific fuel consumption, specific thrust, etc. are numerically the same in both the c.g.s. and f.p.s. systems.

In my opinion any logical system of units should be based on fundamental constants of nature such as the speed of light, the universal constant of gravitation, the mass of a neutron—and so on (though there is no certainty that these constants will maintain their 'fundamentality').

As the reader will find, I have made generous use of non-dimensional methods which, of course, are independent of any system of units.

I cannot end without acknowledging that I owe much to the extensive engineering training I received in the Royal Air Force—3 years as an Aircraft Apprentice in No. 4 Apprentices' Wing, RAF Cranwell, 2 years as a Flight Cadet at the Royal Air Force College, also at Cranwell, 2 years at the Officer's School of Engineering, RAF Henlow, and 2 years at Cambridge University taking the Mechanical Sciences Tripos, followed by a post graduate year for aerodynamic research and jet engine design—a total of 10 years. Also the experience I gained as a fighter pilot, flying instructor and float plane test pilot during the four years of General Duties following my graduation from Cranwell proved invaluable. I also owe much to the brilliant team of engineers (mostly younger than myself) that I succeeded in recruiting after I had been placed on the Special Duty List and loaned to Power Jets Ltd. as Honorary Chief Engineer.

Finally, this work owes much to the encouragement I received from Professor Andrew A. Pouring, Commander Marle D. Hewett, Chairman of the Aero-Space Department of the U.S. Naval Academy, and the help of Commander Hewett's staff, especially Mrs. Morva Hamaleinan.

Postscript

Since the foregoing was written the original of this work has been printed in limited numbers in report form for the U.S. Naval Academy as "Engineering & Weapons Report EW-7-79" dated June 1979. This book, however, differs from the original work in that Sections 3, 4, 5 and 17 have been substantially amended or rewritten with simplification as the main objective.

SECTION 1
Fundamentals

Notation and Units (for this section)

P	Absolute Pressure	lb/ft² (p.s.f.)
p	Pressure Ratio	
V	Specific Volume	ft³/lb
T	Absolute Temperature	°Kelvin (°K)
t	Temperature Ratio	
ρ	Density	lb/ft³
σ	Density Ratio	
E	Internal Energy	C.H.U/lb or ft lb/lb
H	Heat Added	C.H.U (Centigrade Heat Units)
h	Enthalpy	C.H.U/lb or kj/kg
s	Entropy	C.H.U/°K
W	Work/unit mass	C.H.U or Ft lbs/lb
C_v	Specific Heat at Constant Volume in thermal units	(C.H.U/lb/°K)
C_p	Specific Heat at Constant Pressure in thermal units	(C.H.U/lb/°K)
K_v	Specific Heat at Constant Volume in mechanical units	(ft lb/lb/°K)
K_p	Specific Heat at Constant Pressure in mechanical units	(ft lb/lb/°K)
γ	Ratio of Specific Heats $\equiv \dfrac{C_p}{C_v} \equiv \dfrac{K_p}{K_v}$	
R	Gas Constant per unit mass $\equiv K_p - K_v$	ft lb/lb/°K
H_R	Heat Added Reversibly	C.H.U.

Note:— One Centigrade Heat Unit (C.H.U.) is the amount of heat required to raise the temperature of one pound of water by 1.0°C: i.e. 1.0 C.H.U. = 1.8 B.T.U.s or 1400 ft.lbs.

Basic Definitions and Laws

All that follows (unless otherwise stated) is based on the assumption that air behaves as a perfect gas of constant specific heat, and on the use of three definitions, two thermodynamic laws and Newton's Laws of Motion. Steady flow is also assumed.

The *Definitions* are:
1) $E \equiv C_v T$ or $E \equiv K_v T$
2) $h \equiv E + PV$
3) $ds \equiv \dfrac{dH_R}{T}$

The *Thermodynamic Laws* are:

1) The First Law of Thermodynamics, namely $\Delta H = W + \Delta E$ i.e. Heat added equals work done plus the increase in Internal Energy. (This is the thermodynamic form of the Law of Conservation of Energy)

2) That for a *Perfect Gas*: $PV = RT$ or $P = \rho RT$ – This is known as the "Gas Equation"

(Note:—the Second Law of Thermodynamics is not used herein—except incidentally in the section on shock waves. It is variously stated; it is the thermodynamic equivalent of the axiom that energy cannot be created out of nothing. One form of stating it is:—"It is impossible for a self-acting machine, *unaided by any external agency*, to convey heat from one body to another at a higher temperature")

A Note on Entropy

Entropy is a concept which many students have difficulty in assimilating. It is a somewhat intangible quantity which, being defined as a differential, has no absolute value. One can only assign a specific value for entropy by choosing an arbitrary zero such as 0°C (as in steam tables). It is a very valuable concept for thermodynamic calculations concerning vapours and non-perfect gases, and many text book writers make generous use of temp-entropy diagrams, etc. to illustrate thermal cycles with a perfect gas (or mixture of perfect gases) as the working fluid. The writer has never found this to be necessary and so there are no such diagrams herein except for Fig. 1e of Section 6 as one of eight methods of representing thermal cycles, four of which use entropy as one of the basic parameters, merely to illustrate the several alternatives for thermal diagrams.

The word "isentropic" is frequently used, however, to indicate an adiabatic change of state which is reversible i.e. without loss. At one time the word "adiabatic" was used for this but was distinctly misleading since "adiabatic" covers *any* change of state where heat is neither added nor lost. Thus a heat insulated throttling process is adiabatic but not isentropic. Indeed, any adiabatic process in which utilisable energy is converted into non-useable internal energy is non-isentropic because entropy increases. An increase of entropy is analogous to the loss of energy in a waterfall where the original potential energy is converted to unusable energy of turbulence. Again an increase of entropy may be likened to an increase of the randomness of molecular motion at the expense of the energy of orderly motion i.e. the conversion of 'order into chaos.' There can never be a decrease of entropy in an adiabatic process—which is one way of expressing the second law of Thermodynamics. The most one can do in aero-thermodynamics is to minimize the increase of entropy. This is one of the main objectives in the design of aero-thermodynamic machinery.

Isentropic Compression or Expansion of a Perfect Gas

In isentropic compression or expansion, a perfect gas obeys the law $PV^\gamma = $ const. $= P_o V_o^\gamma$ where P_o and V_o are the pressure and specific volume of a reference state. Alternatively, writing $\rho \equiv \frac{1}{V}$ and $\rho_o \equiv \frac{1}{V_o}$, $\frac{P}{\rho^\gamma} = \frac{P_o}{\rho_o^\gamma}$. This law is of key importance in all that follows. It is derived from the first law of Thermodynamics

$$\Delta H = W + \Delta E \qquad (1\text{-}1)$$

using the Gas Equation:— $PV = RT$ (1-2)

the definition of internal Energy:— $E = K_v T$ (1-3)

and the definition of Entropy:— $ds = \frac{dH_R}{T}$ (1-4)

Since no heat is added or removed, $\Delta H = 0$ ∴ from (1)

$$W + \Delta E = 0 \qquad (1\text{-}5)$$

If $\Delta H = 0$ $dH_R = 0$ ∴ $ds = 0$ ∴ s is constant which confirms that the process is isentropic.

Fig. 1 is a Pressure-Specific Volume (P-V) diagram drawn to illustrate the con-

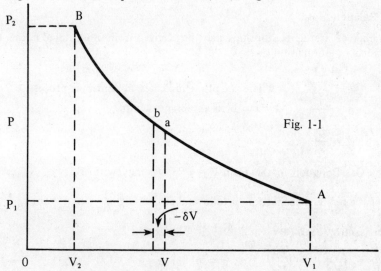

Fig. 1-1

nection between W. P and V. AB shows the relationship between P and V as a gas is compressed from $P_1 V_1$ to $P_2 V_2$. It is not necessarily isentropic. ab represents a part of the compression through a decrease of volume $-\delta V$. The work done on the gas $-\delta W$ in compressing from a to b is seen to be given by $-\delta W = -P\delta V$ (by convention work done in compression is taken to be negative and to be positive in expansion). In the limit

$$dW = PdV \qquad (1\text{-}6)$$

From equ.(5), for isentropic compression or expansion $dW = -dE$ and from equ. (3) $dE = K_v dT$ ∴ $dW = -K_v dT$ (1-7) therefore from (6) & (7) $\quad PdV = -K_v dT$ (1-8)

From which $\quad dT = -\dfrac{PdV}{K_v}$ (1-9)

From the Gas Equation $PV = RT$ we have $PdV + VdP = RdT$

or $\quad PdV + VdP = K_p dT - K_v dT$ (1-10)

Substituting for PdV from (8), (10) becomes $\quad VdP = K_p dT$ (1-11)

Substituting for dT from (9) $\quad VdP = -\dfrac{K_p}{K_v} PdV$ or, since $\dfrac{K_p}{K_v} = \gamma$, $VdP + \gamma PdV = 0$

or $\quad \dfrac{dP}{P} + \gamma \dfrac{dV}{V} = 0$ from which $\log P + \gamma \log V = \text{Const.}$

or

$$PV^\gamma = \text{Const.} \quad (1\text{-}12)$$

An important form of (11) is $\quad \dfrac{dP}{\rho} = K_p dT$ (1-11a)

Note that (11a) can be derived from the First Law when $H = 0$, Fig. 1, and the Gas Equation and is true whether or not the process is isentropic as long as it is adiabatic.

Important Relationships

A number of useful relationships may be derived from those stated in the foregoing using additional notation as follows:

$\theta = \Delta T$
$\theta' = \Delta T'$ where the prime indicates an isentropic process.
$\phi = \dfrac{\theta}{T_1}$ where T_1 is a reference temp.
$\phi' = \dfrac{\theta'}{T_1}$

From the Gas Equation in the form $P = \rho RT$ we may write $P_1 = \rho_1 RT_1$, where the suffix 1 means a reference condition, it follows that $\dfrac{P}{P_1} = \dfrac{\rho}{\rho_1} \dfrac{T}{T_1}$ or $p = \sigma t$ (1-13)

From the isentropic law $PV^\gamma = P_1 V_1^\gamma$ it follows that

$$\dfrac{P}{P_1} = \left(\dfrac{V_1}{V}\right)^\gamma \equiv \left(\dfrac{\rho}{\rho_1}\right)^\gamma \quad \text{or } p = \sigma^\gamma \text{ or } \sigma = p^{\frac{1}{\gamma}} \quad (1\text{-}14)$$

By combining (13) and (14)

$$\dfrac{P}{P_1} = \left(\dfrac{T}{T_1}\right)^{\frac{\gamma}{\gamma-1}} \quad \text{or } p = t^{\frac{\gamma}{\gamma-1}} \text{ or } t = p^{\frac{\gamma-1}{\gamma}} \quad (1\text{-}15)$$

$$\dfrac{\rho}{\rho_1} = \left(\dfrac{T}{T_1}\right)^{\frac{1}{\gamma-1}} \quad \text{or } \sigma = t^{\frac{1}{\gamma-1}} \text{ or } t = \sigma^{\gamma-1} \quad (1\text{-}16)$$

For an isentropic temp rise of θ' from temp T_1

$$t = 1 + \frac{\theta'}{T_1} \quad \text{or} \quad t = 1 + \phi' \tag{1-17}$$

hence $\quad \sigma = (1 + \frac{\theta'}{T_1})^{\frac{1}{\gamma-1}} \quad \text{or} \quad \sigma = (1 + \phi')^{\frac{1}{\gamma-1}}$ (1-18)

and $\quad p = (1 + \frac{\theta'}{T_1})^{\frac{\gamma}{\gamma-1}} \quad \text{or} \quad p = (1 + \phi')^{\frac{\gamma}{\gamma-1}}$ (1-19)

From the definition of enthalpy $h \equiv E + PV$ and the Gas Equation $PV = RT$, $dh = dE + PdV + VdP \equiv dE + RdT \equiv dE + K_p dT - K_v dT$ and since $dE \equiv K_v dT$ it follows that

$$dh = K_p dT \quad \text{or} \quad dh = \frac{dP}{\rho} \quad \text{(from 11a)} \tag{1-20}$$

and therefore *for const. specific heat* $\quad h = K_p T$ (1-21)

hence $\quad \frac{h}{E} = \frac{K_p T}{K_v T} = \gamma$

The Momentum Equation for linear Compressible Flow

Using additional notation as follows:—

    u   Velocity ft/sec
    S   Section Area sq.ft.
    m   Mass lbs
    F   Force lbs

and with the aid of Fig. 2, the very important Momentum Equation will be derived. Fig. 2 represents a small section of a stream tube of sections S at AA' and S+δS at

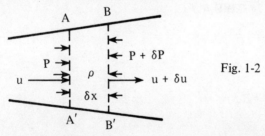

Fig. 1-2

BB' with pressures and velocities as shown for a very small element of length δx and mean density ρ.

The net accelerating force acting on the element ABB'A' is $PS - (P+\delta P)(S+\delta S)$ plus the mean side wall pressure $P + \frac{\delta P}{2}$ acting on the area δS i.e. $\delta F = PS - (P + \delta P)(S + \delta S) + (P + \frac{\delta P}{2}) \delta S$ which, neglecting the products of infinitesimals, reduces to $\delta F = -S\delta P$.

Again neglecting the products of infinitesimals, the mass of the element $\delta m = \rho S \delta x$ and the acceleration is $u \frac{\delta u}{\delta x}$, therefore, since force equals mass times acceleration,

$$\delta F = -S\delta P = \frac{\delta m}{g} u \frac{\delta u}{\delta x} = \frac{\rho S \delta x}{g} u \frac{du}{\delta x} \quad \text{from which it readily follows that} \quad \frac{dP}{\rho} + \frac{du^2}{2g} = 0 \tag{1-22}$$

It may be seen that (22) is the differential form of Bernouilli's Equation for incompressible flow, namely:— $P + \dfrac{\rho u^2}{2g} = \text{const.}$

The Energy Equation for Compressible Flow

This is very simply derived from (22). As we have seen, (Equation (11a)), $\dfrac{dP}{\rho} = K_p dT$, so, substituting for $\dfrac{dP}{\rho}$ in (22), we have $K_p dT + \dfrac{du^2}{2g} = 0$ or $dT + \dfrac{du^2}{2gK_p} = 0 \qquad (1\text{-}23)$

from which $\dfrac{u^2}{2gK_p} + T = T_T$ where T_T is total or 'stagnation' temp. $\qquad (1\text{-}24)$

Writing $\dfrac{u^2}{2gK_p} = \theta_u$ where θ_u is the 'temp equivalent of velocity' then
$$T + \theta_u = T_T \qquad (1\text{-}25)$$
i.e., in words "Total (or stagnation) Temp. equals the Static Temp. plus the Temp. equivalent of velocity"

Equation 25 is a very simple and useful form of the Energy Equation and is true whether the energy conversion is isentropic or not as long as it is adiabatic. It may be described as the 'compressible version' of Bernouilli's Equation. It will be much used in all that follows. *It is emphasised, however, that it is based on the assumption that K_p is constant,* otherwise Equation 23 must be used with K_p as a function of T.

Quantities etc. used for Air treated as a Perfect Gas of Constant Specific Heat

K_p = 336 ft lbs/lb/°C

K_v = 240 ft lbs/lb/°C

R = 96 ft lbs/lb/°C

C_p = 0.24 C.H.U./lb/°C

C_v = 0.171 C.H.U./lb/°C

$\gamma \equiv \dfrac{K_p}{K_v} \equiv \dfrac{C_p}{C_v} = 1.4$

$\dfrac{1}{\gamma-1} = \dfrac{K_v}{R} = 2.5$

$\dfrac{\gamma}{\gamma-1} = \dfrac{K_p}{R} = 3.5$

$\dfrac{\gamma+1}{2} = 1.2$

u_c (the accoustic speed at stalic temp. T) = $\sqrt{gR\gamma T} = 65.786 \sqrt{T}$

$\sqrt{2gK_p} = 147.1$

Note: Since 1.0°C = 1.0°K it is useful to use °K for absolute temps and °C for temp. differences.

In most of this work the numerical quantities above will be used instead of symbols to simplify formulae e.g. $\dfrac{P}{P_1} = \left(\dfrac{T}{T_1}\right)^{\frac{\gamma}{\gamma-1}}$ becomes $p = t^{3.5}$; $\dfrac{\rho}{\rho_1} = \left(\dfrac{T}{T_1}\right)^{\frac{1}{\gamma-1}}$ becomes $\sigma = t^{2.5}$; $u = 147.1 \sqrt{\theta_u}$ etc.

SECTION 2
Flux Density in Isentropic Compressible Flow

Additional Notation

- Q   Mass Flow Rate   lb/sec
- ψ Flux Density   lb/sec/sq ft
- ψ_c Maximum Flux Density
- T_c Static Temp when $\psi = \psi_c$
- u_c Velocity when $\psi = \psi_c$
- θ_c Temp. Equivalent of $u_c = \dfrac{u_c^2}{2gK_p} \equiv T_T - T_c$
- $t_c = \dfrac{T_T}{T_c}$
- ρ_c Density when $\psi = \psi_c$
- $\alpha \equiv \dfrac{1}{\gamma - 1}$ (= 2.5 for air)
- $k = \sqrt{2gK_p}$ (= 147.1 for air)
- S_c Section Area when $\psi = \psi_c$

The suffix T implies total (or 'stagnation') temp., pressure, and density.

Flux Density Relationships

It will be obvious that $\psi = \rho u = \dfrac{Q}{S}$ (2.1)

ψ will be a maximum i.e. will equal ψ_c when $d\psi = 0$ i.e. when $d(\rho u) = 0$ or

$$\rho_c \, du + u_c \, d\rho = 0 \quad \text{or} \quad \frac{du}{u_c} + \frac{d\rho}{\rho_c} = 0 \tag{2-2}$$

Since, for isentropic flow, $\rho = \rho_T \, t^{-\alpha}$, $d\rho = -\alpha \rho_T \, t^{-\alpha-1} \, dt$ $\therefore \frac{d\rho}{\rho} = -\frac{\alpha \, dt}{t}$ (2-3)

and, since $u = k \sqrt{\theta_u} = k\sqrt{T_T - T} = k\sqrt{T_T \left(1 - \frac{1}{t}\right)}$

$$du = k\sqrt{T_T} \, d(1 - t^{-1})^{\frac{1}{2}} = k\sqrt{T_T} \times \frac{1}{2}(1 - t^{-1})^{-\frac{1}{2}} \times t^{-2} \, dt$$

or

$$du = \frac{k\sqrt{T_T}}{2t^2 \sqrt{1 - \frac{1}{t}}} \, dt \quad \text{Using } u = k\sqrt{T_T\left(1 - \frac{1}{t}\right)} \text{ from above}$$

then $\quad \dfrac{du}{u} = \dfrac{dt}{2t^2 - 2t} \tag{2-4}$

hence substituting for $\frac{d\rho}{\rho}$ from (3) and for $\frac{du}{u}$ from (4)

(2) becomes $\quad -\dfrac{\alpha \, dt}{t_c} + \dfrac{dt}{2t_c^2 - 2t_c} = 0$

which reduces to $\dfrac{1}{t_c - 1} = 2\alpha = \dfrac{2}{\gamma - 1}$

$\therefore t_c - 1 = \dfrac{\gamma - 1}{2}$ from which $\underline{t_c = \dfrac{\gamma + 1}{2}}$ (2-5)

i.e. $\psi = \psi_c$ when $\dfrac{T_T}{T_c} = \dfrac{\gamma + 1}{2}$ (= 1.2 for air)

From $\theta_c = T_T - T_c = T_T\left(1 - \dfrac{1}{t_c}\right) = T_T\left(1 - \dfrac{2}{\gamma+1}\right) = T_T\left(\dfrac{\gamma-1}{\gamma+1}\right)$ it follows that, for air,

$\theta_c = \dfrac{T_T}{6}$ or $\dfrac{T_c}{5}$ and, since $u_c = \sqrt{2gK_p \theta_c} = \sqrt{2gK_p \dfrac{\gamma-1}{2} T_c}$ then using $K_p = R\dfrac{\gamma}{\gamma-1}$;

$u_c = \sqrt{g\gamma RT_c}$ which is the acoustic velocity at temp. T_c — thus flux density is a maximum when velocity equals the local speed of sound.

For the 'critical' temp ratio t_c (total to static) the corresponding value of p_c (total to static) $= t_c^{\frac{\gamma}{\gamma-1}}$ and the critical density ratio $\sigma_c = t_c^{\frac{1}{\gamma-1}}$ $\therefore \dfrac{p_T}{p_c} = \dfrac{\gamma+1}{2}^{\frac{1}{\gamma-1}}$

or, since $\rho_T = \dfrac{P_T}{RT_T}$ (from the Gas Equation); $\rho_c = \dfrac{P_T}{RT_T \sigma_c}$. Since $\psi = \rho u$, $\psi_c = \rho_c u_c$ and, as

we have seen, $u_c = \sqrt{g\gamma RT_c}$ it follows that $\psi_c = \dfrac{P_T}{RT_T \sigma_c} \sqrt{g\gamma RT}$ which, using $T_c = \dfrac{T_T}{t_c}$ and

$\sigma_c = t_c^{\frac{1}{\gamma-1}}$, reduces to $\psi_c = \dfrac{P_T}{\sqrt{T_T}} \, t_c^{-\frac{\gamma+1}{2\gamma-2}} \sqrt{\dfrac{g\gamma}{R}}$ (2-6)

For air $t_c = 1.2$ $\frac{\gamma+1}{2\gamma-2} = 3$ and $\sqrt{\frac{g\gamma}{R}} = .6853$

$\therefore$ for air $\psi_c = \frac{.6853}{1.2^3} \frac{P_T}{\sqrt{T_T}} = .3966 \frac{P_T}{\sqrt{T_T}}$ (2-7)

Also, for air $p_c = t_c^{3.5} = 1.2^{3.5} = 1.8929$

and $\sigma_c = t_c^{2.5} = 1.2^{2.5} = 1.5774$

$u_c = 147.1 \sqrt{\frac{T_T}{6}} = 60.053 \sqrt{T_T}$ (2-8)

or $u_c = 147.1 \sqrt{\frac{T_c}{5}} = 65.785 \sqrt{T_c}$

Thus it may be seen that ψ_c is proportional to total pressure and inversely proportional to $\sqrt{\text{total temp.}}$—a fact of much importance in many areas of aero-thermodynamics.

For flux density other than ψ_c, since $\psi = \rho u$; $u = 147.1 \sqrt{T_T} \sqrt{1-\frac{1}{t}}$ and $\rho = \frac{\rho_T}{t^{2.5}} = \frac{P_T}{RT_T} t^{-2.5}$ it follows that $\psi = 147.1 \sqrt{T_T} \sqrt{\frac{t-1}{t}} \times \frac{P_T}{RT_T} t^{-2.5}$ which reduces to

$\psi = 1.532 \frac{P_T}{\sqrt{T_T}} \frac{\sqrt{t-1}}{t^3}$ (2-9)

From (7) $\psi_c = .3966 \frac{P_T}{\sqrt{T_T}}$ $\therefore$ $\frac{\psi}{\psi_c} = 3.863 \frac{\sqrt{t-1}}{t^3}$ (2-10)

The table below gives the values of ψ/ψ_c, u/u_c, ρ/ρ_c and P/P_c for the temp. ratio (total to static) range $1.0 - 3.4$

t	ψ/ψ_c	u/u_c	ρ/ρ_c	P/P_c
1.00	0	0	1.577	1.893
1.05	0.747	0.534	1.396	1.596
1.10	0.917	0.738	1.243	1.356
1.15	0.985	0.885	1.112	1.161
1.20	1.000	1.000	1.000	1.000
1.25	0.990	1.095	0.903	0.867
1.30	0.965	1.177	0.819	0.756
1.35	0.930	1.247	0.745	0.662
1.40	0.892	1.309	0.680	0.583
1.45	0.851	1.365	0.623	0.516
1.50	0.811	1.414	0.572	0.458
1.60	0.731	1.500	0.437	0.365
1.70	0.659	1.572	0.419	0.296
1.80	0.593	1.633	0.363	0.242
1.90	0.534	1.686	0.317	0.200
2.00	0.484	1.732	0.279	0.167
2.20	0.398	1.809	0.220	0.120
2.40	0.331	1.870	0.177	0.0884
2.60	0.278	1.922	0.145	0.0688

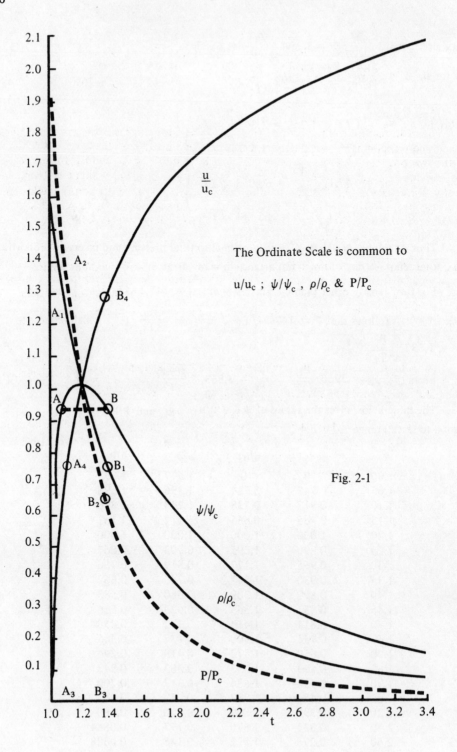

Fig. 2-1

2.80	0.236	1.964	0.120	0.0515
3.00	0.202	2.000	0.101	0.0405
3.20	0.175	2.031	0.086	0.0322
3.40	0.152	2.058	0.074	0.0261

These figures are plotted in Fig. 1.

The main feature to note is that below ψ_{max} ($\psi = \psi_c$) there are two possible values of ψ and, correspondingly, two each of t, p/ρ_c, P/P_c and u/u_c. Thus for $\psi/\psi_c = 0.92$ (A and B) the two possible values of ρ/ρ_c are at A_1 and B_1; for P/P_c at A_2 and B_2; for t at A_3 and B_3; and for u/u_c at A_4 and B_4. It would therefore be possible, without violating continuity, for a sudden 'jump' in pressure from B_2 to A_2 to occur with a corresponding jump in density from B_1 to A_1 and a sudden reduction of velocity from B_4 to A_4. This, of course, is the basis of shock wave phenomena. Unfortunately, as shown in Section 4, these sudden changes are not isentropic.

Another point to notice from the ψ/ψ_c curve is that, near the maximum, in the range, say, t = 1.15 to t = 1.25 the change of ψ/ψ_c is not more than 1.5% of the maximum, so that, in this range (which occurs frequently in turbo machinery) there would be little error in assuming that mass flow rate Q depends only on total temp. and pressure, i.e. that

$$Q \alpha \frac{P_T}{\sqrt{T_T}}.$$

It should be noted that u/u_c is synonymous with Mach Number, i.e. $\frac{\text{actual velocity}}{\text{acoustic velocity}}$ where the acoustic velocity is that for the static temp. T in $u_c = \sqrt{g\gamma RT}$.

It is desirable to emphasize that all the relationships derived in the foregoing discussion of flux density are true only for isentropic flow, however they are a close approximation if the departure from isentropy is minor.

A point to note is that, having magnitude and direction, flux density is a vector quantity.

SECTION 3
Radial Equilibrium in Whirling Isentropic Compressible Flow

This is a subject of considerable importance in turbo machinery blade design, yet, until the mid 1930s, the effect of centrifugal force on the radial pressure gradient in modifying the velocity distribution across the annular whirling flow issuing from, say, a turbine nozzle ring was ignored. It was assumed that the flow from each pair of nozzle blades was 'straight' and that the velocity was uniform across the annulus. This was a legacy from the days of 'partial admission' when a few widely spaced individual nozzles were used too far apart for their individual jets to coalesce into a whirling flow pattern. This non recognition of radial equilibrium or vortex flow was not important when blades were very short relative to mean diameter. For longer blades it was customary to allow for increase of blade speed with radius and blades were 'twisted' accordingly, but no allowance was made for the equally important reduction of whirl velocity with increase of radius with the result that the blade twist was only about half of what it should have been. Moreover calculations of end thrust on a rotor were grossly inaccurate for long bladed turbines.

Wherever a flow has curvature there has to be a pressure gradient normal to the streamlines due to centrifugal force in addition to any tangential pressure gradient there may be due to acceleration along the streamlines. Further if there is a radial component of velocity (as in radial flow machinery, e.g. centrifugal compressors), any variation i.e. acceleration, will be superimposed on the radial pressure gradient due to centrifugal force. Forces due to radial accelerations are, however, normally small compared with centrifugal force when whirl velocities are in the range 800 - 2500 ft/sec as they usually are at entry to turbine rotor blading or exit from axial flow and centrifugal compressor rotors.

In this section the theory of radial equilibrium in whirling flow is dealt with on the assumptions that changes in temp., pressure, velocity etc. take place isentropically and that stagnation (i.e. total) temp and pressure are uniform.

If the flow has tangential velocity (i.e., whirl velocity) u_ω; axial velocity u_a and radial velocity u_r at a given point in space, then, since these components are mutually at right angles, the resultant velocity u is given by

$$u^2 = u_\omega{}^2 + u_a{}^2 + u_r{}^2 \tag{3-1}$$

$$\text{hence} \quad \theta_u = \theta_\omega + \theta_a + \theta_r \tag{3-2}$$

It follows that the static temp. T, at the point concerned is given by $T = T_T - \theta_u \equiv T_T - \theta_\omega - \theta_a - \theta_r$ (3-3)

from which $dT = -d\theta_u \equiv -d\theta_\omega - d\theta_a - d\theta_r$

$$\text{or} \quad dT + d\theta_\omega + d\theta_a + d\theta_r = 0 \tag{3-3a}$$

The diagram of fig. 1 represents a section of whirling flow where the whirl velocity is u_ω at radius r and the density is ρ. If the pressure is P at r and P + δP at r + δr, then,

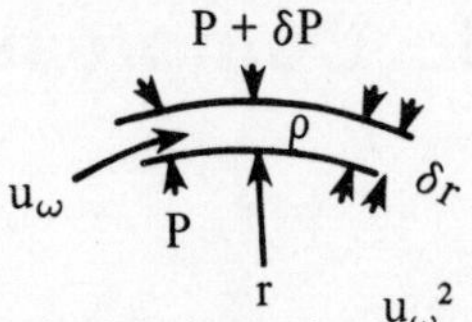

Fig. 3-1

since the acceleration towards the center is $\frac{u_\omega{}^2}{r}$, one may see that

$$\delta P = \frac{\rho}{g} \frac{u_\omega{}^2}{r} \delta r \quad \text{or} \quad \frac{dP}{\rho} = \frac{u_\omega{}^2}{g} \frac{dr}{r} \tag{3-4}$$

Since $\frac{dp}{\rho} = K_p dT$ (equ. 11a of Section 1) and

$$\frac{u_\omega{}^2}{2gK_p} = \theta_\omega \quad \text{it follows that} \quad 2\theta_\omega \frac{dr}{r} = dT \tag{3-5}$$

Substituting for dT from (3a) above

$$2\theta_\omega \frac{dr}{r} = -d\vartheta_\omega - d\theta_a - d\theta_r$$

$$\text{or} \quad 2\theta_\omega \frac{dr}{r} + d\vartheta_\omega + d\vartheta_a + d\theta_r = 0$$

$$\text{or} \quad 2\frac{dr}{r} + \frac{d\theta_\omega}{\vartheta_\omega} + \frac{d\theta_a}{\theta_\omega} + \frac{d\theta_r}{\theta_\omega} = 0 \tag{3-6}$$

From equation (6), if the relationship between r and any two of any two of the three velocity components is known (or decided) the relationship between the third and r may be found. E.g. if $u_r = 0$ or is constant $d\theta_r = 0$ and if u_a is constant $d\theta_a = 0$, hence $\frac{2dr}{r} + \frac{d\theta_\omega}{\theta_\omega} = 0$ from which $r^2 \theta_\omega$ = const. or ru_ω = const. i.e. whirl velocity is inversely proportional to radius. This is the 'law' of the 'free vortex.'

Since ru_ω = const., the angular momentum of a free vortex is the same at all radii.

It should be noted that Equation (6) above was arrived at on the assumption that the only acceleration was that due to centrifugal force. It does not take into account rates of change of u_ω u_a or u_r if present. Except within blade passages these accelerations are usually less than 2% of the centrifugal acceleration in turbo machinery.

There are several relationships between the velocity components and radius which permit easy solutions of Equation (6). Two cases are dealt with below.

Case 1 $u_r = 0$ and $\frac{u_\omega}{u_a}$ = Const = tan α where α is the angle of flow relative to the axis. (This is representative of the flow issuing from a ring of nozzle blades of constant exit angle.)

Since $d\theta_r = 0$ and $u_a \tan\alpha = u_\omega$, $u_a^2 \tan^2\alpha = u_\omega^2$ i.e. $\theta_a \tan^2\alpha = \theta_\omega$ ∴

$\tan^2\alpha \, d\theta_a = d\theta_\omega$ or $d\theta_a = \dfrac{d\theta_\omega}{\tan^2\alpha}$, hence substituting for $d\theta_a$ in equation (6) we have:—

$$\frac{2dr}{r} + \frac{d\theta_\omega}{\theta_\omega} + \frac{d\theta_\omega}{\theta_\omega \tan^2\alpha} = 0$$

or $\dfrac{2dr}{r} + \dfrac{\tan^2\alpha + 1}{\tan^2\alpha} \dfrac{d\theta_\omega}{\theta_\omega} = 0$ or, since $\dfrac{\tan^2\alpha + 1}{\tan^2\alpha} = \text{cosec}^2\alpha$,

$\dfrac{2dr}{r} + \text{cosec}^2\alpha \dfrac{d\theta_\omega}{\theta_\omega} = 0$ from which $2 \ln r + \text{cosec}^2\alpha \ln \theta_\omega = 0$

or $r^2 \theta_\omega^{\text{cosec}^2\alpha} = \text{Const.}$

Case 2 $u_r = 0$ and $u_\omega = \text{Const}$ i.e. $\theta_\omega = \text{Const.}$

Equ. (6) now becomes $\dfrac{2dr}{r} + \dfrac{d\theta_a}{\theta_\omega} = 0$ from which $\ln r^2 + \theta_a/\theta_\omega = \text{const.}$

If, at r_m, θ_a is θ_{am} then $\ln r^2 + \theta_a/\theta_\omega = \ln r_m^2 + \theta_{am}/\theta_\omega$

from which $2 \ln \dfrac{r}{r_m} = \dfrac{1}{\theta_\omega}(\theta_{am} - \theta_a)$

or $2\theta_\omega \ln \dfrac{r}{r_m} = \theta_{am} - \theta_a$ or $\theta_a = \theta_{am} - 2\theta_\omega \ln \dfrac{r}{r_m}$

which is, perhaps, the most convenient equation to use.

Example 3-1.

If, in a free vortex flow in an annulus of inner radius 1.0 ft. and outer radius 1.6 ft., the whirl velocity at mean radius r_m is 1500′/sec and the (constant) axial velocity is 600′/sec. What are the whirl velocities at inner and outer radii and what are the flow angles at inner, mean, and outer radii?

Since, as shown above, $r u_\omega = r_m u_{\omega m}$ $u_\omega = u_{\omega m} \dfrac{r_m}{r}$. Since $r_m = 1.3$, the whirl velocity at the inner radius is $1500 \left(\dfrac{1.3}{1.0}\right) = 1950'$/sec and the flow angle is $\tan^{-1} \dfrac{1950}{600} = 72.9°$. At the outer radius, $u_\omega = 1500 \left(\dfrac{1.3}{1.6}\right) = 1219'$/sec and the flow angle is $\tan^{-1} \dfrac{1219}{600} = 63.8°$. The flow angle at mean radius is $\tan^{-1} \dfrac{1500}{600} = 68.2°$.

Equation (6) has some useful variants. It may be written (since $\theta \propto u^2$):—

$$\frac{2dr}{r} + \frac{du_\omega^2}{u_\omega^2} + \frac{du_a^2}{u_\omega^2} = 0 \quad \text{(neglecting } u_r\text{)} \tag{3-7}$$

or

$$\frac{dr}{r} + \frac{u_\omega \, du_\omega}{u_\omega^2} + \frac{u_a \, du_a}{u_\omega^2} = 0 \qquad (3\text{-}7a)$$

or

$$\frac{dr}{r} + \frac{du_\omega}{u_\omega} + \frac{u_a \, du_a}{u_\omega^2} = 0 \qquad (3\text{-}7b)$$

or

$$u_\omega \frac{dr}{r} + du_\omega + \frac{u_a}{u_\omega} du_a = 0 \qquad (3\text{-}7c)$$

or

$$\frac{u_\omega}{u_{\omega m}} \frac{d\frac{r}{r_m}}{\frac{r}{r_m}} + d\frac{u_\omega}{u_{\omega m}} + \frac{u_a u_{\omega m}}{u_{am} u_\omega} \, d\frac{u_a}{u_{am}} = 0 \qquad (3\text{-}7d)$$

where the suffix m indicates u_a and u_ω at reference radius r_m; usually the mean radius.

Equation 7d is, of course 'doubly non-dimensional'.

From 7b:—

$$\text{Ln. } r + \text{Ln. } u_\omega + \int \frac{u_a}{u_\omega^2} \, du_a = \text{const.} \qquad (3\text{-}8)$$

or

$$\text{Ln. } ru_\omega + \int \frac{u_a}{u_\omega^2} \, du_a = \text{const.} \qquad (3\text{-}8a)$$

This last may be useful for step by step integration where the relationship between u_ω and u_a is known but cannot be put into a form where $\int \frac{u_a}{u_\omega^2} u_{aD}$ can be integrated algebraically. Thus, for example, a turbine nozzle ring designed to produce a free vortex of constant axial velocity at a design mass flow rate Q, may need to be examined to find out what happens if Q is increased or decreased. For such a purpose equ. 8a is probably best used in the form

$$\frac{\delta ru_\omega}{ru_\omega} + \frac{u_a}{u_\omega^2} \, du_a = 0 \qquad (3\text{-}8b)$$

Assuming the leaving flow conforms to the exit angles at all radii and that these conform to the condition that $\tan\alpha = \frac{u_\omega}{u_{aD}}$ where $u_\omega = \frac{r_m}{r} u_{\omega mD}$ and $u_a = \text{const.} = u_{aD}$ (the suffix D implying the design condition). The relationship between u_ω and u_a for the off design condition becomes $u_\omega = u_a \tan\alpha = u_a \frac{r_m}{r} \frac{u_{\omega mD}}{u_{aD}}$ whence one may substitute either for u_ω or u_a in equ. (8b).

Writing $\phi_a = \frac{\theta_a}{T_T}$ and $\phi_\omega = \frac{\theta_\omega}{T_T}$ another useful variant of Equation (6) is:—

$$\frac{2d\frac{r}{r_m}}{\frac{r}{r_m}} + \frac{d\phi_\omega}{\phi_\omega} + \frac{d\phi_a}{\phi_\omega} + \frac{d\phi_r}{\phi_\omega} = 0 \qquad (3\text{-}9)$$

which again is doubly non dimensional.

The static temp. T_s at any radius is $T_s = T_T - \theta_\omega - \theta_a - \theta_r$ (3-10)

or $\dfrac{T_s}{T_T} = 1 - \phi_\omega - \phi_a - \phi_r$ or $t = \dfrac{1}{1-\phi_\omega-\phi_a-\phi_r}$ (3-10a)

where t is the total to static temp. ratio.

If P_s is the static pressure at any radius, then $\dfrac{P_T}{P_s} = t^{3.5}$ or pressure ratio $p = t^{3.5}$
i.e. using Equ. 10a.

$$p = \left(\dfrac{1}{1-\phi_\omega-\phi_a-\phi_r}\right)^{3.5} \qquad (3\text{-}11)$$

Similarly the density ratio σ (total to static) is given by $\sigma = \left(\dfrac{1}{1-\phi_\omega-\phi_a-\phi_r}\right)^{2.5}$ (3-12)

If p_m is the pressure ratio at reference radius r_m then from (11)

$$p_m = \left(\dfrac{1}{1-\phi_{\omega m}-\phi_{am}-\phi_{rm}}\right)^{3.5} \text{ so } \dfrac{p}{p_m} = \left(\dfrac{1-\phi_{\omega m}-\phi_{am}-\phi_{rm}}{1-\phi_\omega-\phi_a-\phi_r}\right)^{3.5} \qquad (3\text{-}13)$$

and $\dfrac{\sigma}{\sigma_m} = \left(\dfrac{1-\phi_{\omega m}-\phi_{am}-\phi_{rm}}{1-\phi_\omega-\phi_a-\phi_r}\right)^{2.5}$ (3-14)

$\left(\text{It presumably being obvious that } t_m = \dfrac{1}{1-\phi_{\omega m}-\phi_{am}-\phi_{rm}} \text{ and that}\right.$

$\left.\dfrac{t}{t_m} = \dfrac{1-\phi_{\omega m}-\phi_{am}-\phi_{rm}}{1-\phi_\omega-\phi_a-\phi_r} \text{ and } \therefore t = \dfrac{t_m}{(1-\phi_\omega-\phi_a-\phi_r)}\right)$

The above doubly non dimensional formulae enable one to draw generalised curves for the variation of temp. ratio, pressure ratio and density ratio with radius for given values of t_m, $\phi_{\omega m}$, ϕ_{am} and ϕ_{rm} at reference radius r_m.

For cases like a free vortex of constant axial velocity or where $\dfrac{u_a}{u_\omega} = $ const (ie constant exit angle in the case of a nozzle ring) where the differential equations (6), (7), (9) and (10) or their variants are readily solved, it is a relatively simple matter to find how temp. ratio, pressure ratio and density ratio vary with radius—as may be seen in the following example:—

Example 3-2

A free vortex nozzle ring has a mean radius r_m of $1.5\, r_i$ where r_i is the root radius (i.e. $r_i = 1.0$) and at r_m, $u_a = 0.4 u_\omega$ and $\phi_{\omega m} = 0.7$, how do the temp., pressure and density ratios vary from $r/r_i = 1.0$ to $r/r_i = 2.0$?

u_ω varies inversely as the radius, ie $u_\omega r = u_{\omega m} r_m$ from which $\dfrac{u_\omega}{u_{\omega m}} = \dfrac{r_m}{r}$ ∴ $\dfrac{\phi_\omega}{\phi_{\omega m}} = \left(\dfrac{r_m}{r}\right)^2$ or $\phi_\omega = \phi_{\omega m} \left(\dfrac{1.5}{r}\right)^2 = \dfrac{.07 \times 2.25}{r^2} = \dfrac{.1575}{r^2}$; ϕ_a = const. = $0.4^2 \phi_{\omega m} = 0.16 \phi_{\omega m}$ = .16 × .07 = .0112

$$t = \dfrac{1}{1-\phi_\omega-\phi_a} = \dfrac{1}{1-\phi_\omega-.0112} = \dfrac{1}{.9888-\phi_\omega}$$

so, since (from above) $\phi_\omega = \dfrac{.1575}{r^2}$ $t = \dfrac{1}{0.9888 - \dfrac{.1575}{r^2}}$

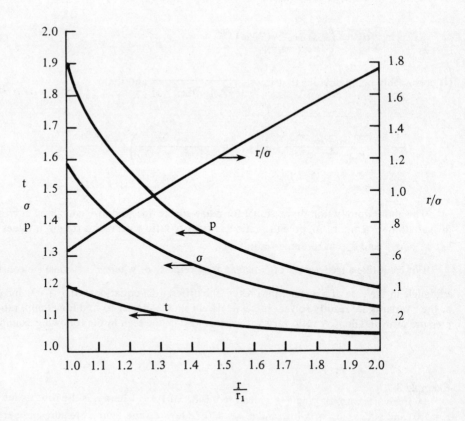

Fig. 3-2

Using $\sigma = t^{2.5}$ and $p = t^{3.5}$ the following table results:—

r	t	σ	p	$\frac{r}{\sigma}$
1.0	1.203	1.587	1.909	0.630
1.2	1.137	1.379	1.568	0.870
1.4	1.101	1.271	1.399	1.101
1.6	1.078	1.208	1.302	1.324
1.8	1.064	1.167	1.241	1.542
2.0	1.053	1.139	1.199	1.756

These figures are plotted in Fig 2. The reader will note the striking variation in p in particular. Thus, for example if the total pressure were, say, 12,000 p.s.f. the static pressure in the vortex would increase from 6286 at r_i to 10,000 p.s.f. at $r/r_1 = 2.0$

The significance of including r/σ in the above table is that the mass flow rate Q is given by:—

$$Q = 2\pi \rho_T \int_{r_1}^{r_2} \frac{r u_a}{\sigma} \, dr \quad \text{where } \rho_T \text{ is the stagnation density.}$$

If, as in the free vortex, u_a is const then

$$Q = 2\pi \rho_T \, u_a \int_{r_1}^{r_2} \frac{r}{\sigma} \, dr$$

As may be seen from Fig. 2, $\frac{r}{\sigma}$ is virtually a straight line for a free vortex, so there would be little error in writing

$$Q = 2\pi \rho_T \, u_a \frac{r_m}{\sigma_m} (r_2 - r_1)$$

SECTION 4
A Method of Dealing with Shock Waves in Air

Part 1 — Normal Shock Waves

Introductory

With the advent of flight at supersonic speeds and the increasing trend towards supersonic relative velocities in turbo machinery blading, some knowledge of shock wave phenomena is essential.

It is not intended herein to prove that *very small* pressure disturbances in a compressible fluid travel at the acoustic speed for the fluid. It will be assumed that the reader is familiar with the fact that the acoustic speed u_c is given by $u_c = \sqrt{g\gamma RT}$ which, for air, becomes $u_c = 65.79\sqrt{T}$ where T is the static temperature in °K and u_c is in ft/sec. Thus at 288°K $u_c = 1116.5$ ft/sec.

With other than small pressure disturbances the speed of propagation exceeds the acoustic speed as is obvious from the fact that the portion of the shock wave immediately ahead of the nose (i.e. a 'detached shock') of a body travelling at supersonic speed must be travelling at the same speed as the body. Fig. 1 shows the kind of shock wave ahead of a blunt nosed supersonic missile, and Fig. 2 shows the kind of shock wave ahead of, say, a pitot tube of a supersonic aircraft.

When concerned with supersonic flight in the stratosphere, the temperatures involved are (except at Mach numbers well above 3.0) within the range where there is negligible error in assuming that the specific heat is constant. For shock waves in turbines operating at much higher temperatures, however, accuracy would require that γ, K_p, etc., would need to be adjusted accordingly, but for present purposes it will be assumed that K_p is constant at 336 ft lbs/lb/°K and $\gamma = 1.4$; $\dfrac{\gamma}{\gamma - 1} = 3.5$, etc.

The velocity u relative to the body is not affected until the shock front is crossed.

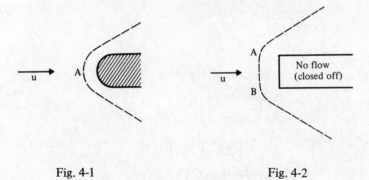

Fig. 4-1 Fig. 4-2

At the point A in Fig. 1 and from A to B in Fig. 2 the shock wave is normal to the axis of the body. At points further from the axis, the wave curves as shown and (if the body of Fig. 1 is a solid of revolution and that of Fig. 2 is a closed off parallel tube—or solid cylinder) becomes an 'oblique shock' which tends to a conical shock front, eventually becoming a conical shock front, when the pressure differential across it becomes very small. The half angle of the cone will then be the 'Mach angle' α_M where $\alpha_M = \text{cosec}^{-1} M$ where M is the Mach number u/u_c. Thus for $M = 2$, i.e. $u = 2u_c$, $\alpha = \sin^{-1} 0.5 = 30°$.

It is not the writer's intention to cover the whole range of shock wave phenomena but only to indicate a method of dealing with them.

Fig. 3 represents part AB of a normal shock which is stationary relative to an observer and across which velocity drops sharply from u to zu (where z is less than unity) and there is an associated rise in static pressure, static temperature and density. The change will be adiabatic but not isentropic. However, total temperature will remain unchanged.

Static Temp. = T_{s0}
Static Press. = P_{s0}
Density = ρ_0
Total Temp. = T

$\theta_0 = \dfrac{u_0^2}{2gK_p} = \left(\dfrac{u}{147.1}\right)^2$

$\phi_0 = \dfrac{\theta_0}{T_{s0}}$

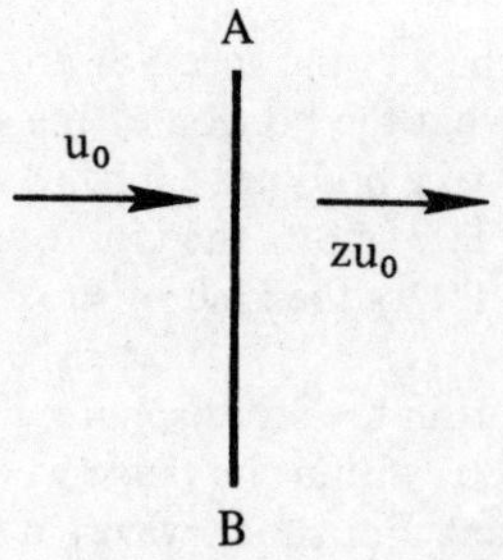

Static Temp. = T_{s1}
Static Press. = P_{s1}
Density = ρ_1
Total Temp. = T

$\theta_1 = \dfrac{z^2 u_0^2}{2gK_p} = z^2 \theta_0 = z^2 \left(\dfrac{u}{147.1}\right)^2$

$\phi_1 = \dfrac{\theta_1}{T_{s0}} = z^2 \phi_0$

Fig 4-3

The notation used and the more simple relationships are indicated in Fig. 3. (*It should be noted that ϕ is here used as a multiple of T_{s0} and not as a multiple of total temperature T as hitherto*).

The increase of entropy which occurs is best explained by visualising AB as a very thin diaphragm across which high speed molecules can 'escape' from right to left and, by collision, raise the internal energy immediately ahead of AB, and vice versa. Also radiation from the higher temperature fluid to the right can increase the internal energy to the left and vice versa (probably negligible at temperatures of the order of 300°K).

For the solution of this problem, four equations are necessary, namely those for 1) continuity, 2) energy, 3) the gas equation and 4) the momentum equation.

From Continuity: The flux density is unchanged, i.e. $u_0 \rho_0 = z u_0 \rho_1$ $\therefore$ $\dfrac{\rho_1}{\rho_0} = \dfrac{1}{z}$ (4-1)

From Energy: $T = T_{s0} + \theta_0 = T_{s1} + \theta_1 = T_{s1} + z^2 \theta_0$ (4-2)

from which $1 + \phi_0 = \dfrac{T_{s1}}{T_{s0}} + z^2 \phi_0$ or $\dfrac{T_{s1}}{T_{s0}} = 1 + \phi_0(1 - z^2)$ (4-3)

From the Gas Equation: $\dfrac{P_{s1}}{P_{s0}} = \dfrac{\rho_{s1}}{\rho_{s0}} \dfrac{T_{s1}}{T_{s0}}$ or, by substitution from (1) and (3)

$$\dfrac{P_{s1}}{P_{s0}} = \dfrac{1}{z}[1 + \phi_0(1 - z^2)]$$ (4-4)

From Momentum: $P_{s1} - P_{s0} = \dfrac{\psi}{g} u_0(1 - z)$ or, since $\psi = \rho_0 u_0 = \dfrac{P_{s0} u_0}{RT_{s0}}$

$$P_{s1} - P_{s0} = \dfrac{P_{s0} u_0^2}{R g T_{s0}}(1 - z) = P_{s0}\dfrac{2K_p u_0^2}{2gRK_p T_{s0}}(1 - z)$$

$$= P_{s0}\dfrac{2K_p}{R}\dfrac{\theta_0}{T_{s0}}(1 - z)$$

or, since $\dfrac{2K_p}{R} = 7$ $\left(\text{from } \dfrac{K_p}{R} = \dfrac{\gamma}{\gamma - 1} = 3.5\right)$ and $\dfrac{\theta_0}{T_{s0}} = \phi_0$

$$P_{s1} - P_{s0} = 7 P_{s0} \phi_0 (1 - z) \quad \text{or} \quad \dfrac{P_{s1}}{P_{s0}} - 1 = 7\phi_0(1 - z)$$

or

$$\frac{P_{s1}}{P_{so}} = 1 + 7\phi_0 (1-z) \tag{4-5}$$

∴ from equations (4) and (5), $\frac{1}{z} [(1 + \phi_0 (1-z^2))] = 1 + 7\phi_0 (1-z)$

or

$$1 + \phi_0 (1-z^2) = z[1 + 7\phi_0 (1-z)] \tag{4-6}$$

Equation (6) reduces to $\phi_0 = \frac{1}{6z-1}$ \hfill (4-6a)

It should be noted that when z = unity $\phi_0 = \frac{1}{5}$, i.e. $\frac{\theta_0}{T_{so}} = \frac{1}{5}$ which is the condition for $u_0 = u_c$. (Note that if $z > 1.0$, ϕ_0 is less than $1/5$, i.e. u_0 is subsonic and no shock wave will exist.)

Alternatively $z = \frac{1}{6}\left(1 + \frac{1}{\phi_0}\right)$ or $\frac{1}{z} = \frac{6\phi_0}{1 + \phi_0}$ \hfill (4-6b)

Hence by substitution in equations (1), (3), and (4) or (5) $\frac{\rho_1}{\rho_0}; \frac{T_{s1}}{T_{so}}$ and $\frac{P_{s1}}{P_{so}}$ may be found in terms of ϕ_0.

Thus, substituting for z from (6b), equation (1) becomes:

$$\frac{\rho_1}{\rho_0} = \frac{1}{z} = \frac{6\phi_0}{1 + \phi_0} \tag{4-7}$$

Equation (3) becomes:

$$\frac{T_{s1}}{T_{so}} = 1 + \phi_0 (1-z^2) = 1 + \phi_0 \left[1 - \frac{1}{36}\left(\frac{\phi_0 + 1}{\phi_0}\right)^2\right]$$

∴ $$\frac{T_{s1}}{T_{so}} = 1 + \phi_0 - \frac{(\phi_0 + 1)^2}{36\phi_0} \tag{4-8}$$

Equation (5) becomes:

$$\frac{P_{s1}}{P_{so}} = 1 + 7\phi_0 (1-z) = 1 + 7\phi_0 \left[1 - \frac{1}{6}\left(1 + \frac{1}{\phi_0}\right)\right]$$

or

$$\frac{P_{s1}}{P_{so}} = 1 + 7\phi_0 \left(1 - \frac{\phi_0 + 1}{6\phi_0}\right) = 1 + 7\phi_0 - \frac{7}{6}\phi_0 - \frac{7}{6}$$

or

$$\frac{P_{s1}}{P_{so}} = \frac{1}{6}(35\phi_0 - 1) \tag{4-9}$$

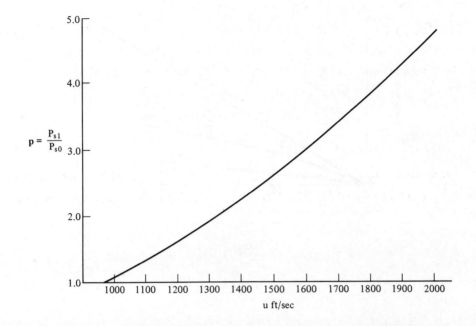

Fig. 4-4

Since $\phi_0 \equiv \dfrac{\theta_0}{T_{s0}} \equiv \dfrac{1}{T_{s0}} \left(\dfrac{u_0}{147.1}\right)^2$, equations (7), (8) and (9) provide a ready connection between density ratio, temperature ratio and pressure ratio and velocity u_0. In fact, since from the Gas Equation, $\dfrac{P_{s1}}{P_{s0}} = \dfrac{\rho_{s1}}{\rho_{s0}} \dfrac{T_{s1}}{T_{s0}}$, any two of the ratios can be used to find the third.

Fig. 4 shows the relationship between the pressure ratio across a normal shock and flight speed u for an air temperature of 220°K. This curve is, of course, readily calculated from equation (9).

Fig. 5 shows curves for pressure, density, and temperature ratios in non-dimensional form over the range $\phi = 0.2$ (Mach 1.0) to 1.0.

Example 4-1

What are the velocity, pressure, temperature and density changes across a normal shock ahead of a bluff body travelling at 2000 ft/sec through air having a static temperature of 220°K (i.e. in the stratosphere)?

For $u = 2000$, $\theta_0 = \left(\dfrac{2000}{147.1}\right)^2 = 184.9°C$ $\therefore$ $\phi_0 = \dfrac{184.9}{220} = 0.84$

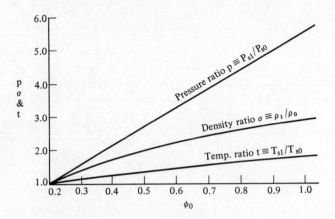

Fig. 4-5

$$\therefore \quad z = \frac{1}{6}\left(1 + \frac{1}{.84}\right) = .365$$

∴ the velocity after the shock is zu = .365 × 2000 = 730.2 ft/sec.

The total temp. T_T relative to the body is $T_T = 220 + 184.9 = 404.9°K$. T_T is the same before and after the shock. The kinetic temp. after the shock $\theta_1 = \left(\frac{730.2}{147.1}\right)^2 = 24.6°C$.
∴ the static temp. $T_{s2} = T_T - \theta_1 = 404.9 - 24.6 = 380.3$ ∴ the static temp. rise $\Delta T_s = 380.3 - 220 = 160.3°C$ and $\frac{T_{s1}}{T_{s0}} = \frac{380.3}{220} = 1.728$.

From equation (1) $\frac{\rho_1}{\rho_0} = \frac{1}{z} = \frac{1}{.365} = 2.74$

$$\frac{P_{s1}}{P_{s0}} = \frac{\rho_1}{\rho_0}\frac{T_{s1}}{T_{s0}} = 2.74 \times 1.728 = 4.734.$$

Referring to Fig. 5, it is obvious from equation (9) that there is a straight line relationship between p and ϕ_0. Also, as may be seen, there is a near straight line relationship between t and ϕ_0. Even when t is drawn at 10 times the scale the curve is very nearly straight at values of ϕ_0 greater than 0.6.

Example 4-2

If the pressure ratio across a normal shock is 10.0, what is the wave velocity u_0 when $T_{so} = 220°K$?

This problem is very simply solved by re-arranging equation (9) into the form

$$\phi_0 = \frac{6p + 1}{35} \quad \text{from which} \quad \phi_0 = \frac{61}{35} \equiv 1.7429$$

∴ $\theta_0 = 1.7429 \times 220 = 383.4°C$ for which $u_0 = 147.1\sqrt{383.4} = 2880$ ft/sec, i.e. nearly three times acoustic speed at 220°K.

In Fig. 6, z is plotted against ϕ_0 over the range 0.2 to 1.0 and brings out very clearly the very large reduction of kinetic energy which can occur across a normal shock wave. Thus, at $\phi_0 = 1.0$, $z = \frac{1}{3}$, i.e. the velocity after the shock is 33.3% of u_0. This corresponds

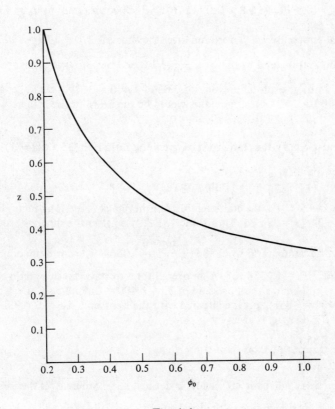

Fig. 4-6

to a reduction of kinetic energy of 89% approximately. This, however, entails a considerable conversion of k.e. into internal energy; i.e. a loss. This matter will be discussed below.

From $z = \dfrac{1 + \phi_0}{6\phi_0}$ it is evident that the asymptotic limit of z is $\dfrac{1}{6}$ with the assumption of constant K_p. In practice, at such high values of ϕ_0 one would be dealing with plasmas rather than a 'perfect gas.'

The Efficiency of Shock Compression

The efficiency of engines, gears, etc. is easily defined as $\dfrac{\text{energy out}}{\text{energy in}}$, but the efficiency of shock-compression is not so easily defined, and there is some inconsistency among engineers in definition. Some talk of 'pressure efficiency,' others of 'adiabatic efficiency' and so on. The difficulty is, perhaps, best illustrated by an example.

Suppose $\phi_0 = 0.8$ and T_0, the static temperature before the shock, is 220°K and P_0, the static pressure before the shock, is 400 psf, then we have the following figures:

From $\phi_0 = 0.8$, $\theta_{uo} = 0.8 \times 220 = 176°C$ which corresponds to $u_0 = 1951.5$ ft/sec.

The total temperature T before and after the shock is $220 + 176 = 396°K$. ∴ the isentropic temp. ratio (total to static) is $\dfrac{396}{220} = 1.8$ and the corresponding pressure ratio is $1.8^{3.5} = 7.8244$ to give a total pressure of $7.8244 \times 400 = 3129.8$ p.s.f., i.e. a pressure rise of $3129.8 - 400 = 2729.8$ p.s.f., i.e. this would be the pressure rise if there were no losses.

From equation (9), the static to static pressure ratio is $\dfrac{1}{6}(35 \times 0.8 - 1) = 4.5$ and from equation (7) $z = \dfrac{1.8}{4.8} = 0.375$ to give $u_1 = 0.375 \times 1951.5 = 731.8$ ft/sec which is equivalent to 24.75°C, i.e. the static temp. after the shock is the total temp. less 24.75°, namely $396 - 24.75 = 371.25$. Thus if the 731.8 ft/sec after the shock is brought to rest isentropically, the temp. ratio would be $\dfrac{396}{371.25} = 1.0667$. The corresponding pressure ratio would be $1.0667^{3.5} = 1.2534$ to give an overall total to static pressure ratio of $1.2534 \times 4.5 = 5.64$ and a total pressure of $5.64 \times 400 = 2256.2$ p.s.f., i.e. a pressure rise of $2256.2 - 400 = 1856.2$ p.s.f. compared with the isentropic rise of 2729.8 p.s.f. Thus the pressure efficiency η_p is given by

$$\eta_p = \dfrac{1856.2}{2729.8} = 0.68 \quad \text{or} \quad 68\%.$$

In compressors 'adiabatic efficiency' is defined as $\dfrac{\theta'_c}{\theta_c}$ where θ'_c is the isentropic temperature rise corresponding to the total to static pressure ratio and θ_c is the actual temperature rise static to total. This definition is imperfect in that a proportion of the losses (converted into internal energy) are recoverable on re-expansion. Alternatively, in a thermal cycle, the heat which must be added to achieve a given maximum cycle temperature is reduced by the amount of internal energy resulting from losses. Nevertheless, 'adiabatic efficiency' is a very convenient definition for cycle calculations.

Since shock compression is of greatest interest from the point of view of supersonic aircraft intakes where it contributes substantially to the overall compression, it seems desirable to use adiabatic efficiency for shock compression. Thus, in the above example the actual pressure ratio was found to be 5.64. The corresponding isentropic temperature ratio t' is $5.64^{0.2857} = 1.639$. $\therefore \theta' = 1.639 \times 220 - 220 = 140.63°C$. The actual temperature rise θ was found to be $176°C$. $\therefore$ the adiabatic efficiency η_c is

$$\eta_c = \frac{140.63}{176} = 0.799 \text{ or } 79.9\%,$$ i.e. a much higher figure than pressure efficiency.

The conversion of kinetic energy into internal energy on passage through the shock is clearly $176 - 140.63 = 35.87°C$ which is 20% of the initial kinetic energy.

Example 4-3

If $T_0 = 220°K$ and $\phi_0 = 0.4$ what is η_c if there are no further losses after a normal shock wave?

$\theta_{u0} = .4 \times 220 = 88°C$. This is the actual temperature rise static to total. From equation (9), the static to static pressure ratio is $\frac{1}{6}(35 \times 0.4 - 1) = 2.1667$. The initial velocity $u_0 = 147.1\sqrt{88} = 1{,}379.9$ ft/sec.

From equation (7), $z = \frac{1.4}{2.4} = 0.5833$ $\therefore$ the velocity $zu_0 = .5833 \times 1379.9 = 804.9$ ft/sec the temperature equivalent of which is $29.94°C$ $\therefore$ the isentropic temperature ratio involved in reducing zu_0 to zero is $\frac{220 + 88}{220 + 88 - 29.94} = 1.1077$ for which the corresponding pressure ratio is $1.1071^{3.5} = 1.4304$ to give a total pressure ratio of $1.4304 \times 2.1667 = 3.0993$. The corresponding value of t' is $3.0938^{0.2857} = 1.3815$ to give a θ'_c of $.3815 \times 220 = 83.93°C$.

$$\therefore \eta_c = \frac{83.93}{88} = 0.954 \text{ or } 95.4\%.$$

It may thus be seen that the greater the 'strength' of a shock wave the lower the adiabatic efficiency. In fact if ϕ_0 is only a little above its lowest limit of 0.2 the adiabatic efficiency is virtually 100%.

It should be noted that, since $\phi_0 = \frac{1}{5}$ corresponds to the acoustic speed, ϕ_0 cannot be less than 0.2 for a shock wave to exist. It follows from equation (6a) that z cannot be greater than unity (corresponding to $\phi_0 = 0.2$) therefore a rarefaction shock wave cannot exist. (This point is usually made by invoking the second law of thermodynamics. It can also be made by pointing out that if, in Fig. 3, the arrows were reversed, though the continuity, energy, gas and momentum equations would still be satisfied, there would be a decrease in entropy which is impossible in an adiabatic process—a statement which amounts to "invoking the second law . . . .")

Summary

$z \equiv \dfrac{u_1}{u_0}$; $\theta_{u0} \equiv \dfrac{u_0^2}{2gK_p}$; $\phi_0 \equiv \dfrac{\theta_{u0}}{T_{s0}}$; P_{s0}, ρ_0 and T_{s0} are the static pressure, density and temperature ahead of the shock wave. P_{s1}, ρ_1 and T_{s1} are the static pressure, density and temperature after the shock wave. T is the total temperature.

$T = T_{s0} + \theta_{u0} = T_{s1} + \theta_{u1}$

$\sigma \equiv \rho_1/\rho_0 = 1/z$ or, eliminating z, $\sigma = \dfrac{6\phi_0}{1 + \phi_0}$

$t \equiv T_{s1}/T_{s0} = 1 + \phi_0 (1 - z^2)$ or, eliminating z, $t = 1 + \phi_0 - \dfrac{(\phi_0 + 1)^2}{36\phi_0}$

$p \equiv P_{s1}/P_{s0} = 1 + 7\phi_0 (1 - z)$ or, eliminating z, $p = \dfrac{1}{6}(35\phi_0 - 1)$

$\phi_0 = \dfrac{1}{6z - 1}$ or $z = \dfrac{1 + \phi_0}{6\phi_0}$ or, eliminating z, $\phi_0 = \dfrac{6p + 1}{35}$

Part 2. Oblique Shock Waves

Introductory

The discussion herein is limited to two dimensional cases.

The key to the theory of oblique shock waves is that an oblique shock is a normal shock travelling in a direction perpendicular to the oblique shock.

Discussion

Fig. 7 illustrates a sharp wedge of angle β with its upper surface parallel to the direction of u_0 and generating an oblique shock AB, the angle between AB and the under surface being α.

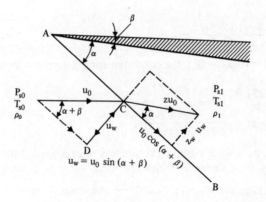

Fig 4-7

Treating AB as a normal shock travelling in the direction CD with velocity u_w one may see that $u_w = u_0 \sin(\alpha + \beta)$ (4-10)

and that $\dfrac{z_w u_w}{u_0 \cos(\alpha + \beta)} = \tan \alpha$ (4-11)

From equation (6b) one may also see that $z_w = \dfrac{1}{6}\left(1 + \dfrac{1}{\phi_w}\right)$

or

$$z_w = \frac{\phi_w + 1}{6\phi_w} \qquad (4\text{-}12)$$

By using these three equations the relationship between ϕ_0, α and β may be found as follows:

From (10), $\left(\dfrac{u_w}{u_0}\right)^2 = \dfrac{\phi_w}{\phi_0} = \sin^2(\alpha + \beta) \therefore \phi_w = \phi_0 \sin^2(\alpha + \beta)$ (4-13)

From (10) and (11), $z_w \tan(\alpha + \beta) = \tan \alpha$,

i.e. $z_w = \dfrac{\tan \alpha}{\tan(\alpha + \beta)}$ (4-14)

From (12) and (14), $\dfrac{\phi_w + 1}{6\phi_w} = \dfrac{\tan \alpha}{\tan(\alpha + \beta)}$

or

$6\phi_w \tan \alpha = \phi_w \tan(\alpha + \beta) + \tan(\alpha + \beta)$

from which $\phi_w(6 \tan \alpha - \tan(\alpha + \beta)) = \tan(\alpha + \beta)$

or

$$\phi_w = \frac{\tan(\alpha+\beta)}{6\tan\alpha - \tan(\alpha+\beta)} \tag{4-15}$$

(Note that if $\beta = 0$ $\phi_w = 0.2$, the condition for the acoustic speed.)

Hence, using (13),

$$\phi_0 \sin^2(\alpha+\beta) = \frac{\tan(\alpha+\beta)}{6\tan\alpha - \tan(\alpha+\beta)} \tag{4-16}$$

Given α and β, ϕ_0 is readily found though some 'juggling' is still necessary to find α if ϕ_0 and β are given. (The writer finds that, for small values of β, then if an initial guess at α is a little larger than the Mach angle $\sin^{-1}\frac{u_c}{u_0}$ then two other trials are sufficient to find the value of α which satisfies (16).)

Equation (16) can, of course, be manipulated into many forms but it seems very doubtful that any of them show any advantage over (16).

For the density, temperature, and pressure ratios across an oblique shock it remains to adapt equations (7), (8), (9), etc., by substituting ϕ_w for ϕ_0.

One could substitute further for ϕ_w from (15). This will clearly lead to some distinctly clumsy equations of doubtful value and so will not be attempted herein. One would clearly prefer to determine ϕ_w from (15) and use it in equations (7), (8), (9), etc., instead of ϕ_0.

For ease of reference the oblique shock wave relationships are given below.

$$T = T_{s0} + \theta_{u0} = T_{s1} + \theta_{u1} \tag{4-17}$$

$$z_w = \frac{1+\phi_w}{6\phi_w} \tag{4-18}$$

From Fig. 7 $\quad z = \dfrac{\cos(\alpha+\beta)}{\cos\alpha}$ \hfill (4-19)

$$\sigma \equiv \rho_1/\rho_0 = \frac{1}{z_w} = \frac{6\phi_w}{1+\phi_w} \tag{4-20}$$

$$t \equiv T_{s1}/T_{s0} = 1 + \phi_w(1 - z_w^2) = 1 + \phi_w - \frac{(\phi_w+1)^2}{36\phi_w} \tag{4-21}$$

$$p \equiv \frac{P_{s1}}{P_{s0}} = 1 + 7\phi_w(1-z_w) = \frac{1}{6}(35\phi_w - 1) \tag{4-22}$$

Example 4-4

If $\beta = 5°$ what must u_0 be to give $\alpha = 30°$ assuming $T_{s0} = 220°K$?

From (16) $\phi_0 \sin^2 35 = \dfrac{\tan 35}{6 \tan 30 - \tan 35}$

$\sin 35 = .5736 \quad \tan 30 = .5774 \quad \text{and} \quad \tan 35 = .7002$

$\therefore \phi_0 (.5736)^2 = \dfrac{.7002}{6 \times .5774 - .7002} = .2533$

$\therefore \phi_0 = \dfrac{.2533}{.5736^2} = 0.77$

$\therefore \theta_{uo} = .77 \times 220 = 169.4°C \quad \therefore u_0 = 147.1 \sqrt{169.4} = 1914.5 \text{ ft/sec}.$

Example 4-5

Using the values of α, β and T_{so} of Example 4 what are the values of σ, t and p?

From equation (15) $\phi_w = \dfrac{.7002}{6 \times .5774 - .7002} = .2533$

From equation (22) $p = \dfrac{1}{6}(35 \times .2533 - 1) = 1.3109$

From equation (20) $\sigma = \dfrac{6 \times .2533}{1.2533} = 1.2126$

$t = \dfrac{p}{\sigma} = \dfrac{1.3109}{1.2126} = 1.081$

Example 4-6

Using the data of Examples 4 and 5 what is the velocity reduction through the shock wave?

From equation (19) $z = \dfrac{\cos 35}{\cos 30} = \dfrac{.8192}{.866} = .9459 \quad \therefore$, since in Example 4, u_0 was found to be 1914.5 ft/sec $u_1 = .9459 \times 1914.5 = 1810.9$ ft/sec, i.e. a reduction of 103.6 ft/sec.

Example 4-7

Using the figures of the foregoing examples what is the loss (i.e. conversion to internal energy) as a percentage of the kinetic energy reduction?

In Example 5 the pressure ratio (static to static) was found to be 1.3109 for which the corresponding isentropic temperature ratio is $1.3109^{0.2857} = 1.0804$ to give an isentropic temperature increase of $0.0804 \times 220 = 17.69°C$. The actual temperature ratio was found to be 1.081, to give an actual temperature increase of $0.081 \times 220 = 17.82°C$. ∴ the conversion of part of the kinetic energy to internal energy is only $17.82 - 17.69 = 0.13°C$.

The velocity reduction was found to be from 1914.5 ft/sec ($\theta_{u0} = 169.4°C$) to 1810.9 ft/sec ($\theta_{u1} = 151.55°C$), i.e. the temperature equivalent of the kinetic energy reduction is $169.4 - 151.55 = 17.85°C$. Thus, the proportion of conversion to internal energy is $\frac{0.13}{17.85} = .0073$ or 0.73%. So, in the very unlikely event that subsequent conversion would entail negligible loss, the adiabatic efficiency of the shock compression is over 99%. This points to the obvious conclusion that a succession of weak oblique shocks can give high compression efficiency. Indeed, as will be demonstrated below, it should be possible to approximate very closely to isentropic compression at least until the velocity is reduced to the acoustic velocity corresponding to $\frac{T°}{6}$ C. (T being total temperature.)

Fig. 8 illustrates a sequence of oblique shocks from a two-dimensional surface ABCDEF for a speed u = 2000 ft/sec. with a static temperature of 220°K at station 0.

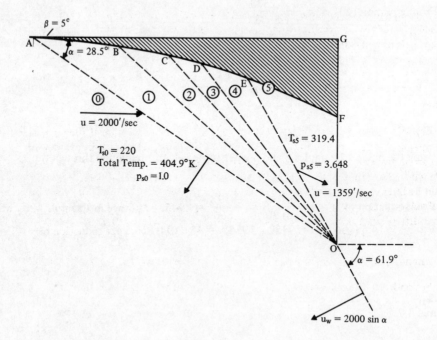

Fig. 4-8

(Total temperature $= 220 + \left(\frac{2000}{147.1}\right)^2 = 404.9°K$.) The initial wedge angle at A is 5° and there are 5° increments at BCD & E. Between E and F increments of slope are not specified but would need to be very much less than 5° for the shocks to pass through 0.

It is no coincidence that all waves focus at 0. The distances AB, BC, etc., were chosen to achieve this and avoid the complexities of wave junctions at other points.

If there were no bounding surface, all waves coalescing at O would combine into a single shock as shown.

The pressure, temperature ratios across shocks OA, OB, etc., are readily calculated by the methods indicated above. The overall pressure ratio and static temp. after OE are indicated in Fig. 8. The efficiency of compression from A to E is found to be very high. The isentropic temperature ratio corresponding to $P_{ss} = 3.648$ is $3.648^{0.2857} = 1.4474$ to give an isentropic temperature increase of $0.4474 \times 220 = 98.42°C$. The actual temperature rise is $319.4 - 220 = 99.4$ $\therefore$ the loss to internal energy is $0.98°C$. The kinetic energy conversion is $\left(\frac{2000}{147.1}\right)^2 - \left(\frac{1359}{147.1}\right)^2 = 184.86 - 85.35 = 99.51$ so that the loss to internal energy is less than 1.0%. The efficiency across the combined shock beyond O, however, would be distinctly lower.

Fig. 8 represents an idealised case in that it takes no account of the inevitable boundary layer which viscosity would cause along the surface AF and which would be subjected to a substantial adverse pressure jump at each 'kink'. This boundary layer would be very thin and, though it would mean a small change in the effective profile of AF, the tendency to reverse flow at the surface, i.e. 'breakaway,' would be countered by the high velocities in the outer layers of the boundary layer. Nevertheless AF should really represent the 'surface' of the boundary layer film.

It will be obvious from Fig. 8 that AF could be a smooth curve. If such were the case one may visualise an infinite number of oblique shocks each of infinitesimally small strength except at A (and even at A if the curve had zero slope at A). In reality there would be smooth rise of pressure downstream of AO and the 'rays' BO, CO, etc., would be lines of equal pressure, i.e. isobars, of equal temperature, equal density and equal velocity. It seems appropriate that they be called 'isostats' (there being no probability of confusion with its geological use in 'isostatic', 'isostasy', etc.) implying lines of equal state since every property of the fluid, including entropy and enthalpy will be uniform along them. It follows that $\frac{dP_s}{P_s} \propto \frac{dT_s}{T_s} \propto \frac{d\rho}{\rho}$ etc., from which it further follows that beyond AO, the compression is isentropic, at least until u falls to the local value of u_c ($\equiv u_{cF}$).

The geometry of the situation is represented in Fig. 9.

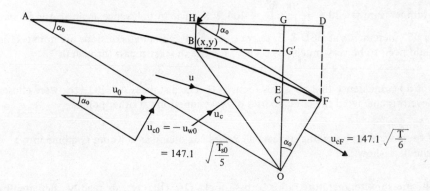

T_{so}

$T = T_{so} + \left(\dfrac{u_c}{147.1}\right)^2$

ρ_0

$\sin \alpha_0 = \dfrac{u_{c0}}{u_0}$

$AG = x_0$

$GO = y_0$

$AO = r_0 = \dfrac{x_0}{\cos \alpha_0}$

$OF = r_F = \left(\dfrac{T_{so}}{T_{sF}}\right)^3 r_0$

$GD = CF = r_F \sin \alpha_0$

$OB = r$

At OF, $T_{sF} = \dfrac{5}{6} T$ and $\rho_F = \left(\dfrac{T_{sF}}{T_{so}}\right)^{2.5}$

Fig. 4-9

Neglecting the boundary layer, AF may be described as an 'isentropic surface.'

Taking A as the origin of co-ordinates, HB ≡ y and AH ≡ x, so the isostat OB of length r ends on the isostatic surface at xy. So, to locate B, one must either find x and y, or, using polar co-ordinates, the length r and the angle BOA.

Since the compression beyond AO is assumed isentropic the relationships of isentropic compression apply, i.e. $\rho/\rho_0 = \left(\dfrac{T_s}{T_{so}}\right)^{2.5}$; $\dfrac{P_s}{P_{so}} = \left(\dfrac{T_s}{T_{so}}\right)^{3.5}$ etc.

The scale of the diagram is one of choice. x_0, for example, may be set at ten units of length.

Assuming that $\beta_0 = 0$ then α_0 is the Mach angle for u_0, i.e. $\sin^{-1} \frac{u_{co}}{u_0}$

where $\frac{u_{co}}{u_0} = \sqrt{\frac{\theta_{co}}{\theta_{uo}}} = \sqrt{\frac{T_{so}}{5\theta_{uo}}} = \sqrt{\frac{1}{5\phi_0}}$ from which r_0, the point O and y_0 are readily determined. Evidently these are dependent on ϕ_0 which means that the isentropic surface varies with ϕ_0.

One may see that $x_0 = r_0 \cos \alpha_0$ and $y_0 = r_0 \sin \alpha_0$. These determine the general frame of the diagram.

Information can be obtained from the requirements of continuity. Thus the flow through each isostat is the same as that flowing through AO, namely $\rho_0 u_0 y_0$ or $\rho_0 u_{co} r_0$. It follows that for any isostat of length r

$$\rho_0 u_{co} r_0 = \rho u_c r \qquad (4\text{-}23)$$

Using the fact that the total temperature T is constant and given by $T = T_{so} + \theta_{co}$, one may see that the static temperature T_s along an isostat such as OB may be obtained from $\theta_u = T_{so} + \theta_{uo} - T_s$ where θ_u is the temperature of the (uniform) velocity u across OB. (Note that the direction of u is parallel to the tangent at B.)

Equation (23) may be manipulated as follows:

$$\frac{r}{r_0} = \frac{\rho_0}{\rho} \frac{u_{co}}{u_c} = \left(\frac{T_{so}}{T_s}\right)^{2.5} \sqrt{\frac{\theta_{co}}{\theta_c}} = \left(\frac{T_{so}}{T_s}\right)^{2.5} \sqrt{\frac{T_{so}}{5} \times \frac{5}{T_s}} = \left(\frac{T_{so}}{T_s}\right)^3 \qquad (4\text{-}24)$$

(Remembering that the temperature equivalent of the critical velocity is $0.2 \times T_s$.)

Hence, for given values of r_0 and T_{so} there is a very simple relationship between r and T_s. For any selected value of T_s the length r is readily determined but, unfortunately, not its direction except for OE and OF.

For some time the writer supposed that Fig. 9 contained enough information to find the shape of the isostatic surface but, after many fruitless hours, was forced to the conclusion that it is indeterminate. It must, however, satisfy the following conditions:

1) If $\beta = 0$ then AG is the tangent at A.
2) The slope must continually increase from A to F.
3) The tangent at F must be parallel to AO.
4) The length r_F of OF must be consistent with (24).

For the last, T_{sF}, the static temperature at OF, is determined by the fact that the critical velocity u_{cF} across OF coincides with the actual velocity

$$\therefore \theta_{ucF} = \frac{T}{6}, \text{ hence } T_{sF} = T_0 + \theta_{uc} - \frac{T}{6} \text{ or } T_{sF} = \frac{5}{6} T$$

Example 4-8

What is the length of OF given that x_0 is 100 units of length, $T_{s0} = 220°K$ and $u_0 = 1600$ ft/sec?

The temperature equivalent of 1600 ft/sec is $\left(\dfrac{1600}{147.1}\right)^2 = 118.3°C$

∴ the total temperature $T = 220 + 118.3 = 338.3°K$

$$\text{Sin } \alpha_0 = \frac{u_{c0}}{u_0} \text{ and } \theta_{uc0} = \frac{T_{s0}}{5} = 44°C \therefore u_{c0} = 147.1\sqrt{44} = 975.8 \text{ ft/sec}$$

$$\therefore \text{Sin } \alpha_0 = \frac{975.8}{1600} = 0.6098 \text{ and Cos } \alpha_0 = 0.7925$$

Since $\dfrac{\alpha_0}{r_0} = \text{Cos } \alpha_0, r_0 = \dfrac{x_0}{\text{Cos } \alpha_0} = \dfrac{100}{0.7925} = 126.18$ units

At OF, $T_{sF} = \dfrac{5}{6}T = \dfrac{5}{6} \times 338.8 = 281.7°K$

$\therefore$ from (24) (using $r_F \equiv OF$) $\dfrac{r_F}{T_0} = \left(\dfrac{T_{s0}}{T_{sF}}\right)^3 = \left(\dfrac{220}{281.7}\right)^3 = 0.4765$

$\therefore r_F = 0.4765 \ r_0 = 0.4765 \times 126.18 = 60.12$ units of length.

To make the isentropic surface determinate it is necessary to specify a relationship between two of the several variables. For example one might decide that T_s should rise in proportion to x. More simply it might be specified that $\dfrac{dy}{dx} = kx$. This, of course, defines the parabola $y = \dfrac{k}{2} x^2$. AF thus becomes a segment of a parabola where $\dfrac{dy}{dx}$ at F is Tan α_0.

At F, $x = x_0 + r_F \text{ Sin } \alpha_0$

$\therefore$ at F, $\dfrac{dy}{dx} \equiv \text{Tan } \alpha_0 = k(x_0 + r_F \text{ Sin } \alpha_0)$, which gives the value of k, namely

$$k = \frac{\text{Tan } \alpha_0}{x_0 + r_F \text{ Sin } \alpha_0}$$

Example 4-9

Using figures from the preceding example and assuming that AF is a segment of the parabola $y = \dfrac{1}{2}\left(\dfrac{\text{Tan } \alpha_0}{x_0 + r_F \text{ Sin } \alpha_0}\right)x^2$, what are the values at B of r; T_s; u; u_c; ρ/ρ_0 and P/P_0 if, in Fig. 9, the value of x at B is 60 units of length?

$$\text{Tan } \alpha_0 = \frac{\text{Sin } \alpha_0}{\text{Cos } \alpha_0} = \frac{0.6098}{0.7925} = 0.7695 \quad \therefore \text{ at B,}$$

$$y = \frac{3600}{2}\left(\frac{.7695}{100 + 60.12 \times 0.6098}\right) \text{ from which } y = 10.13. \text{ Now } y_0 = \sqrt{r_0^2 - x_0^2} = 76.95$$

$\therefore$ the length of $OG' = 76.95 - 10.13 = 66.82$ units and $BG' = 40$ units $\therefore$ the length of OB is given by $r = \sqrt{66.82^2 + 1600} = 77.88$ units.

Having found r, then, from (24) T_s is obtained from $\dfrac{r}{r_0} = \left(\dfrac{220}{T_s}\right)^3$

$$\therefore \quad \sqrt[3]{\frac{77.88}{126.18}} = \frac{220}{T_s} = .8514, \text{ from which } T_s = 258.4°K$$

All the other required quantities readily follow.

From $\theta_u = T - T_s = 338.3 - 258.4 = 79.91°C, u = 147.1\sqrt{79.91} = 1315$ ft/sec.

From $\theta_{uc} = \dfrac{T_s}{5}, u_c = 147.1\sqrt{\dfrac{258.4}{5}} = 1057.5$ ft/sec

From $\dfrac{\rho}{\rho_0} = \left(\dfrac{T_s}{T_{s0}}\right)^{2.5}, \quad \dfrac{\rho}{\rho_0}\left(\dfrac{258.4}{220}\right)^{2.5} = 1.4951$

From $\dfrac{P}{P_0} = \dfrac{\rho T_s}{\rho_0 T_{s0}}, \quad \dfrac{P}{P_0} = 1.4951 \times \dfrac{258.4}{220} = 1.7561$

Check

If the above figures are correct then, from continuity

$$u_{c0}r_0 = \frac{\rho}{\rho_0}u_c r$$

For the left hand side, $975.8 \times 126.18 = 123.126$.
For the right hand side, $1.4951 \times 1057.5 \times 77.88 = 123.134$. This is an exceptionally close agreement in view of the numbers of significant figures used in the calculations.

We have thus verified that between A and F the equation for the isentropic surface is

$y = \dfrac{x^2}{2}\left(\dfrac{\tan \alpha_0}{x_0 + r_F \sin \alpha_0}\right)$. This of course is only one of several possible surfaces, but r_F, T_{sF} and α_0 would be the same for all for the same values of x_0, T_{s0}, and u_0.

In practice a value of $\beta = 0$ at A would be impracticable so for $\beta > 0$ AO would in fact be a weak oblique shock entailing a very modest loss. Nevertheless, even allowing for boundary layer loss, the adiabatic efficiency as far as the throat OF should be over 98%.

In the case on which Fig. 9 was based ($u_0 = 2000$ ft/sec and $T_{s0} = 220°K$), $u_{cF} = 1208.3$ ft/sec which represents 36.5% of the initial kinetic energy. Further reduction of velocity beyond OF would require an increase of section, i.e. a subsonic diffuser in which, without special devices, one might expect a further loss of the order of 10% of the residual kinetic energy at OF.

It should be noted that if β_0 is greater than $0°$, then the relevant values of u_0, T_{s0}, etc., for calculating the isentropic surface are those immediately *after* the oblique shock, and that the initial isostat is some distance from the nose of the wedge at the under surface. If A' denotes the leading edge and A denotes the initial point of the first isostat then A'A is straight conforming to the wedge angle β. Also if the oblique shock and the initial isostat are to meet at O then the length of A'A is given by

$$\text{Cos A'OA} = \frac{r_s^2 + r_0^2 - (A'A)^2}{2 r_s r_0}$$ where r_s is the length of the oblique shock. The isentropic surface begins at A where $\frac{dy}{dx} = \tan \beta_0$.

The question arises, does the isentropic surface have any practical use? From the point of view of an intake for a supersonic aircraft engine, the direction of u_{cF} is undesirable. However, this might be dealt with by accepting a 'reflected' oblique shock from O, of such strength as to change the direction of flow as desired and accept the comparatively small loss which would result (because the velocity through it would not be all that much above u_{cF}). Again, a 'cascade' of curves similar to AF might have some useful function as, for example, in a supersonic compressor.

A more interesting case is that of successive reflections as shown in Fig. 10 where from A to E the slope is constant at $5°$ and BH is parallel to AO.

The effect of the initial shock AB is to deflect the flow (as shown in broken lines) parallel to AC. The reflected shock BC must be such that the flow between BC and CD is parallel to BD. The second reflection CD must be such that the flow becomes parallel to CE. The reflection DE must once more cause the flow to be parallel to BDF.

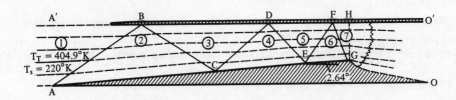

Fig. 4-10

At E and F the calculation becomes 'delicate'. Referring to (16) it is found that a curve for $\phi \sin^2 (\alpha + \beta)$ does not intercept a curve for $\dfrac{\tan (\alpha + \beta)}{6 \tan \alpha - \tan (\alpha + \beta)}$ if the slope EG is maintained at 5°. It was found necessary to reduce it to 2.64° to obtain the reflection EF to give the final reflection FG. The figures for station 7 in Fig. 10 are therefore somewhat uncertain.

Station	1	2	3	4	5	6	7
ϕ	0.840	0.699	0.575	0.465	0.365	0.268	0.201
$\alpha°$	33.4	36.9	41.15	46.9	55.9	69.4	$90_{\text{approx.}}$
$\alpha - \beta°$	28.4	31.9	36.15	41.9	50.9	66.76	$0_{\text{approx.}}$
$T_s °K$	220.0	238.2	257.1	276.4	296.6	319.3	$337_{\text{approx.}}$
$\theta_u °C$	184.9	166.5	147.7	128.5	108.3	85.6	67.9
u'/sec.	2000	1898	1788	1667	1531	1360.7	1212
p/stage	←1.3273→	←1.3012→	←1.2858→	←1.2781→	←1.2922→	←1.2058→	
σ/stage	←1.2241→	←1.2062→	←1.9662→	←1.1910→	←1.2003→	←1.1425→	
t/stage	←1.0843→	←1.0788→	←1.0749→	←1.0731→	←1.0756→	←1.0554→	
p	1.000	1.3273	1.7272	2.2209	2.8385	3.6680	4.4228

p ≡ cumulative static pressure ratio

At HG the velocity is the local acoustic velocity and so if a further velocity reduction is desired the channel must diverge to become a subsonic diffuser. But, as pointed out in Section 5, when a divergence follows a choking throat, over-expansion with acceleration occurs followed by sudden recompression through a normal shock. The lower the value of ϕ ahead of this shock the lower the loss through it, so it seems desirable to 'induce' it as near to the beginning of the divergence as possible by making the initial divergence much larger than that after the recompression shock—as indicated in Fig. 10 between G and O. The efficiency of compression from AA' to HG is remarkably high—over 99%—for an energy conversion of about 63%. So, with a small wedge angle, there would be no great error in assuming isentropic compression from AA' to HG.

Obviously the pattern of shocks shown in Fig. 10 is particular to the initial value of ϕ, i.e. to u and T_s at station 1. Moreover, if Fig. 10 represents the engine intake of a supersonic aircraft, it would be very sensitive to any change of direction of u at station 1, i.e. to yaw.

Further, some very careful design would be necessary if reflection DE is to coincide exactly with the sudden reduction of slope at E. This difficulty might be alleviated by avoiding such a sudden change of slope.

If diffusion between G' and OO' were reasonably efficient the velocity at OO' would be about 500 ft/sec. which is a 'comfortable' figure for the velocity ahead of an engine compressor intake. It is emphasised, however, that Fig. 10 is a two dimensional diagram while a compressor entry is annular.

If the line A'O' in Fig. 10 were regarded as a mirror, the solid boundary removed, and the system below it inverted above A'O as a mirror image then one would have a diagram of what happens at oblique shock wave interceptions.

Though, as we have seen, a rarefaction shock wave is impossible, the information that the pressure along a surface is dropping must be transmitted at sonic speed. Here again, the concept of a system of rays or isostats is useful. These, however, must diverge because as the pressure drops the velocity increases and the static temperature drops. The latter means a reduction of the acoustic velocity u_c, therefore as u increases and u_c decreases the Mach angle $\sin^{-1} \frac{u_c}{u}$ decreases.

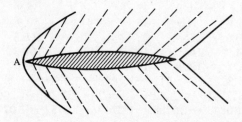

Fig. 4-11

A guess at what happens in the case of a two dimensional body is shown in Fig. 11. Being wholly convex there can be no shock waves along its length. A single nose shock would exist which would be normal at A and the point of highest pressure would be at the nose. Thereafter the pressure drops from nose to tail. There would be a tail shock of recompression to the initial ambient pressure before A.

At this point it should be emphasised that though a convex surface cannot generate a rarefaction shock, it *can* reflect a compression shock (as at the point E in Fig. 10) weakening it in the process.

The rarefaction isostats will intercept the front and rear shocks and progressively reduce their strength and therefore their angle relative to the axis of the body. Also, owing to losses across the front shock, at a sufficient distance from the body the velocity ahead of the rear shock will be less than that ahead of the front shock. It would seem, therefore, that the nose and tail shocks must eventually merge into a single oblique shock. This is contrary to the general belief that the 'double boom' often heard from supersonic aircraft is due to separate nose and tail shocks. In the opinion of the writer the second boom is more likely to be a ground reflection of the first. If this belief is correct then the time interval between the booms would depend on the observer's height from the ground, the slope of the latter, etc.

SECTION 5
Isentropic Flow Through Nozzles

The discussion of flux density in Section 2 clears the way for a brief discussion of flow through convergent and convergent-divergent nozzles neglecting losses due to wall friction (for the time being).

For simplicity, only the formulae applicable to air acting as a perfect fluid will be used.

It will be obvious that the flow dQ through a small section δS is given by $dQ = \psi \delta S$, and that if the flow is uniform across a section S, then $Q = \psi S$. Further, that the maximum value of Q cannot exceed $\psi_c S_c$ where S_c is the minimum section, and will, in fact, be less than this if the flow at S_c is non-uniform due to curvature.

Fig. 1a represents a convergent nozzle of exit area S discharging isentropically into a region, e.g. the atmosphere, where the ambient pressure is P_0, from a reservoir where total temp. and total pressure are T_T and P_T (assumed maintained constant by continuous replacement).

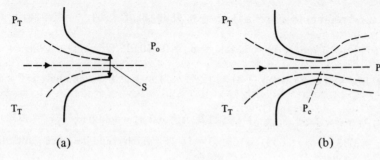

Fig. 5-1

If it be assumed that the effect of flow curvature at nozzle exit is negligible, i.e. that pressure and velocity is uniform across S, then, if the pressure ratio is less than the critical, i.e. that $\frac{P_T}{P_0} < 1.893$, one may assume that the pressure at S is P_0 and that $t = p^{\frac{\gamma-1}{\gamma}} = \left(\frac{P_T}{P_0}\right)^{\frac{1}{3.5}}$, i.e. $t < 1.2$. The flux density at exit from equation (2-9) is given by $\psi = 1.532 \frac{P_T}{\sqrt{T_T}} \frac{\sqrt{t-1}}{t^3}$, so the discharge flow is $Q = 1.532 \ S \frac{P_T}{\sqrt{T_T}} \frac{\sqrt{t-1}}{t^3}$ lb/sec.

Example 5-1

Air is discharged through a convergent nozzle of exit area of 0.5 ft² from a reservoir, where $P_T = 3000$ p.s.f. and $T_T = 800°K$, to atmosphere where ambient pressure $P_0 = 2116$ p.s.f. What are the values at exit of a) ψ_e; b) $\frac{\psi_e}{\psi_c}$; c) u_c; d) $\frac{u_e}{u_c}$; e) T_e; f) ρ_e; and g) Q? h) What should be the value of P_T for max. possible Q assuming the same T_T and that pressure at S_e is still 2116 p.s.f.?

The pressure ratio $\frac{P_T}{P_0} = \frac{3000}{2116} = 1.418$ ∴ $t = \frac{T_T}{T_e} = 1.418^{.286} = 1.105$ and $\frac{\rho_T}{\rho_e} = 1.105^{2.5} = 1.284$. $\rho_T = \frac{P_T}{RT_T} = \frac{3000}{96 \times 800} = .0391$ lb/ft.³ ∴ $\rho_e = \frac{.0391}{1.284} = .0304$ lb/ft.³

$T_e = \frac{800}{1.105} = 724°K$ ∴ $\theta_{ue} = 800 - 724 = 76°C$ ∴ $u_e = 147.1\sqrt{76} = 1282'$/sec. ∴ $\psi_e = u_e \rho_e = 1282 \times .0304 = 38.98$ lb/sec./ft.² ∴ $Q = \psi_e S_e = 19.49$ lb/sec.

$u_c = 147.1\sqrt{\frac{800}{6}} = 1698.6$ ∴ $u_e/u_c = \frac{1282}{1698.6} = .755$. $\rho_c = \frac{\rho_T}{1.2^{2.5}} = .0248$

∴ $\psi_c = \rho_c u_c = 42.1$ lb/sec/ft.² ∴ $\psi/\psi_c = \frac{38.98}{42.1} = .926$.

Hence the answers for a) to g) are:– a) $\psi_e = 38.98$ lb/sec./ft.²; b) $\frac{\psi_e}{\psi_c} = 0.926$; c) $u_e = 1282'$/sec.; d) $\frac{u_e}{u_c} = 0.755$; e) $T_e = 724°K$; f) $\rho_e = 0.0304$ lb/ft.³; g) Q = 19.49 lb/sec. For h) the max. flow will occur when $\frac{P_T}{2116} = 1.2^{3.5} = 1.893$ ∴ P_T would be $1.893 \times 2116 = 4005$ p.s.f.; ρ_T would be $\frac{4005}{96 \times 800} = .0521$ lb/ft.³ and ρ_e would be $\frac{.0521}{1.2^{2.5}} = .0331$ lb/ft.³ ∴ $\psi_e = \psi_c = .0331 \times u_c = .0331 \times 1698.6 = 56.15$ (which could have been obtained from equation (2-7) to give a Q_{max} of 28.075 lb/sec.)

It is of interest to note that in the sub critical case of the above example (i.e. $P_T = 3000$ p.s.f., etc.) the reactive thrust on the reservoir would be $\frac{Q}{g} \times u_e$, i.e. $\frac{19.49}{32.2} \times 1282 = 776$ lbs. while ΔP acting on $S_e = .5(3000 - 2116) = 442$ lbs. The difference, of course, is due to the reduction of internal pressure as the air accelerates into the nozzle. The proportion of thrust accounted for by this is $\frac{776 - 442}{776} = .43$ or 43%. In the critical case where $P_T = 4005$ p.s.f., $u_c = 1698.6$ and $Q_{max} = 28.075$, the reactive thrust is $\frac{28.075}{32.2} \times 1698.6 = 1481$ lbs. while $0.5(4005 - 2116) = 944.5$ lbs., so the reduction of internal pressure accounts for a proportion equal to $\frac{1481 - 944.5}{1481} = .362$, i.e. 36.2% of the thrust. When the pressure ratio to a convergent nozzle exit is the value corresponding to max. flow, i.e. 1.893, the nozzle is said to be 'choking'. Then, as we have already seen, $Q_{max} = .397\, S. \frac{P_T}{\sqrt{T_T}}$ where S is the minimum section.

Fig. 1b illustrates what happens in a convergent nozzle when the pressure ratio $\frac{P_T}{P_o}$ exceeds 1.893 (i.e. the temp ratio exceeds 1.2). The pressure at nozzle exit plane is $\frac{P_T}{1.893}$ and further expansion to P_o occurs external to the nozzle exit as shown. For complete expansion within a nozzle when $\frac{P_T}{P_o}$ exceeds 1.893 there must be a divergent section beyond the throat, though without a divergent section and for pressure ratios not greatly above the critical, the velocity a short distance downwind from the exit would not differ greatly from that corresponding to complete expansion.

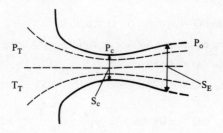

Fig. 5-2

Fig. 2 illustrates a 'con-di' nozzle for complete expansion from P_T to P_o where $\frac{P_T}{P_o}$ is well in excess of 1.893.

The size of S_E can readily be calculated from the fact that continuity requires that

$\psi_c S_c = \psi_E S_E$, i.e. that $\dfrac{S_c}{S_E} = \dfrac{\psi_E}{\psi_c}$ and, as we have seen (equation 2-10), $\dfrac{\psi_E}{\psi_c} = \dfrac{3.864\sqrt{t-1}}{t^3}$

$\therefore \dfrac{S_c}{S_E} = \dfrac{3.864\sqrt{t-1}}{t^3}$. Thus *exit area/throat area is determined by the value of t only (and vice versa).*

Example 5-2

If P_T is $4P_o$ and $T_T = 1000°K$ what is the exit area of a con-di nozzle of throat section $S_c = 0.5$ ft.^{2} for complete expansion. And if $P_o = 2116$ p.s.f. what are the values of a) u_c; b) u_e; c) ρ_c; d) ρ_e; and e) Q.?

For $p = 4$ $t = 4^{.286} = 1.486$ $\therefore \dfrac{S_E}{S_c} = \dfrac{1.486^3}{3.864\sqrt{.486}} = 1.218$

$\therefore S_E = .609$ ft.^{2}

for a) $u_c = 147.1\sqrt{\dfrac{1000}{6}} = 1899'/\text{sec.}$

for b) $u_e = 147.1\sqrt{1000\left(\dfrac{t-1}{t}\right)} = 147.1\sqrt{1000\left(\dfrac{.486}{1.486}\right)} = 2660'/\text{sec.}$

for c) $\rho_c = \dfrac{\rho_T}{1.577}$ and $\rho_T = \dfrac{P_T}{RT_T} = \dfrac{4 \times 2116}{96 \times 1000} = .08817$

$\therefore \rho_c = \dfrac{.08817}{1.577} = .0559$ lb/ft.^{3}

for d) $\rho_e = \dfrac{\rho_T}{t^{2.5}} = \dfrac{.08817}{1.486^{2.5}} = .0328$ lb/ft.^{3}

for e) One may use either $Q = \rho_c u_c S_c$ or $Q = \rho_e u_e S_e$
or $Q = \psi_e S_e$ using equation (2-9) for ψ, or $Q = \psi_c S_c$ using equation (2-7) for ψ_c.
From $Q = \rho_c u_c S_c$ $Q = .0559 \times 1899 \times .5 = 53.08$ lb/sec.
From $Q = \rho_e u_e S_e$ $Q = .0328 \times 2660 \times .609 = 53.13$ lb/sec.

From $Q = \psi_e S_e$ $Q = .609 \times 1.532 \dfrac{P_T}{\sqrt{T_T}} \dfrac{\sqrt{t-1}}{t^3} = .933 \left(\dfrac{8464}{\sqrt{1000}}\right) \dfrac{\sqrt{.486}}{1.486^3}$
$= 53.05$ lb/sec.

From $Q = \psi_c S_c$ $Q = .5 \times .397 \times \dfrac{8464}{\sqrt{1000}} = 53.13$ lb/sec.

(The minor differences are, of course, due to using a limited number of significant figures.)

The first two alternatives are to be preferred since they involve the memorisation of fewer constants.

It is of interest to note what would happen to the reactive thrust if the divergent portion were omitted. The pressure at S_c is $\frac{8464}{1.893} = 4471$ p.s.f. and the velocity $u_c = 1899'/\text{sec}$. Thus there would be a pressure thrust of $0.5\,(4471 - 2116) = 1177.5$ lb. and a reaction thrust of $53.1 \times \frac{1899}{32.2} = 3131.6$ lb. to give a total thrust of 4309.1 lb., whereas, with complete expansion the reactive thrust $= 53.1 \times \frac{2660}{32.2} = 4386.5$ lb.; i.e. an increase of 1.8%.

In practice, flow through nozzles is not quite isentropic, though nearly so if correctly designed. The inevitable boundary layer due to surface friction reduces the effective pressure drop and, though very thin in a sharply descending pressure gradient, reduces the effective flow area. Also, if wall curvature is appreciable at the throat, one cannot assume uniform flux density because of the pressure gradient across the section due to centrifugal force. In a symmetrical nozzle with a straight axis ψ_c can occur only on the axis so that ψ between the axis and the wall is less than ψ_c. If the axis were curved this effect would be even more pronounced and the streamline for ψ_c would not be on the axis (except for one special case).

If, in the case of a con-di nozzle designed for a pressure ratio well above the critical ratio 1.893, the pressure ratio (total to static) is reduced, either by increasing exit static pressure or reducing entry total pressure, so is the temp. ratio, and the nozzle no longer conforms to the requirement that $\frac{S_c}{S_E} = \frac{3.864\sqrt{t-1}}{t^3}$. There will be a section between S_c and S_E which does conform and beyond which the divergent portion no longer runs full. Referring to this intermediate section as S_i one would expect that there would be recompression from S_c to S_i until $\frac{S_c}{S_i} = \frac{3.864\sqrt{t-1}}{t^3}$ and that at S_i the flow would break away with the static pressure equal to the ambient pressure at discharge. But in practice this does not happen. For a short distance beyond S_c there is over expansion to a pressure well below exit static P_o and then sudden recompression to P_o across a normal shock wave much as shown in Fig. 3.

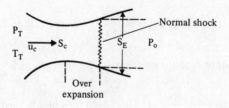

Fig. 5-3

Recompression through a shock wave entails loss not only in the compression across it but also in the turbulence in the space between the discharge flow and the nozzle wall downstream of the shock. Moreover, the effect of the sudden pressure rise on the boundary layer tends to cause a fluctuating flow breakdown where the shock contacts the wall and this in turn acts as an effective reduction in section, the net result being a complex interaction which causes a rapid fluctuation in the position of the shock.

Even if $\frac{P_T}{P_o}$ were reduced to the critical ratio 1.893 this over expansion and shock recompression would still occur, but if reduced below 1.893 the throat would cease to choke, the shock would disappear and the divergent portion would become a diffuser. Recompression in a diffuser, however, is inefficient even if the divergence is very gradual because of the adverse pressure gradient effect on the boundary layer, so the flow would be far from isentropic beyond S_c. The pressure ratio would need to be appreciably below the critical for the divergent portion to become fully established as a subsonic diffuser. When the flow through S_c becomes 'stably' subsonic then S_E becomes the flow controlling section.

Nozzles in Series

If two or more nozzles are in series and the flow is isentropic (i.e. no friction losses) and there is no power extraction between them then clearly the flow will be determined by the smaller (or smallest) throat. But if there is some loss of total pressure between them this may not be so if the difference in throat sections is small.

Fig. 4 represents such a situation in which S_2 is slightly larger than S_1. The 'grid' AB symbolises a cause of total pressure loss from P_{T1} to P_{T2} where P_{T2} is only a little less than P_{T1}. There will be no change of total temperature T_T if the flow is adiabatic.

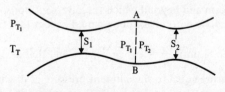

Fig. 5-4

For the throat S_1, if choking, $\psi_{c1} = k\frac{P_{T1}}{\sqrt{T_T}}$. For the throat S_2, if choking, $\psi_{c2} = k\frac{P_{T2}}{\sqrt{T_T}}$, so, for simultaneous choking $\psi_{c1} S_1 = \psi_{c2} S_2$, i.e. $\frac{P_{T1}}{\sqrt{T_T}} S_1 = \frac{P_{T2}}{\sqrt{T_T}} S_2$ (for continuity) from which $\frac{P_{T1} S_1}{P_{T2} S_2} = 1$. This could be a highly unstable state of affairs if $P_{T1} - P_{T2}$ varied according to which nozzle was choking. Choking might oscillate rapidly between the nozzles and cause a rapid fluctuation in mass flow Q. To avoid this $\frac{P_{T1}S_1}{P_{T2}S_2}$ would need to be significantly larger or smaller than unity to ensure steady flow.

We will now proceed to examine nozzle combinations which are very relevant to later portions of this work.

In the diagrams a vertical axis is chosen for the various arrangements. This makes it easier to include many of the relevant relationships on the diagrams rather than in the text.

Case 1. Two Choking Nozzles in Series with Energy Extraction between them

This is analagous to the situation in a simple jet engine. The turbine driving the compressor may be, in effect, a choking nozzle beyond which total temperature and total pressure are reduced. The final exhaust includes the second choking nozzle.

The situation is represented diagrammatically in Fig. 5.

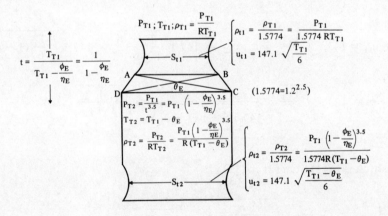

Fig. 5-5

ABCD represents a turbine or other means of energy extraction θ_E with efficiency η_E which reduces the total temperature from T_{T1} to T_{T2}, and the total pressure from P_{T1} to P_{T2}. The corresponding 'total' densities are ρ_{T1} and ρ_{T2}.

The suffixes t1 and t2 pertain to conditions at the throats S_1 and S_2.

With efficiency η_E, the isentropic energy drop $\theta'_E = \dfrac{\theta_E}{\eta_E}$; hence, as shown, the total to total temperature ratio t across S_1 and ABCD is given by

$$t = \frac{1}{1 - \dfrac{\phi_E}{\eta_E}} \quad \left(\text{where } \phi_E = \frac{\theta_E}{T_{T1}}\right)$$

Many of the relevant relationships are indicated in Fig. 5.

For continuity, $\dfrac{S_{t2}}{S_{t1}} = \dfrac{\rho_{t1} \, u_{t1}}{\rho_{t2} \, u_{t2}}$. \hfill (5-1)

From Fig. 5 we see that $\rho_{t1} = \dfrac{P_{T1}}{1.5774 \, RT_{T1}}$ and $\rho_{t2} = \dfrac{P_{T1}\left(1 - \dfrac{\phi_E}{\eta_E}\right)^{3.5}}{1.5774 \, R \, (T_{T1} - \theta_E)}$

$$\therefore \frac{\rho_{t1}}{\rho_{t2}} = \frac{P_{T1}}{1.5774 \, RT_{T1}} \times \frac{1.5774 \, RT_1 \, (1 - \phi_E)}{P_{T1}\left(1 - \dfrac{\phi_E}{\eta_E}\right)^{3.5}} = t^{3.5} \, (1 - \phi_E) \quad (5\text{-}2)$$

Also $u_{t1} = 147.1 \sqrt{\dfrac{T_{T1}}{6}}$ and $u_{t2} = 147.1 \sqrt{\dfrac{T_1 - \theta_E}{6}} = 147.1 \sqrt{\dfrac{T_1}{6} (1 - \phi_E)}$

$$\therefore \frac{u_{t1}}{u_{t2}} = 147.1 \sqrt{\frac{T_{T1}}{6}} \times \frac{1}{147.1 \sqrt{\dfrac{T_1}{6} (1 - \phi_E)}} = \sqrt{\frac{1}{1 - \phi_E}} \quad (5\text{-}3)$$

Substituting in (1) from (2) and (3)

$$\frac{S_{t2}}{S_{t1}} = t^{3.5} \, (1 - \phi_E) \sqrt{\frac{1}{1 - \phi_E}} = t^{3.5} \sqrt{1 - \phi_E} \quad (5\text{-}4)$$

Note that if $\eta_E = 1.0$

$$t = \frac{1}{1 - \phi_E} \quad \text{so (4) becomes} \quad \frac{S_{t2}}{S_{t1}} = t^3. \quad (5\text{-}5)$$

This result is relevant to relatively moving nozzles in Case 6 on page 60.

Numerical Check

Using assumptions as follows: $T_{T1} = 1000°K$, $P_{T1} = 12{,}000$ p.s.i.;

$\theta_E = 300$ (so that $\phi_E = 0.3$) and $\eta_E = 0.87$, we have:

$$\rho_{T1} = \frac{12000}{96 \times 1000} = 0.125 \quad \text{and} \quad \rho_{t1} = \frac{0.125}{1.5774} = .07924$$

$$u_{t1} = 147.1\sqrt{\frac{1000}{6}} = 1899.1 \text{ ft/sec.}$$

$$T_{T2} = 1000 - 300 = 700; \quad t = \frac{1}{1 - \frac{0.3}{.87}} = 1.5263,$$

$$\therefore P_{T2} = \frac{12.000}{(1.5263)^{3.5}} = 2731.7 \quad \therefore \rho_{T2} = \frac{2731.7}{96 \times 700} = .04065,$$

$$\therefore \rho_{t2} = \frac{.04065}{1.5774} = .02577$$

$$u_{t2} = 147.1\sqrt{\frac{700}{6}} = 1588.9 \text{ ft/sec.}$$

$$\therefore \frac{\rho_{t1}}{\rho_{t2}} \frac{u_{t1}}{u_{t2}} = \frac{S_{t2}}{S_{t1}} = \frac{.07924}{.02577} \times \frac{1899.1}{1588.9} = 3.675.$$

From equation (4) $\dfrac{S_{t2}}{S_{t1}} = 1.5263^{3.5}\sqrt{1-0.3} = 3.675$ Q.E.D.

Note that if there had been no less, i.e. $\eta_T = 1.0$, then t would be $\dfrac{1}{0.7} = 1.4286$ which, from (5), would make $\dfrac{S_{t2}}{S_{t1}} = t^3 = 1.4286^3 = 2.9155$. Thus it may be seen that losses in energy extraction have an appreciable effect on the throat area ratio.

Equation (4) brings out three very important facts, namely, that if $\dfrac{S_{t2}}{S_{t1}}$ is fixed and η_E may be regarded as constant then i) *ϕ_E is fixed, i.e. θ_E is proportional to T_{T1}* ii) *t is also fixed*, iii) *T_{T2} is proportional to T_{T1}*. Thus if one wishes to vary ϕ_E and t it is necessary to vary $\dfrac{S_{t2}}{S_{t1}}$. If S_{t1} represents turbine nozzle throats in parallel and S_{t2} represents the final nozzle of a jet engine, it is no problem to vary S_{t2} but, up to now, S_{t1} cannot be varied (though attempts to achieve this are probably in progress).

Case 2. Three Choking Nozzles in Series with Energy Extraction after the First and Second

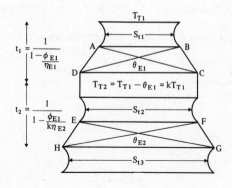

Fig. 5-6

This, shown diagrammatically in Fig. 6, is relevant to a two-spool jet engine with a choking final nozzle preceded by two choking turbines in series.

For the expansion across S_{t1} and ABCD, the situation is the same as Case 1,

$$\therefore \frac{S_{t2}}{S_{t1}} = t_1^{3.5} \sqrt{1 - \phi_{E1}} \tag{5-6}$$

and the total temperature after CD is proportional to T_{T1}, so we can write $T_{T2} = kT_{T1}$.

For an observer at EF the situation is also the same as Case 1 except that kT_{T1} replaces T_{T1} and the temperature ratio t_2 across S_{t2} and EFGH becomes

$$t_2 = \frac{kT_{T1}}{kT_{T1} - \frac{\theta_{E2}}{\eta_{E2}}} = \frac{1}{1 - \frac{\phi_{E2}}{k\eta_{E2}}}$$

as shown on Fig. 6.

Hence $\dfrac{S_{t3}}{S_{t2}} = t_2^{3.5} \sqrt{1 - \dfrac{\phi_{E2}}{k}}$ (5-7)

Multiplying (6) and (7) we obtain

$$\frac{S_{t3}}{S_{t1}} = (t_1 t_2)^{3.5} \sqrt{(1 - \phi_{E1})\left(1 - \frac{\phi_{E2}}{k}\right)} \tag{5-8}$$

As with Case 1, θ_{E2} is proportional to T_{T2} which in turn is proportional to T_{T1}, and, since θ_{E1} is also proportional to T_{T1} it follows that if $\dfrac{S_{t3}}{S_{t1}}$ is fixed, so is $\dfrac{\theta_{E2}}{\theta_{E1}}$ (so long as η_{E1} and η_{E2} may be regarded as constants). Thus *there is a 'thermodynamic lock' between ABCD and EFGH* which may be 'broken' by variation of the throat S_{t3}.

These 'fixed links' in expansion imposed by choking nozzles are very useful for calculating the part load performance of jet engines over the range where η_{E1} and η_{E2} may be assumed constant.

If there is an energy loss between BC and EF it may be allowed for by adjusting η_{E1} to include it. Similarly if there is an energy loss between GH and S_{t3} it may be allowed for by adjusting η_{E2} to include it.

In summary, if S_{t1}, S_{t2}, and S_{t3} are fixed so are $\dfrac{T_{T1}}{T_{T2}}, \dfrac{\theta_{E1}}{\theta_{E2}}, \dfrac{\theta_{E1}}{T_{T1}}, \dfrac{\theta_{E2}}{T_{T1}}$, etc. Also t_1 and t_2 are fixed as long as η_{E1} and η_{E2} may be assumed constant.

Finally, if $\eta_{E1} = \eta_{E2} = 1.0$ $\quad t_1 = \dfrac{1}{1 - \phi_{E1}}$ and $\quad t_2 = \dfrac{1}{1 - \dfrac{\phi_{E2}}{k}}$

So, substituting in (8) we get

$$\frac{S_{t3}}{S_{t1}} = (t_1 t_2)^3. \tag{5-9}$$

This result is relevant to relatively moving nozzles (see below).

Case 3. A Non-Choking Nozzle in Series with a Choking Nozzle with Energy Extraction between them

This is relevant to a gas turbine with open exhaust or a jet engine with a variable final nozzle opened up to a non-choking condition for operational reasons.

In the cases considered in the foregoing, the temperature ratio across the final nozzle did not enter into the picture but it must be taken into account if the temperature ratio (total to static) is less than the critical value (= 1.2) at all points from entry to exit of the final nozzle. The static pressure of final discharge must also be taken into account.

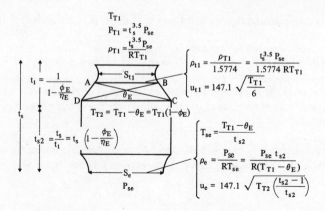

Fig. 5-7

Figure 7 is a diagrammatic representation of the system. In it the suffix s implies static quantities or total to static ratios; thus t_s is the overall total to static temperature ratio and P_{se} is the static pressure at S_e.

This case is somewhat more complex than the two already considered above.

Again, for continuity $\dfrac{S_e}{S_1} = \dfrac{\rho_{t1} u_{t1}}{\rho_e u_e}$.

Using the relationships shown on Fig. 7 and the fact that $T_{T2} = T_{T1}(1 - \phi_E)$

$$\frac{\rho_{t1}}{\rho_e} = \frac{t_s^{3.5} P_{se}}{1.5774 \, RT_{T1}} \times \frac{RT_{T1}(1-\phi_E)}{P_{se} \, t_{s2}} = \frac{t_s^{3.5}}{t_s^2}\left(\frac{1-\phi_E}{1.5774}\right) \tag{5-10}$$

$$\frac{u_{t1}}{u_e} = 147.1\sqrt{\frac{T_{T1}}{6}} \times \frac{1}{147.1\sqrt{T_{T1}(1-\phi_E)\left(\dfrac{t_{s2}-1}{t_{s2}}\right)}} = \sqrt{\frac{t_{s2}}{6(1-\phi_E)(t_{s2}-1)}} \tag{5-11}$$

Multiplying (10) by (11) gives

$$\frac{S_e}{S_1} = \frac{1}{1.5774}\frac{t_s^{3.5}}{t_{s2}}\sqrt{\frac{t_{s2}(1-\phi_E)}{6(t_{s2}-1)}} \quad \text{or, since } t_s = t_1 t_{s2}$$

$$\frac{S_e}{S_1} = \frac{1}{1.5774} t_1^{3.5} t_{s2}^3 \sqrt{\frac{1-\phi_E}{6(t_{s2}-1)}} \quad \text{or, since } t_1 = \frac{1}{1-\dfrac{\phi_E}{\eta_E}}$$

$$\frac{S_e}{S_1} = \frac{1}{1.5774} \left(\frac{1}{1 - \frac{\phi_E}{\eta_E}}\right)^{3.5} t_{s2}^3 \sqrt{\frac{1 - \phi_E}{6(t_{s2} - 1)}}. \tag{5-12}$$

Numerical Check

Using assumptions as follows: $T_{T_1} = 1000°K$, $\theta_E = 300°C$, $\eta_E = 0.87$, $P_{se} = 2000$ p.s.f. and $t_{s2} = 1.15$, we have: $t_1 = \dfrac{1}{1 - \dfrac{0.3}{0.87}} = 1.5263$

$\therefore\ t_s = t_1 \times t_{s2} = 1.5263 \times 1.15 = 1.7553 \therefore P_{T_1} = 1.7553^{3.5} \times 2000 = 14{,}329$ p.s.f.

$\therefore\ \rho_{T_1} = \dfrac{14\,329}{96 \times 1000} = 0.1493$ and $\rho_{t_1} = \dfrac{0.1493}{1.5774} = .09463$

Since $T_{T_2} = 1000 - 300$, $T_{se} = \dfrac{700}{1.15} = 608.7 \therefore \rho_e = \dfrac{2000}{96 \times 608.7} = .03423$

$\therefore\ \dfrac{\rho_{t_1}}{P_a} = \dfrac{.09463}{.03423} = 2.7649$

$u_{t_1} = 147.1 \sqrt{\dfrac{1000}{6}} = 1899.1$ ft/sec and $u_e = 147.1 \sqrt{700 \times \dfrac{0.15}{1.15}} = 1405.6$ ft/sec.

$\therefore\ \dfrac{u_{t_1}}{u_e} = \dfrac{1899.1}{1405.6} = 1.3511 \therefore \dfrac{S_e}{S_1} = 2.7649 \times 1.3511 = 3.736$

Using equation (12) we have $\dfrac{S_e}{S_1} = \dfrac{1}{1.5774} \left(\dfrac{1}{1 - \dfrac{0.3}{0.87}}\right)^{3.5} \times 1.15^3 \times \sqrt{\dfrac{1 - 0.3}{6 \times 0.15}} = 3.736$ Q.E.D.

If $\eta_E = 1.0$ (12) becomes $\dfrac{S_e}{S_1} = \dfrac{1}{1.5774} \left(\dfrac{1}{1 - \phi_E}\right)^{3.5} t_{s2}^3 \sqrt{\dfrac{1 - \phi_E}{6(t_{s2} - 1)}} =$

$$\frac{1}{1.5774} \left(\frac{t_{s2}}{1 - \phi_E}\right)^3 \sqrt{\frac{1}{6(t_{s2} - 1)}} \tag{5-13}$$

This is relevant to relatively moving nozzles (see over).

Case 4. Choking Nozzle in Series with a Non-Choking Nozzle with Energy Extraction between them

This is representative of a turbine with non-choking blading followed by another turbine or final nozzle across which the total to static temperature ratio is above the critical ratio 1.2.

In this case the situation reverts to that in which the system is unaffected by the static pressure at exit from the choking nozzle.

The system is represented in Fig. 8.

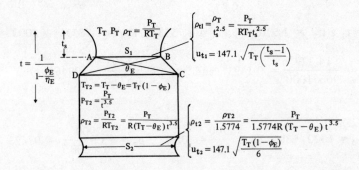

Fig. 5-8

Using the relationships shown on Fig. 8, we have:

$$\frac{\rho_{t1}}{\rho_{t2}} = \frac{P_T}{RT_T\, t_s^{2.5}} \times \frac{1.5774\, RT_T\, (1-\phi_E) t^{3.5}}{P_T} = 1.5774\, (1-\phi_E) \frac{t^{3.5}}{t_s^{2.5}} \qquad (5\text{-}14)$$

$$\frac{u_{t1}}{u_{t2}} = 147.1 \sqrt{T_T\left(\frac{t_s-1}{t_s}\right)} \times \frac{1}{147.1}\sqrt{\frac{6}{T_T(1-\phi_E)}} = \sqrt{\frac{6(t_s-1)}{t_s(1-\phi_E)}} \qquad (5\text{-}15)$$

Hence, multiplying (14) by (15)

$$\frac{S_2}{S_1} = 1.5774\,(1-\phi_E)\frac{t^{3.5}}{t_s^{2.5}} \times \sqrt{\frac{6(t_s-1)}{t_s(1-\phi_E)}} = 1.5774\,\frac{t^{3.5}}{t_s^{2.5}}\sqrt{\frac{6(t_s-1)(1-\phi_E)}{t_s}}$$

(5-16)

or, using $t = \dfrac{1}{1-\dfrac{\phi_E}{\eta_E}}$, $\dfrac{S_2}{S_1} = 1.5774\left(\dfrac{1}{1-\dfrac{\phi_E}{\eta_E}}\right)^{3.5} t_s^{-2.5}\sqrt{\dfrac{6(t_s-1)(1-\phi_E)}{t_s}}$

or, $\quad \dfrac{S_2}{S_1} = \dfrac{1.5774}{t_s^3} \left(\dfrac{1}{1 - \dfrac{\phi_E}{\eta_E}} \right)^{3.5} \sqrt{6(t_s - 1)(1 - \phi_E)}$ (5-17)

from which it is more readily seen that, for a given value of η_E, if any two of $\dfrac{S_2}{S_1}$, t_s or ϕ_E are fixed, so is the third.

Numerical Check

Assuming that $T_T = 1000°K$; $P_T = 12000$ p.s.f.; $\theta_E = 300$; $\eta_E = 0.87$ and $t_s = 1.15$;

we have: $\rho_T = \dfrac{12000}{96 \times 1000} = 0.125$; $\rho_{t1} = \dfrac{0.125}{1.15^{2.5}} = .08814$;

$t = \dfrac{1}{1 - \dfrac{0.3}{0.87}} = 1.5263 \therefore P_{T2} = \dfrac{12000}{1.5263^{3.5}} = 2731.7$ p.s.f.,

$\therefore \rho_{T2} = \dfrac{2731.7}{96 \times 700} = .04065$

$\therefore \rho_{t2} = \dfrac{.04065}{1.5774} = .02577 \therefore \dfrac{\rho_{t1}}{\rho_{t2}} = \dfrac{.08814}{.02577} = 3.4203$

$u_{t1} = 147.1 \sqrt{1000 \left(\dfrac{0.15}{1.15} \right)} = 1680$ ft/sec, $\quad u_{t2} = 147.1 \sqrt{\dfrac{700}{6}} = 1588.9$ ft/sec

$\therefore \dfrac{u_{t1}}{u_{t2}} = \dfrac{1680}{1588.9} = 1.0574 \therefore \dfrac{S_2}{S_1} = 1.0574 \times 3.4203 = 3.616$.

From (17) $\dfrac{S_2}{S_1} = \dfrac{1.5774}{1.15^3} \left(\dfrac{1}{1 - \dfrac{0.3}{0.87}} \right)^{3.5} \sqrt{6 \times 0.15 \times 0.7} = 3.616$ Q.E.D.

If, in (17), $\eta_E = 1.0$ the equation becomes:

$$\dfrac{S_2}{S_1} = \dfrac{1.5774 \sqrt{6(t_s - 1)}}{[t_s (1 - \phi_E)]^3} \quad (5\text{-}18)$$

which is again relevant to relatively moving nozzles (see over).

Case 5. Two Non-Choking Nozzles in Series with Energy Extraction Between them

This is relevant to a two stage turbine in which neither stage is choking and there is no choking restriction in the exhaust.

In this case the exhaust static pressure has to be taken into account once more.

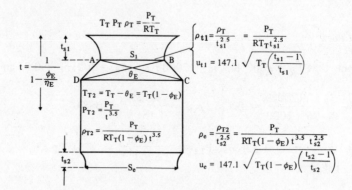

Fig. 5-9

The situation is represented in Fig. 9 on which as before several of the major relationships are indicated.

For continuity, $\dfrac{S_E}{S_1} = \dfrac{\rho_{t1} \, u_{t1}}{\rho_e \, u_e}$.

Using the relationships shown on Fig. 9:

$$\dfrac{\rho_{t1}}{\rho_e} = \dfrac{P_T}{RT_T t_{s1}^{2.5}} \times \dfrac{RT_T(1-\phi_E)t_{s2}^{2.5}}{P_T} = (1-\phi_E)t^{3.5}\left(\dfrac{t_{s2}}{t_{s1}}\right)^{2.5} \qquad (5\text{-}19)$$

$$\dfrac{u_{t1}}{u_e} = \dfrac{147.1\sqrt{T_T\left(\dfrac{t_{s1}-1}{t_{s1}}\right)}}{147.1\sqrt{T_T(1-\phi_E)\left(\dfrac{t_{s2}-1}{t_{s2}}\right)}} =$$

$$\sqrt{\dfrac{t_{s2}}{t_{s1}}\dfrac{(t_{s1}-1)}{(t_{s2}-1)}(1-\phi_E)} \qquad (5\text{-}20)$$

Multiplying (19) by (20):

$$\dfrac{S_e}{S_1} = (1-\phi_E)t^{3.5}\left(\dfrac{t_{s2}}{t_{s1}}\right)^{2.5}\sqrt{\dfrac{t_{s2}}{t_{s1}}\dfrac{(t_{s1}-1)}{(t_{s2}-1)}(1-\phi_E)} = \left(\dfrac{t_{s2}}{t_{s1}}\right)^{3} t^{3.5}\sqrt{(1-\phi_E)\left(\dfrac{t_{s1}-1}{t_{s2}-1}\right)} \qquad (5\text{-}21)$$

Numerical Check

Assuming that $T_T = 1000°K$; $P_T = 12,000$ psi; $\theta_E = 300$; $\eta_E = 0.87$
$t_{s1} = 1.17$ and $t_{s2} = 1.15$ then:

$$\rho_T = \frac{12000}{96 \times 1000} = 0.125; \quad \rho_{t1} = \frac{0.125}{1.17^{2.5}} = .08442$$

$$t = \frac{1}{1 - \frac{0.3}{0.87}} = 1.5263 \quad \therefore P_{T2} = \frac{12000}{1.5263^{3.5}} = 2731.7 \text{ p.s.f.}$$

$$\therefore \rho_{T2} = \frac{2731.7}{96 \times 700} = .04065$$

$$\therefore \rho_e = \frac{.04065}{1.15^{2.5}} = .02866 \quad \therefore \frac{\rho_{t1}}{\rho_e} = \frac{.08442}{..02866} = 2.9456$$

$$u_{t1} = 147.1\sqrt{1000 \times \frac{0.17}{1.17}} = 1773.1 \text{ ft/sec, and } u_e = 147.1\sqrt{700 \times \frac{0.15}{1.15}} = 1405.6 \text{ ft/sec.}$$

$$\therefore \frac{u_{t1}}{u_e} = 1.2615 \quad \therefore \frac{S_e}{S_1} = 1.2615 \times 2.9456 = 3.715.$$

Using equation (21) we have:

$$\left(\frac{1.15}{1.17}\right)^3 \left(\frac{1}{1 - \frac{0.3}{0.87}}\right)^{3.5} \sqrt{0.7 \times \frac{0.17}{0.15}} = 3.715 \text{ Q.E.D.}$$

At this point, a word of caution is necessary. In the numerical check, with $\theta_E = 300°$, $t = 1.5263$ which is well above the critical ratio 1.2, so that if the energy extractor were a turbine it would need to have at least two stages to avoid choking in either the stationary or moving blading

Referring to equation (21), if $\eta_E = 1.0$, $t^{3.5} = \left(\frac{1}{1 - \phi_E}\right)^{3.5}$ and (21) then becomes

$$\frac{S_e}{S_1} = \left(\frac{t_{s2} \, t}{t_{s1}}\right)^3 \sqrt{\frac{t_{s1} - 1}{t_{s2} - 1}} \tag{5-22}$$

which is again relevant to relatively moving nozzles (see over).

Case 6. Relative Movement between Nozzles

Figure 10 represents a nozzle B receding from fixed nozzle A with velocity v. The broken lines indicate a confining channel between them.

Improbable as it may seem from the diagram, this is relevant to the stationary and moving blades of a turbine.

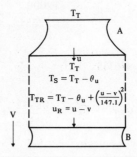

Fig. 5-10

As will now be shown, the relative movement of B away from A is equivalent to energy extraction between them, this energy being without loss if the flow is wholly isentropic as will be assumed.

The key fact is that the static temperature T_s is the same relative to both nozzles in the internozzle space, but the total temperature T_{TR} relative to B is reduced by the relative movement.

Using the relationships shown on the diagram, the reduction of total temperature is

$$T_T - \left\{ T_T - \theta_u + \left(\frac{u-v}{147.1}\right)^2 \right\} = \theta_u - \left(\frac{u-v}{147.1}\right)^2.$$

This reduction is equivalent to energy extraction θ_E, so one may write

$$\theta_E = \theta_u - \left(\frac{u-v}{147.1}\right)^2 = \theta_u - \left(\frac{u^2}{147.1^2} - \frac{2uv}{147.1^2} + \frac{v^2}{147.1^2}\right)$$

or

$$\theta_E = \theta_u - \left(\theta_u - \frac{2u^2 \frac{v}{u}}{147.1^2} + \theta_v\right) = 2\theta_u \frac{v}{u} - \theta_v$$

or, since $\dfrac{v}{u} = \sqrt{\dfrac{\theta_v}{\theta_u}}$,

$$\theta_E = 2\theta_u \sqrt{\dfrac{\theta_v}{\theta_u}} - \theta_v = 2\sqrt{\theta_u \theta_v} - \theta_v.$$

Dividing throughout by T_T we have

$$\phi_E = 2\sqrt{\phi_u \phi_v} - \phi_v. \tag{5-22}$$

Hence in Cases 1 to 5 above, if relatively moving nozzles are substituted for energy extraction between non-relatively moving nozzles with $\eta_E = 1.0$, then the 'presumptive' value of t $\left(= \dfrac{1}{1-\phi_E}\right) = \dfrac{1}{1 - 2\sqrt{\phi_u \phi_v} + \phi_v}$. Thus in Case 1, for $\eta = 1.0$ it was found that $\dfrac{S_2}{S_1} = t^3$ (equation (5)) so, for a second choking nozzle receding from a first choking nozzle with velocity v, (5) becomes $\dfrac{S_2}{S_1} = \left(\dfrac{1}{1 - 2\sqrt{\phi_u \phi_v} + \phi_v}\right)^3$ or $1 - 2\sqrt{\phi_u \phi_v} + \phi_v = \sqrt[3]{\dfrac{S_1}{S_2}}$ from which it may be seen that if $\dfrac{S_1}{S_2}$ is fixed ϕ_v is a function of ϕ_u.

Similar substitutions for t and/or ϕ_E may be made in equations (9), (13), (18) and (22).

It is not recommended that the reader attempts to memorise the numerous formulae of this Section. The writer's main purpose has been to demonstrate the methods to be used for finding them and to bring out the 'links' and 'locks' which nozzles in series can impose, especially in the cases involving choking nozzles.

One final point: it is emphasised that only non-choking convergent nozzles can run full in off design conditions. For choking con-di nozzles to run full, as has been shown, the temperature ratio across them must conform to the exit/throat area ratio and vice versa. If this conformity does not exist the flow tends to fluctuate because in the instant following 'detachment' from the wall of the divergent portion the rapidly moving detached flow tends to pump away the relatively stagnant fluid between it and the wall and to re-attach at one point while becoming detached from another. Thus there would be a rapid fluctuation in direction superimposed on a fluctuation in flow rate. One may surmise that this phenomenon would be a serious source of noise.

SECTION 6
Representation of Thermal Cycles for Perfect Gases

The state of a perfect gas is determined by any two of the six quantities Pressure P, Specific Volume V (or $1/\rho$), Temp. T, Internal Energy E, Enthalpy h, and Entropy s. Hence changes of state can be represented on diagrams using any pair for coordinates or even pairs of combinations.

Figs. 1a–1h show eight of many possible ways of representing an *ideal* constant pressure cycle for a heat engine (or heat pump) using air as working fluid, it being assumed that the specific heats C_v and C_p (or K_v and K_p) are constant. Each diagram is for the same cycle–the well known 'Brayton Cycle'–in which, in a heat engine, air is compressed isentropically from A to B, heated at constant pressure from B to C, expanded isentropically from C to D, and cooled at constant pressure from D to A. This is the basic cycle of most modern gas turbines, though, in practice, except for 'closed cycle' engines, the 'spent' fluid is exhausted to atmosphere at D.

None of the diagrams involve enthalpy or internal energy because, with constant specific heats they would be similar to those with T as one of the ordinates but to a different scale. The same applies to the product PV. Even allowing for increase of specific heat with temperature, P–h, P–E, h–s diagrams etc., would be quite similar to their constant specific heat counterparts.

Any of the diagrams can be represented in non-dimensional form, thus the p–v diagram of Fig. 1d is the non-dimensional version of the P–V diagram of Fig. 1a.

Entropy, by reason of its definition as a differential, has no absolute value. So, where it is used, it is necessary to stipulate an arbitrary datum. For the purpose of the diagrams involving s (Figs. 1c, 1e, 1f and 1g) the datum chosen was s = 0 when T = 200°K.

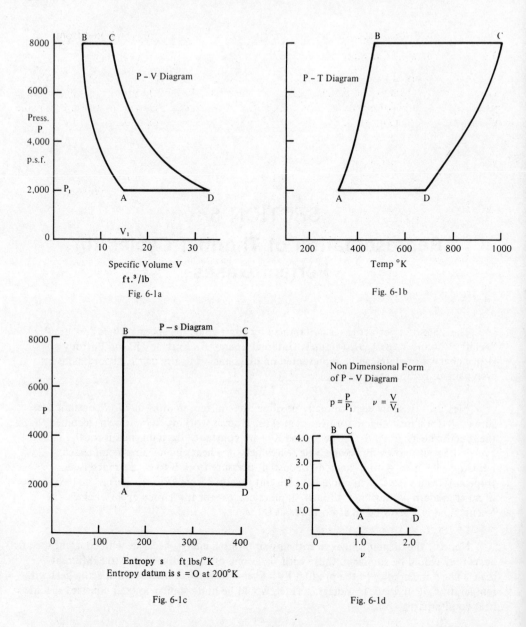

Fig. 6-1a

Fig. 6-1b

Fig. 6-1c

Fig. 6-1d

Of the diagrams shown, the P–V diagram (Fig. 1a) and the T–s diagram (Fig. 1e) are the most frequently used in text books. The P–V diagram because areas on it are a measure of mechanical energy, e.g. the area below AB, i.e. $\int_A^B PdV$; is the work done in (ideal) compression from A to B and $\int_C^D PdV$ is the ideal work of expansion from C to D. The area ABCD is the net work of the cycle. On the T–s diagram, areas on it are a measure of heat.

Thus $\int_B^C Tds$ equals the heat added between B and C and $\int_D^A Tds$ is the heat rejected from D to A. The area ABCD represents the net heat converted to mechanical energy.

Many writers of text books on gas turbines make generous use of T–s diagrams but the writer, in many years of gas turbine design, has never had occasion to use them with a perfect gas as working fluid (or mixtures of perfect gases). The use of entropy is, however, invaluable when dealing with steam and other vapours or mixtures of vapours and gases.

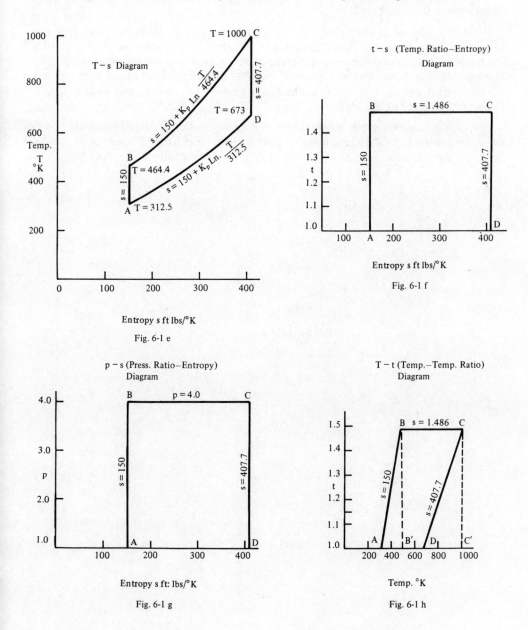

Fig. 6-1 e

Fig. 6-1 f

Fig. 6-1 g

Fig. 6-1 h

The main value of thermal diagrams is to illustrate theoretical propositions.

The P–T diagram of Fig. 1b is rarely used, if at all. But, as may be seen, if set alongside the corresponding P–V diagram (as in Fig. 1) or superimposed on it, one may find temperatures by reading across from one to the other.

Figs. 1c, 1f and 1g are of interest because of their simplicity.

The T–t diagram is of academic interest in that all isentropics radiate from the hypothetical point $T = 0$ $t = 0$.

Another interesting feature of Fig. 1h is that projections on the Temp. scale allow one to read off energy changes. Thus AB' indicates the temp. rise of compression, $B'C'$ ($\equiv BC$) the temp. rise due to heat addition, AD is the temp. equivalent of the heat rejected, $C'D$ is the temp. drop in expansion and $C'D - AB'$ gives the temp. equivalent of net work. A defect of this diagram, as shown, is that an isothermal is a mere point on the temp. scale so the Carnot cycle cannot be depicted.

As will be seen hereafter, the writer evolved methods of cycle calculations based on Fig. 1f using, for constant specific heats, temps. T, temp. ratios t and temp. differences θ. And, for variable specific heat, enthalpy h, pressure ratios p, and enthalpy differences Δh.

SECTION 7

Gas Turbine Cycle Calculations Using Approximate Methods

The approximate methods referred to in the heading are those referred to in the last paragraph of Section 6, i.e., cycles are dealt with almost entirely in terms of absolute temperatures, temperature differences and temperature ratios, it being assumed a) that specific heats are constant at $C_V = 0.171$ C.H.U./lb/°C; $C_P = 0.24$ C.H.U./lb/°C; $K_V = 240$ ft lbs/lb/°C and $K_P = 336$ ft lbs/lb/°C and therefore that $\gamma = 1.4$ and $R = 96$ ft lbs/lb/°C; b) that increase of mass due to fuel addition in combustion may be neglected.

All energy and other quantities are for unit mass flow rate (designated by the word 'specific' where necessary).

The adiabatic efficiency of compression η_C is defined as $\eta_C = \dfrac{\theta'_C}{\theta_C}$ shere θ'_C is the isentropic total temperature rise and θ_C is the actual total temperature rise for the pressure ratio. (total to total).

The adiabatic efficiency of expansion η_E is defined as $\eta_E = \dfrac{\theta_E}{\theta'_E}$ where θ_E is the actual total temperature drop and θ'_E is the isentropic total temperature drop for the pressure ratio concerned (the suffixes T or J may be used where it is necessary to distinguish between expansion through a turbine and expansion through a jet).

The Brayton Cycle With No Losses

Fig. 1a is a representation of the P-V diagram of Fig. 6-1a of Section 6 in diagrammatic form (i.e., not to scale as are all the diagrams of Section 6) with certain additional features. $T_O T_C$ is isentropic compression from P_1 to P_2 and $T_M T_E$ is isentropic expansion from P_2 to P_1. $T_M - T_C$ is heat addition at constant pressure P_2 and $T_E - T_O$ is heat rejection at constant pressure P_1. The shaded area $T_O\ T_C\ P_2\ P_1$ represents the 'negative work' of compression W_C, and is given by $W_C = \displaystyle\int_{P_1}^{P_2} V dP$ or $\displaystyle\int_{P_1}^{P_2} \dfrac{dP}{\rho}$.

As we have seen from Section 1, $\dfrac{dP}{\rho} = K_P dT \therefore W_C = \int_{T_O}^{T_C} K_P dT$, i.e., $W_C = K_P(T_C - T_O)$
Similarly the positive work of expansion W_E, i.e., the shaded area $P_2 T_M T_E P_1$ is given by $W_E = K_P(T_M - T_E)$.

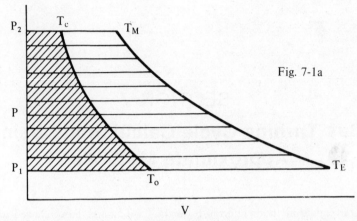

Fig. 7-1a

Since the pressure ratio p is the same for compression and expansion so is the temperature ratio t, i.e., $t = \dfrac{T_C}{T_O} = \dfrac{T_M}{T_E}$ $\left(= \left(\dfrac{P_2}{P_1}\right)^{0.286} \right)$. It follows that $\dfrac{T_M}{T_C} = \dfrac{T_E}{T_O} = K$ where K is the positive/negative work ratio.

Fig. 1b is a modification of Fig. 1a using $T_C = tT_O$ and $T_M = kT_O$ from which it follows that Figs. 1a and 1b can be represented non-dimensionally as in Fig. 1c.

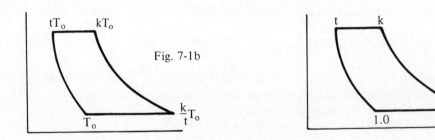

Fig. 7-1b

Fig. 7-1c

In Fig. 1c, the positive work W_e is proportional to $k - \dfrac{k}{t}$ while the negative work W_C is proportional to $t - 1$ ∴ the net work $W_E - W_C$ is proportional to $k\left(\dfrac{t-1}{t}\right) - (t-1)$ or $(k - t)\left(\dfrac{t-1}{t}\right)$. The heat added H is proportional to $k - t$ (the constants of proportionality being the same for W_E, W_C, and H, namely $K_P T_O$ in mechanical units).

The overall efficiency η_O is $\dfrac{\text{net work}}{\text{heat added}}$ (or since net work equals heat added less heat rejected, $\eta_O = \dfrac{\text{heat added} - \text{heat rejected}}{\text{heat added}}$). $\therefore \eta_O = \dfrac{k\left(\dfrac{t-1}{t}\right) - (t-1)}{k-t}$ which reduces to $\eta_O = \dfrac{t-1}{t}$ or $\eta_O = 1 - \dfrac{1}{t}$.

Note that, in the ideal case, i.e., 100% efficiency in compression and expansion, η_O is independent of k and depends only on t.

$1 - \dfrac{1}{t}$ is the 'Air Standard Efficiency'

The specific work W_S, however, depends on k since it is proportional to $\eta_O (k - t)$, i.e.,
$W_S \equiv W_E - W_C = \eta_O K_P T_O (k - t)$ or, since $\eta_O = 1 - \dfrac{1}{t}$, $W_S = K_P T_O \left(1 - \dfrac{1}{t}\right)(k - t)$.

Brayton Cycle With Losses

This is depicted in the P-V diagram of Fig. 2.

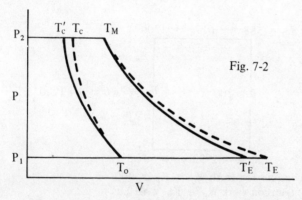

Fig. 7-2

$T_O T'_C$ is isentropic compression and $T_O T_C$ represents the actual compression (assumed adiabatic) in which losses are converted to internal energy thus causing an increase of temperature (and therefore of V) at any given value of P.

$T_M T'_E$ is isentropic expansion and $T_M T_E$ is the actual expansion (also assumed adiabatic) in which again losses take the form of an increase of internal energy (and entropy).

Unfortunately, areas bounded by the broken lines are no longer measures of work done. $T'_C T_C$ and $T'_E T_E$ represent the conversion of losses into heat. And, as may be seen, losses are not totally lost. The heating effect of compression losses reduces the amount of heat needed to raise the temperature to T_M at constant pressure P_2 and the expansion losses are partially recovered in the portion of the expansion following their heating effect. The lower down the expansion curve they occur the less the recovery.

F.W. Representation of the Ideal Brayton Cycle

As indicated in Section 6 the writer has always preferred to think in terms of diagrams similar to Fig. 6-1f of that section, i.e., using vertical lines for isentropics and horizontal lines for constant temperature ratio (and therefore constant pressure or pressure ratio) but further simplifying by omitting ordinands and scales. Thus the ideal Brayton cycle would be represented (or merely imagined) as in Fig. 3a.

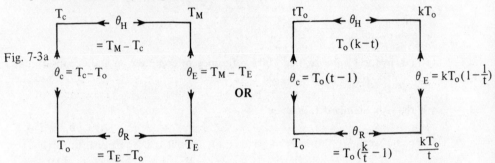

Fig. 7-3a

The non-dimensional form will be obvious from the right hand diagram of Fig. 3a – division by T_o at all points.

Fig. 3b gives a numberical example for $T_O = 288°K$, $T_M = 1200°K$, and $\theta_C = 300°C$,
$$t = \frac{588}{288} = 2.042.$$

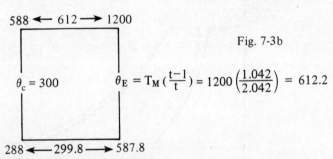

Fig. 7-3b

The specific compression work $W_C = K_P \times 300$
The specific expansion work $W_E = K_P \times 612.2$
The specific net work $W_E - W_C = K_P \times 312.2$

$$\therefore \eta_O = \frac{312.2}{612} = 0.51 \left(\text{note that this equals } 1 - \frac{1}{t}\right)$$

Representation of the Actual Cycle with Losses

This is shown in Fig. 4a. The temperature ratio t of the cycle is now $t = \dfrac{T'_C}{T_o} =$
$$\frac{T_o + \eta_C \theta_C}{T_o} = 1 + \frac{\eta_C \theta_C}{T_o}$$

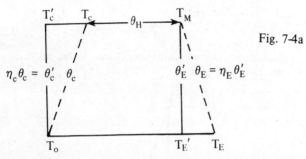

Fig. 7-4a

(Note: The 'slope' of the broken lines is exaggerated to accommodate the symbols, and their length is not representative of their magnitudes thus $T_M T_E = T_M - T_E = \eta_E \theta'_E$ which should be shorter than $T_M T'_E$. (This would be rectified if a diagram similar to Fig. 6-1h of Section 6 were used.)

Figure 4b shows a numerical example for $T_O = 288°K$; $T_M = 1200°K$; $\theta_C = 300°C$; $\eta_C = 0.85$ and $\eta_e = 0.87$.

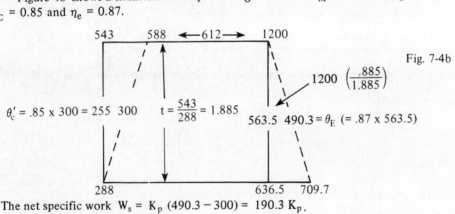

Fig. 7-4b

The net specific work $W_s = K_p (490.3 - 300) = 190.3 \, K_p$.

Heat added $H = K_p (1200 - 588) = 612 \, K_p$ ∴ $\eta_o = \dfrac{190.3}{612} = .311$

Once one is familiar with the layout, it is unnecessary to draw a diagram. The spatial relationship of figures only will serve the purpose, thus the example shown in Fig. 7-4b may be set out thus:

```
543   588  |612|  1200
|255| 300         |563.5| |490.3|  t = 1.885
288               635.5    709.7
```

The temperature differences are distinguised from absolute temperatures by 'boxing'.

For most practical purposes one may omit T'_C, θ'_C, θ'_E and T'_E and reduce the layout to:

```
      588  |612|  1200
      |300|  t = 1.885   |490.3|
288                       709.7
```

since t may be obtained from $t = 1 + \dfrac{\eta_C \theta_C}{T_O}$ and $\theta_E = \eta_E T_M \left(1 - \dfrac{1}{t}\right)$

Fig. 5 is a repeat of Fig. 4a in non-dimensional form where all temperatures and temperature differences are multiples of T_O.

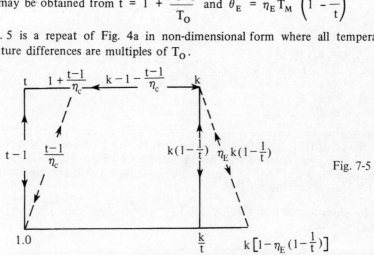

Fig. 7-5

In non-dimensional terms, the positive work of expansion is seen to be $W_E = \eta_E k \left(1 - \dfrac{1}{t}\right)$ and the negative work of compression is seen to be $\dfrac{t-1}{\eta_C}$, hence the net work is given by

$$W_E - W_C = \eta_E k \left(1 - \frac{1}{t}\right) - \frac{t-1}{\eta_C}$$

The heat added is H (non-dimensional) $= k - 1 - \dfrac{t-1}{\eta_C}$ ∴ $\eta_O = \dfrac{\text{net work}}{\text{head added}} =$

$$\frac{\eta_E k \left(\dfrac{t-1}{t}\right) - \dfrac{t-1}{\eta_C}}{k - 1 - \dfrac{t-1}{\eta_C}}$$

which reduces to

$$\eta_O = \left(\frac{t-1}{t}\right)\left(\frac{\eta_C \eta_E k - t}{\eta_C (k-1) - t + 1}\right) \tag{7-1}$$

Since $\dfrac{t-1}{t}$ is the air standard efficiency, the quantity in the second pair of brackets may be described as the 'thermodynamic efficiency,' which, if $\eta_C = \eta_E = 1.0$ reduces to 1.0.

The above formula for η_O shows that η_O is zero if $\eta_C \eta_E \dfrac{k}{t} - 1 = 0$, i.e., if $\eta_C \eta_E \dfrac{k}{t} = 1$. This explains the many failures to evolve practical gas turbines in the early part of this century when component efficiencies were about 70% or below and the maximum temperature of the cycle was severely limited by the materials available.

As may be seen from Figure 5 $\frac{k}{t}$ is the 'positive/negative work ratio' of the ideal cycle.

It should also be fairly obvious that expansion efficiency is relatively more important than compression efficiency and that this relative importance increases with increase of $\frac{k}{t}$.

For any particular values of k, η_C and η_E the optimum value of t can be found from $\frac{d\eta_O}{dt} = 0$. The result is a somewhat clumsy quadratic for t. It is preferable to plot η_O as a function of t to get the optimum and also a picture of the sensitivity of η_O to t, especially if one allows for the normal decrease of η_C and increase of η_E with increase of t (i.e., with pressure ratio). (The increase of η_E will normally more than offset the decrease of η_C.)

SECTION 8
Turbo Gas Generators

Gas turbines operating with the constant pressure (or Brayton) cycle can take many forms. Such engines as the piston engine operating with the constant volume (or 'Otto') cycle or with the Diesel cycle (in which heat is added at substantially constant pressure and rejected at constant volume) are limited in variety by the fact that all the cycle stages – inspiration (or 'suction'), compression, combustion and exhaust, all take place in a single organ, the cylinder, so variations are basically limited to number, size, and arrangement of cylinders. In the gas turbine, on the other hand, each phase of the cycle takes place in separate 'organs' – the compressor or compressors, the combustion chamber or chambers and the turbine or turbines, and this fact makes possible a large number of arrangements. Moreover such devices as heat exchangers, compressor intercoolers (rarely used) add further to the possible range of types of gas turbine power plant. However, one of the commonest features is to use a combination of compressor, combustion chamber/s and compressor turbine to generate a supply of heated and pressurized gas for use in one or more power turbines (or propelling jet, etc.) more or less mechanically independent of the compressor–combustion–compressor turbine combination which is usually known as a 'gas generator.' A gas generator for use with a power turbine is depicted schematically in Figure 1.

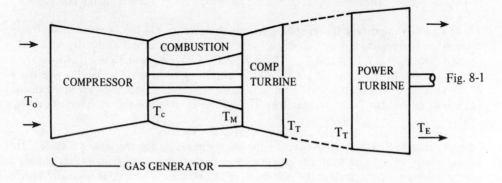

Fig. 8-1

The corresponding P-V diagram is shown in Fig. 2 and the F. W. diagram in Fig. 3a where, for convenience, the relationships between temperature differences, efficiencies and temperature are also shown.

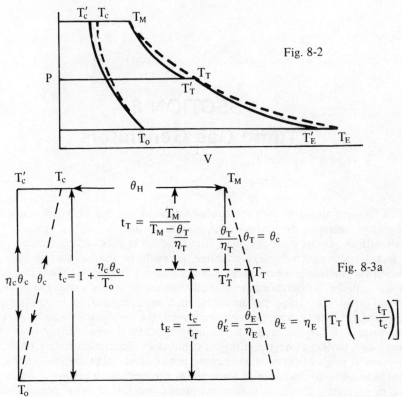

Fig. 8-2

Fig. 8-3a

The compressor turbine usually also drives auxiliaries such as a fuel pump, etc., but the power for such purposes is very small compared with that required by the compressor, so that there is negligible error in assuming that $\theta_T = \theta_C$, i.e., the actual temperature drop in the compressor turbine is equal to the actual temperature increase in the compressor.

Two other important things should be noted, namely: (1) the product of the temperature ratios in expansion must equal the overall temperature ratio, i.e., $t_T \times t_E = t_C$, and (2) that since $T_C = T_O + \theta_C$ and $T_M = T_C + \theta_H = T_O + \theta_C + \theta_H$ and $T_T = T_M - \theta_C = T_O + \theta_C + \theta_H - \theta_C = T_O + \theta_H$ then $\theta_H = T_T - T_O$. (This must be so since, at T_T, the negative work of the cycle has been dealt with, so the total temperature at compressor turbine exhaust (T_T) must exceed the initial temperature T_O by the temperature rise in combustion (θ_H).)

A further point: the exhaust kinetic energy of the compressor turbine is a loss in relation to that turbine but contributes to the energy available for the power turbine. The kinetic energy of exhaust from the power turbine is, however, a total loss (unless made use of in some way). Nevertheless the efficiency of the power turbine η_E is normally higher than that of the compressor turbine η_T, especially if T_M is so high that blade cooling means are necessary in the gas generator turbine (i.e., compressor turbine).

θ_E represents the net specific work of the cycle, i.e., specific power = $K_p \theta_E$,

$$\therefore \eta_O = \frac{\theta_E}{\theta_H}.$$

From Fig. 3a it may be seen that $\theta_E = \eta_E T_T \left(1 - \frac{t_T}{t_C}\right) = \eta_E (T_M - \theta_C) \left(1 - \frac{t_T}{t_C}\right)$
hence, since T_O, T_M, θ_C, η_C, η_T and η_E are normally known (or assumed) θ_E may be readily calculated.

Example 8-1
Fig. 3b shows a numberical example for $T_O = 288°K$; $T_M = 1200°K$; $\theta_C = \theta_T = 300°C$; $\eta_C = 0.85$; $\eta_t = 0.87$ and $\eta_E = 0.90$, the isentropics being omitted.

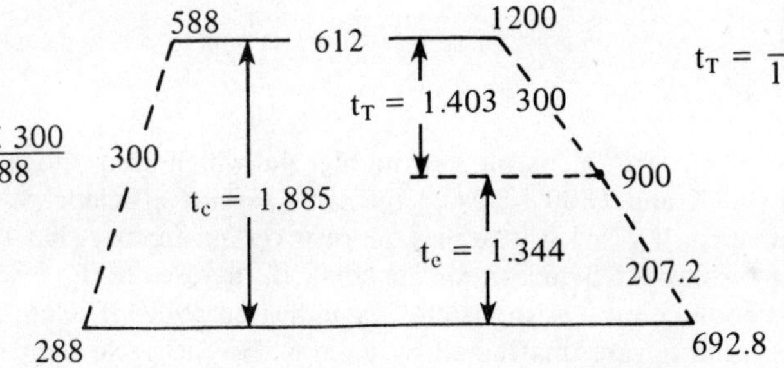

$t_c = 1 + \frac{.85 \times 300}{288}$

$t_c = 1.885$

$t_T = 1.403$

$t_c = 1.344$

$t_T = \frac{1200}{1200 - 300}$
$.87$

$t_e = \frac{1.885}{1.403}$

Fig. 8-3b

$$\theta_E = .9 \times 900 \left(1 - \frac{1}{1.344}\right) = 207.2 \quad \therefore \eta_O = \frac{207.2}{612} = .339$$

The specific fuel consumption f_S for fuel of calorific value of 10,500 CHU/lb is $\therefore f_S = \frac{C_p \theta_H}{10,500} = \frac{0.24 \times 612}{10,500} = 0.014$ lb/sec. The specific power $W_S = 207.2 \times 336 = 69,612$ ft/lbs/sec or 126.6 HP. (It is useful to remember that the multiplier to convert θ_E to HP is 0.611.)

Note that, in the above example, the fuel represents a mass addition of 1.4%.

Figs. 2, 3a and 3b could equally well represent a simple jet engine at *static sea level* operation, in which case the expansion from T_T to T_E would represent the kinetic energy of the jet. The efficiency of this expansion, however, would be much higher than the 90% assumed for the example of Fig. 3b.

With the layouts of Figures 3a and 3b in mind, it is a simple matter to find how η_O, etc., vary with t_C by tabulation. For example the table below shows the effect of varying θ_C and therefore t_C assuming $T_O = 288°K$, $T_M = 1200°K$, $\eta_C = 0.85$, $\eta_T = 0.87$, $\eta_E = 0.90$ (i.e., the same assumptions as for Fig. 8-3b except for the variation of θ_C).

θ_C	t_C	T_C	θ_H	T_T	t_T	t_E	θ_E	η_O	f_S	W_S(HP)
200	1.590	488	712	1000	1.237	1.286	200.0	0.281	0.0163	122.2
250	1.738	538	662	950	1.315	1.322	208.1	0.314	0.0151	127.1
300	1.885	588	612	900	1.403	1.344	207.2	0.339	0.0140	126.6
350	2.033	638	562	850	1.504	1.351	198.9	0.354	0.0128	121.6
400	2.181	688	512	800	1.621	1.345	184.7	0.361	0.0117	112.9
450	2.328	738	462	750	1.758	1.325	165.4	0.358	0.0106	101.1

$$\left(t_C \text{ from } t_C = 1 + \frac{\eta_C \theta_C}{288} \quad T_C = 288 + \theta_C \quad \theta_H = 1200 - T_C \quad T_T = 1200 - \theta_C \right.$$

$$t_T = \frac{1200}{1200 - \frac{\theta_C}{\eta_T}} \quad t_E = \frac{t_C}{t_T} \quad \theta_E = \eta_E T_T \left(1 - \frac{1}{t_E}\right) \quad \eta_O = \frac{\theta_E}{\theta_H} \quad f_S = \frac{0.24 \times \theta_H}{10{,}500}$$

$$\left. W_S = 0.611 \theta_E . \right)$$

The variation of η_O with θ_C is plotted in Fig. 4a which also shows the effect of changing T_M to 1100°K and 1300°K, using the same turbine efficiencies. This assumption means that the curve for T_M = 1300°K may be over optimistic and that for 1100°K a little pessimistic in that turbine efficiencies tend to fall with increase of T_M owing to losses resulting from blade cooling, etc. Also, as already indicated above it is unreasonable to assume that η_C, η_T and η_E are unaffected by θ_C, i.e., by the cycle temperature ratio t_C. This effect will be examined briefly below.

The values of W_S are shown in Fig. 4b.   Fig. 4c shows pressure ratio p v. θ_C.

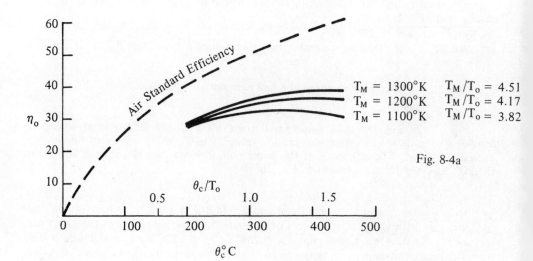

Fig. 8-4a

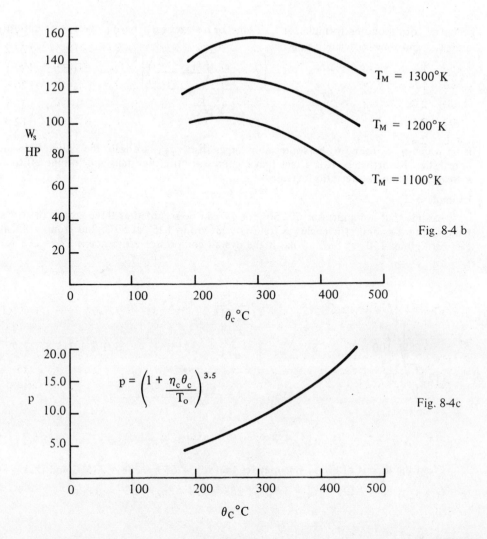

Fig. 8-4 b

Fig. 8-4c

Polytropic Efficiency

It can be shown that if a compressor has n stages and it is assumed that each stage has the same temperature ratio t_s and the same stage efficiency η_{cs} then $\eta_c = \dfrac{t_s^n - 1}{\left(1 + \dfrac{t_s - 1}{\eta_{cs}}\right)^n - 1}$
from which η_c is found to be increasingly less than η_{cs} as n is increased. The proof is not difficult but it is not proposed to include it here because it is unrealistic to assume equal efficiency per stage because of the differences of geometry of successive stages (blade length, etc.) and the fact that, except for the first stage, each stage is receiving the disturbed flow from all the preceding stages.

The corresponding formula for a turbine of n stages each having stage temperature ratio t_{TS} and stage efficiency η_{TS} is

$$\eta_T = \frac{1 - \left[1 - \eta_{TS}\left(\frac{t_{TS} - 1}{t_{TS}}\right)\right]^n}{1 - \frac{1}{t_{TS}^n}}$$

from which η_T is found to be appreciably larger than η_{TS}, but again the assumptions are unrealistic. Nevertheless, it is a fact that compressor efficiency falls and turbine efficiency increases as temperature ratio increases.

Example 8-2

Assume that a compressor of 350° for θ_C can be regarded as three groups of stages of temperature rises and efficiencies as follows: 1st group 125° at 89%; 2nd group 100° at 89%; 3rd group 150° at 88%. What is the overall compressor efficiency if $T_0 = 288°K$?

$$\theta_3' = 132 \quad \begin{array}{c} 663 \\ 150 = \theta_3 \text{ @ }.88 \end{array} \quad t_3 = 1 + \frac{132}{513} = 1.257$$

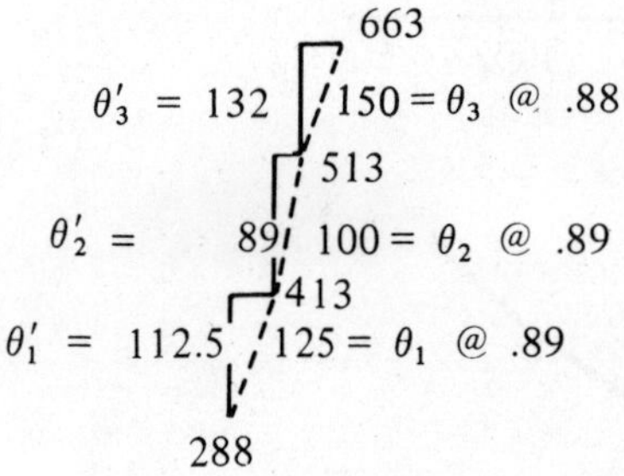

$$\theta_2' = 89 \quad \begin{array}{c} 513 \\ 100 = \theta_2 \text{ @ }.89 \end{array} \quad t_2 = 1 + \frac{89}{413} = 1.215 \quad t_c = 1.391 \times 1.215 \times 1.257 = 2.1$$

$$\theta_1' = 112.5 \quad \begin{array}{c} 413 \\ 125 = \theta_1 \text{ @ }.89 \end{array}$$

$$288 \qquad\qquad t_1 = 1 + \frac{112.5}{288} = 1.391$$

Fig. 8-5

From the layout of Fig 5, one may see that $\theta_C = 663 - 288 = 375°C$ and that $t_C = t_1 \times t_2 \times t_3 = 2.125$ ∴ $\theta_C' = 1.125 \times 288 = 324$ ∴ $\eta_C = \frac{324}{375} = 0.864$.

Example 8-3

It would be possible to drive the compressor of Example 2 with a single stage turbine in certain circumstances, but let it be supposed that with a T_M of 1300°K it is considered desirable to use a three stage turbine (to limit the need for blade cooling to the first stage and for other reasons) and that the temperature drops and stage efficiencies are as follows:

1st stage $\theta_{T1} = 200°C$ at $\eta_{T1} = 0.86$
2nd stage $\theta_{T2} = 90°C$ at $\eta_{T2} = 0.91$
3rd stage $\theta_{T3} = 85°C$ at $\eta_{T3} = 0.91$

What is the value of η_T for the compressor turbine?

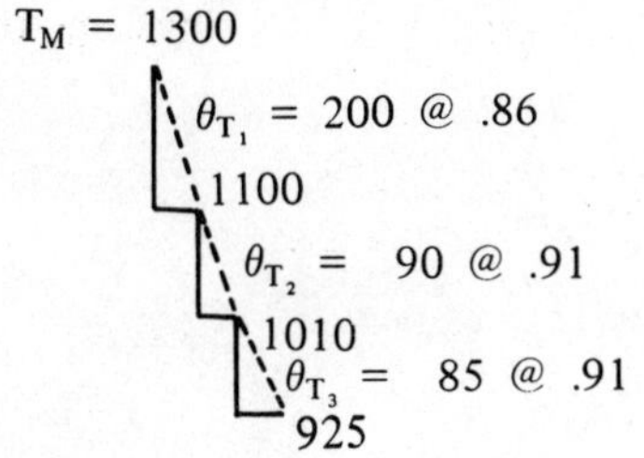

$T_M = 1300$

$\theta_{T_1} = 200$ @ .86 $t_1 = \dfrac{1300}{1300-\dfrac{200}{.86}} = 1.218$

1100

$\theta_{T_2} = 90$ @ .91 $t_2 = \dfrac{1100}{1100-\dfrac{90}{.91}} = 1.099$

1010

$\theta_{T_3} = 85$ @ .91 $t_3 = \dfrac{1010}{1010-\dfrac{85}{.91}} = 1.102$

925

Fig. 8-6

From the layout of Fig. 6, $t_T = 1.218 \times 1.099 \times 1.102 = 1.475$ $\therefore \theta'_T = 1300 \left(\dfrac{0.475}{1.475}\right) =$ 418.6 and since $\theta_T = 200 + 90 + 85 = 375$ $\eta_T = \dfrac{375}{418.6} = 0.896.$

Example 8-4

Using the figures of Examples 2 and 3, and assuming that final expansion efficiency $\eta_E = 0.9$, what is η_O?

Since, (from Example 2), $t_C = 2.125$ and, (from Example 3), $t_T = 1.475$ $t_E = \dfrac{2.125}{1.475} = 1.4407$ $\therefore \theta_E = 0.9 \times 925 \left(\dfrac{0.4407}{1.4407}\right) = 254.7°C.$ From Example 2, $T_C = 663°K$ $\therefore \theta_H = 1300 - 663 = 637°C$ $\therefore \eta_O = \dfrac{\theta_E}{\theta_H} = \dfrac{254.7}{637} = 0.40$

This improvement as compared with the curve for 1300°K in Fig. 4a is, of course, due to the higher values of η_C and η_T.

It may be seen from the tabulation associated with Example 1 that specific power peaks at a much lower value of θ_C than that for maximum efficiency but this is greatly offset by the effect of pressure ratio on the mass flow rate. This is the subject of the following section.

SECTION 9
Mass Flow Rate in Gas Turbines

The thermal cycles discussed in Sections 7 and 8 provide means for calculating directly such quantities as specific fuel consumption, specific power, etc., but for complete performance calculations it is, of course, essential to know the mass flow rate Q.

A simple 'axiom' may be stated here: "What flows in is determined by what can get out." In the limit, if the exhaust (and any other means of escape) is completely blocked off there can be no through flow (as in the case of a pitot tube). One would think that the above stated axiom is self evident, yet one of the longest and most costly patent infringement actions in British history (Rateau vs. Rolls Royce and others) turned on it, in that Rateau claimed that, in jet engines and the like, a divergent intake ahead of a compressor 'controlled' the axial velocity at the compressor inlet, which was tantamount to arguing that the shape of the intake duct was a controlling factor in determining the throughput. In fact, unless it has a choking throat, the only way in which the intake duct can influence the mass flow is by the small effect its efficiency has on overall pressure ratio.

The design value of Q is determined by the designer to meet the power W_D needed for the design purpose. If the design specific power is W_{SD} then $Q_D = \dfrac{W_D}{W_{SD}}$. The value of Q_D thus decided determines the size of the power plant and 'fixes' a number of critical apertures, e.g., turbine nozzle throats, exhaust duct dimensions, etc. These critical areas then determine the value of Q at all running conditions.

In short, Q depends primarily on the geometry of the expansion part of the system. (As will be seen later, there are circumstances where 'variable geometry' may be used to obtain more desirable values of Q in certain modes of operation.)

There is, however, a limiting factor of considerable importance, namely the maximum permissible value of the axial velocity at entry to the first stage compressor rotor blades. In the limit this cannot exceed the local speed of sound, but, in the present state of the art, it normally has to be well below this to keep down the Mach number relative to the moving blades at full speed. The axial velocity at entry to the first compressor rotor is usually designed to be in the range 500–550'/sec. This, coupled with the largest possible tip/hub ratio, i.e., $\dfrac{\text{blade tip diameter}}{\text{blade root diameter}}$ at the entry plane, for the first rotor often determines the diameter of an engine (almost always in the case of aero-engines other than helicopter engines).

It will be obvious that, for geometrically similar engines with the same thermal cycles, the power will be proportional to the square of the diameter while the weight would be proportional to the cube of the diameter, i.e., the 'square-cube law' would apply, so it would seem to pay to use, say, 4 small engines instead of one engine of twice the size, but unfortunately, though the 'aerodynamic' scale down is feasible (at the expense of some loss due to scale effects, i.e., the effect of reduced Reynolds Number) the apparent potential 50% weight saving would not be achieved at reasonable cost because mechanical and manufacturing problems are greatly increased by size reduction. It is in fact, far easier to design efficient high power gas turbines than low power units of acceptable efficiency—hence the failure, to date, of the gas turbine to 'invade' the automotive and light aircraft fields.

Multiple small engines would also greatly complicate control problems, instrumentation, etc. In any case, designers have been very skilful in 'defeating' the square-cube law.

It has been shown in Section 2 that flux density ψ for isentropic flow is given by $\psi = 1.532 \dfrac{P_T}{\sqrt{T_T}} \dfrac{\sqrt{t-1}}{t^3}$ for air (see Eq. 2-9) hence the flow Q through a duct of section S is given by $Q = \psi S = 1.532 \dfrac{S \cdot P_T}{\sqrt{T_T}} \dfrac{\sqrt{t-1}}{t^3}$. It has also been shown that the maximum value of ψ is $\psi_C = 0.397 \dfrac{P_T}{\sqrt{T_T}}$, (see Eq. 2-7) and occurs in choking throats when the temperature ratio from total to static in the throat is $t_C = 1.2$. It has further been shown that $\psi = 0.397 \dfrac{P_T}{\sqrt{T}}$ is a good approximation for total to static temperature ratios in the range 1.15–1.25 which largely covers the values of t which occur in turbine nozzle and rotor blade rings.

Though the flow through a gas turbine is not isentropic due to losses it is nevertheless reasonable to assume that though the constant of proportionality will be affected by this fact, the magnitude of Q will still be proportional to the total pressure at entry to the first turbine stage and inversely proportional to $\sqrt{T_M}$, i.e., to assume that $\dfrac{Q}{Q_D} = \dfrac{P_D}{P} \sqrt{\dfrac{T}{T_D}}$ where the suffix D signifies a reference condition (e.g., the design values of P_{max} and T_{max}).

The value of Q can be affected by the compressor and combustion chamber characteristics in that if the power output is reduced by reduction of T_M, (i.e., cutting back on fuel injection) while maintaining speed of rotation, the resulting increase of mass flow will usually reduce compressor delivery pressure which tends to counteract the effect of reduced

T_M, the net effect being small. It is unusual, however, for a gas generator to maintain speed with reduction of fuel, but it can be done by manipulating the effective 'back pressure' at compressor turbine exhaust, e.g., by opening up the final nozzle in the case of a jet engine (a means used by the writer in the early days of jet engine development to obtain a limited range of compressor characteristics).

Referring to the curves of Figs. 8–4a and 8–4b of Section 8, one may see that for optimum efficiency for the values of T_M the following figures apply:

For T_m = 1100 max η_O occurs at θ_C = 350 and W_s = 92.6 HP/lb/sec (η_O = 0.33)
For T_m = 1200 max η_O occurs at θ_C = 400 and W_s = 112.9 HP/lb/sec (η_O = 0.36)
For T_m = 1300 max η_O occurs at θ_C = 450 and W_s = 134.0 HP/lb/sec (η_O = 0.39)

Thus, over the range considered, there is virtually a linear relationship between T_M and θ_C for optimum efficiency and between either T_M or θ_C and optimum efficiency.

As mentioned above, it is undesirable to exceed 500–550'/sec for the axial velocity at compressor entry and this fact will therefore decide the first stage compressor diameter for a given horse power.

Example 9-1

What is the diameter D of the first stage compressor for a 10,000 HP gas turbine, given T_O = 288°K, P_O = 2116 p.s.f. T_{max} = 1300°K; axial velocity at compressor entry = 550'/sec; θ_C = 450°C, and compressor tip/hub ratio = 2.0?

From above, W_s = 134 HP/lb/sec $\therefore Q = \frac{10,000}{134}$ = 74.63 lb/sec $\therefore$ 74.63 = $\rho u S$ where S is the entry annulus = $\frac{0.75\pi D^2}{4}$. For u = 550'/sec. $\theta_u = \left(\frac{550}{147.1}\right)^2$ = 13.98 $\therefore$ t (total to static) for acceleration from u = 0 to u = 550 is given by t = $\frac{288}{288 - 13.98}$ = 1.051 for which $\sigma = 1.051^{2.5}$ = 1.132. The stagnation density $\rho_O = \frac{2116}{288 \times 96}$ = 0.0765 $\therefore \rho$ at compressor entry = $\frac{0.0765}{1.132}$ = 0.0676. $\therefore$ 74.63 = 0.0676 x 550 x $\frac{0.75\pi D^2}{4}$ from which D^2 = 3.408 and D = 1.85 ft or 22.2".

For T_M = 1100°K and θ_C = 350, D would need to be increased in the ratio $\sqrt{\frac{134}{92.6}}$ = 1.203. Why, therefore, one might ask, might a design be chosen which has a 20% larger diameter and substantially lower efficiency. The answer is that there are certain circumstances in which capital cost is more important than running costs, e.g., 'peak load lopping' in electric power generation where 'boost power' is required for only a fraction of total generating time. The machine with a lower T_{max} though larger in diameter, might be much cheaper by reason of fewer compressor and turbine stages and freedom from the complications of blade cooling, etc.

The above example illustrates the difficulty of scaling down. For a geometrically similar unit of 100 HP, the diameter would be 2-1/4" approximately which would be quite impractical from the manufacturing point of view, and the efficiency would be seriously affected by scale effect. (This is a very hypothetical case and somewhat contradictory in that to obtain the 100 HP with efficiency reduced by scale effect, Q would have to be somewhat larger than is implied.)

The geometry having been 'frozen' for the design condition, Q for conditions other than design is, as stated above, given by $\frac{Q}{Q_D} = \frac{P}{P_D} \sqrt{\frac{T_D}{T}}$ where the suffix D again means the design condition.

Example 9-2

Taking the figures of Example 9-1 above and referring to Figs. 8-4a, 8-4b and 8-4c of Section 8, what would be the effect on power if 'throttling back' reduced θ_C to 400 with $T_M = 1200°K$ and θ_C to 350 with $T_M = 1100°K$? (This example ignores the 'lock' which may exist between T_M and θ_C to be discussed later.)

For 400/1200 (this is an abbreviation for $\theta_c = 400°C$ and $T_M = 1200°K$) the overall pressure is ratio 15.31 as compared with 19.25 for the design case (450/1300)

$$\therefore \frac{Q}{Q_D} = \frac{15.31}{19.25} \sqrt{\frac{1300}{1200}} = 0.828, \text{ while } W_S \text{ is reduced from } 134 \text{ HP/lb/sec}$$

to 107.3 HP/lb/sec $\therefore \frac{W_S}{W_D} \frac{Q}{Q_D} = \frac{107.3}{134} \times 0.828 = 0.6629$ $\therefore$ Total HP is reduced from 10,000 to 6,629 HP.

For 350/1100 the pressure ratio is 11.98 $\therefore \frac{Q}{Q_D} = \frac{11.98}{19.25} \sqrt{\frac{1300}{1100}} = 0.677$. W_S is reduced to 92.6 HP/lb/sec $\therefore$ Power $= 0.677 \times \frac{92.6}{134} \times 10,000 = 4,675$ HP.

This example assumes that, over the range considered, component efficiencies are unchanged. While this may be reasonable for the gas generator for the range covered it is not so acceptable for the power turbine unless its function allows it to 'keep in step' with the gas generator unit. The variation of efficiency with load for the power turbine therefore depends upon its purpose. Thus, if driving an AC generator it must run at constant speed whatever the load. If driving a variable pitch airscrew or propellor the load would depend on pitch setting, aircraft speed, etc.

As will be shown later (and as might be guessed from the fact that temperature changes are measures of kinetic energy changes) θ_C tends to be proportional to the square of rotational speed N, hence in the last example $\frac{N}{N_D}$ for $T_M = 1200°K$ and $\theta_C = 400°C$ would be $\sqrt{\frac{400}{450}} = 0.943$ and, for $T_M = 1100°K$ and $\theta_C = 350°$, would be $\sqrt{\frac{350}{450}} = 0.882$, thus power varies very rapidly with N. Thus, in the last Example, at 88% of full speed the power is 46.7% of full power and at 94.3% of full speed the power is 66.3% of full power; again assuming that the nature of the load allows the efficiency of the power turbine to be maintained.

It will be noted, however, that for the speed range 88% to 100% the variation of $\frac{Q}{Q_D}$ is from 67.7% to 100%. Unfortunately the variation of axial velocity at compressor entry is greater than this because the reduced axial velocity due to reduced Q means an increased density at compressor entry. The effect, however, is small for the range under discussion, and would be much smaller if the design speed axial velocity were more conservative—at the expense of increased diameter.

The foregoing suggests that in cases where gas turbine power plant is used at full power for only a small proportion of total running time it would be preferable to design the components to suit the part load condition which is most used and accept a small loss of component efficiency at full load.

The foregoing discussion does not take into account the sensitivity of compressor delivery pressure and efficiency to mass flow rate, especially in the case of multi stage axial flow compressors. This matter will also be discussed later.

SECTION 10

More on Part Load Performance of Gas Generators

The subject of part load performance has been dealt with in general terms in the preceding section. It is the purpose of this section to take a closer look at the relevant theory.

Since the output of pressurized and heated air (and combustion products) may be used for many different purposes — not necessarily involving power turbines — it is useful to think in terms of the gas power generated, and to use the concept of gas generator efficiency η_{GG}, defined as gas power generated/heat energy of combustion. The overall efficiency of the total power plant is then given by $\eta_O = \eta_{GG}\eta_E$. Thus, referring to Example 9-1 the 'gas horsepower' available would be $\frac{10,000}{0.9} = 11,110$ and η_{GG} would be $\frac{\eta_O}{0.9}$, i.e., $\frac{0.39}{0.9} = 0.433$ or 43.3%.

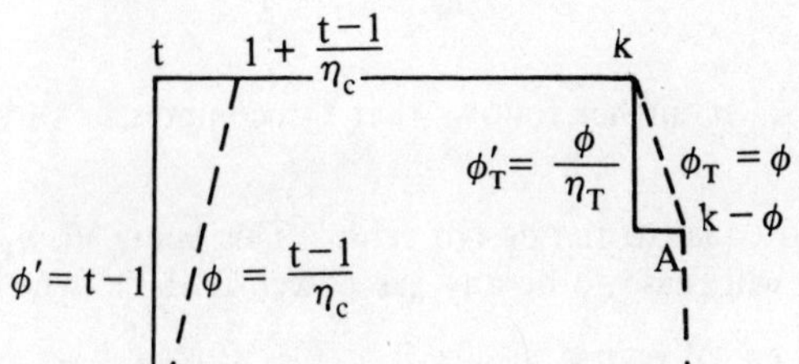

Fig. 10-1

The non-dimensional cycle for a gas generator is shown in the diagram of Fig. 1. All temperatures are shown as multiples of T_o and $\phi \equiv \frac{\theta_C}{T_o}$ etc. Expansion AB after the compressor turbine is unspecified.

If the pressure at k is P then the pressure at A is $\frac{P}{t_T^{3.5}} = P_A$ Q for expansion AB is the same as for the compressor turbine, $\therefore$ if one may assume that, between A and B, there are nozzle rings or equivalent fixed areas such that $Q_{AB} \propto \frac{P_A}{\sqrt{k-\phi}}$ then we have $Q \propto \frac{P}{\sqrt{k}}$ for the compressor turbine. Since $Q = Q_{AB}$ it follows that $K_1 \frac{P}{\sqrt{k}} = K_2 \frac{P_A}{\sqrt{k-\phi}}$ where K_1 and K_2 are constants which include equivalent fixed areas.

Since $P_A = \frac{P}{t_T^{3.5}}$, $K_1 \frac{P}{\sqrt{k}} = K_2 \frac{P}{t_T^{3.5}} \sqrt{k-\phi}$ from which

$$t_T^{3.5} \sqrt{1 - \frac{\phi}{k}} = \frac{K_2}{K_1} = \text{const.} \tag{10-1}$$

From the diagram of Fig. 1, $t_T = \frac{k}{k - \frac{\phi}{\eta_r}}$ or $\frac{1}{1 - \frac{\phi}{\eta_r k}}$

Substituting in (1) for t_T gives

$$\left(\frac{1}{1 - \frac{\phi}{\eta_r k}}\right)^{3.5} = \frac{K_2}{K_1} \frac{1}{\sqrt{1 - \frac{\phi}{k}}} \quad \text{or} \quad \left(1 - \frac{\phi}{\eta_r k}\right)^{3.5} = \frac{K_1}{K_2} \left(1 - \frac{\phi}{k}\right)^{0.5} \tag{10-2}$$

From (2) one may see that, for the range over which η_T is substantially constant, $\frac{\phi}{k} = \text{const.}$ Thus it may be seen that the linear relationship between θ_C and T_M, found in Section 9, was no fluke. Since $\phi = \frac{t-1}{\eta_C}$ it follows that, over the range for which η_C may be considered virtually constant, $k \propto t - 1$. So $\frac{k}{k_D} = \frac{t-1}{t_D - 1}$.

Since $t_T = \frac{k}{k - \frac{\phi}{\eta_T}}$ and $k \propto \phi$, it further follows that the compressor turbine temperature ratio remains constant and equal to the design value. This being so, η_T will be substantially constant, hence there will cease to be any gas power available when t is reduced to t_T, i.e., when $\frac{k}{k - \frac{\phi}{\eta_r}} = t$. Then, of course $\eta_{GG} = 0$.

Summarizing the foregoing, we have the following useful conclusions: For the part load range over which η_C and η_T may be regarded as substantially constant; (a) $\frac{T_M}{\theta_C} = \text{const.} = \frac{T_{MD}}{\theta_{CD}}$ (b) $\frac{t-1}{t_D - 1} = \text{const.} = \frac{T_M}{T_{MD}}$ (c) $t_T = \text{const.} = t_{TD}$.

Example 10-1

If the design conditions of a gas generator unit are $T_O = 288°K$, $\theta_{CD} = 450°C$, $T_{MD} = 1300°K$, $\eta_C = 0.85$, and $\eta_T = 0.87$, at what values of θ_C and T_M would the gas power be nil?

$$k_D = \frac{1300}{288} = 4.514 \quad \phi_D = \frac{450}{288} = 1.5625 \quad \therefore t_T \text{ (at all conditions)} = \frac{4.514}{4.514 - \frac{1.5625}{0.87}} =$$

$1.661 = t = 1 + \eta_C \phi$ at zero gas power $\therefore \phi = \frac{0.661}{0.85} = 0.777 \quad \therefore \theta_C = 0.777 \times 288 = 223.9°C$

$\frac{k}{\phi} = \frac{k_D}{\phi_D} = \frac{4.514}{1.5625} = 2.889 \quad \therefore T_M = 2.889 \times 223.9 = 646.8$. The zero power values of θ_C and T_M are thus seen to be approximately half the design values which means that the speed of rotation would be about 71% of design.

Example 10-2

Using the data of Example 1, how would the zero power mass flow and fuel consumption compare with design?

The design temperature ratio of compression $t_D = 1 + \frac{0.85 \times 450}{288} = 2.328$ $\therefore$ the pressure ratio is $2.328^{3.5} = 19.254$.

At zero power $t = 1.661$ $\therefore$ the corresponding pressure ratio is $1.661^{3.5} = 5.91$. Also at zero power $T_m = 646.8$ $\therefore \frac{Q}{Q_D}$ for $W = 0$ is given by $\frac{Q}{Q_D} = \frac{5.91}{19.254}\sqrt{\frac{1300}{646.8}} = 0.435$.

Since θ_C at $W = 0$ is 223.9; $\theta_H = T_M - T_C = 646.8 - (288 + 223.9) = 134.9$ while $\theta_{HD} = 1300 - (288 + 450) = 562$ $\therefore$ the relative fuel consumption is $\frac{134.9}{562} \times \frac{Q}{Q_D} = 0.24 \times 0.435 = 0.104$, i.e., the fuel consumption at $W = 0$ is about 10% of that at design speed. In practice, whatever follows the gas generator is bound to cause some back pressure unless completely by-passed, which means that fuel consumption, idling speed, etc., would be somewhat greater than those calculated in the above two examples.

In Section 9 and referring to Fig. 4a of Section 8 the apparent linear relationship between η_O and θ_C at the higher values of θ_C was purely fortuitous since, as seen above, $\eta_O = 0$ when $\theta_C = 223.9$. In fact η_O falls off very rapidly below $\theta_C = 350$.

As previously indicated, the efficiency and power of any device which uses the energized gas generated by the gas generator are not necessarily simple multiples of η_{GG} and W_{gg} (i.e., gas power) but the part load behavior of the gas generator will give a useful indication of the characteristics of the complete system.

From the foregoing reasoning, and using $a = \frac{k_D}{\phi_D}$ it is now possible to draw a 'doubly non-dimensional' diagram for the gas generator portion of the cycle as shown in Fig. 2.

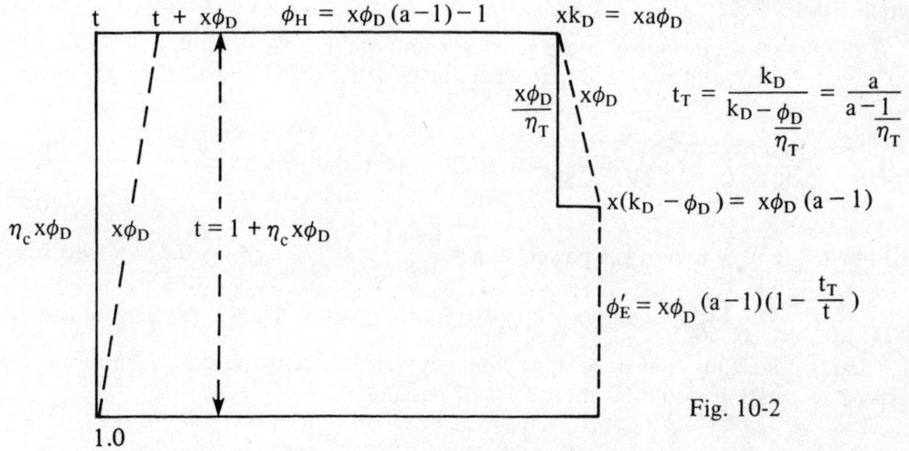

```
t        t + xφ_D     φ_H = xφ_D(a-1)-1      xk_D = xaφ_D
```

$$\frac{x\phi_D}{\eta_T} \quad x\phi_D \qquad t_T = \frac{k_D}{k_D - \frac{\phi_D}{\eta_T}} = \frac{a}{a - \frac{1}{\eta_T}}$$

$$x(k_D - \phi_D) = x\phi_D(a-1)$$

$$\eta_c x\phi_D \quad x\phi_D \quad t = 1 + \eta_c x\phi_D$$

$$\phi'_E = x\phi_D(a-1)\left(1 - \frac{t_T}{t}\right)$$

Fig. 10-2

1.0

ϕ'_E, of course, is the specific gas power available as a multiple of T_0. Writing $\phi'_E = W_S$ then

$$\frac{W_S}{W_{SD}} = \frac{x\left(1 - \frac{t_T}{t}\right)}{1 - \frac{t_T}{t_D}} \qquad (10\text{-}3)$$

where the suffix D again implies the design condition.

For mass flow,

$$\frac{Q}{Q_D} = \frac{P}{P_D}\sqrt{\frac{T_{MD}}{T_M}} = \left(\frac{t}{t_D}\right)^{3.5}\sqrt{\frac{1}{x}} \qquad (10\text{-}4)$$

Hence, multiply (1) and (2) $\quad \dfrac{W}{W_D} = \sqrt{x}\left(\dfrac{t}{t_D}\right)^{3.5}\left(\dfrac{1-\frac{t_T}{t}}{1-\frac{t_T}{t_D}}\right)$ which may be written

$$\frac{W}{W_D} = \sqrt{x}\left(\frac{t}{t_D}\right)^{2.5}\left(\frac{t - t_T}{t_D - t_T}\right) \qquad (10\text{-}5)$$

For given values of ϕ_D and a, t_D and t_T are constants ∴ one may write

$$\frac{W}{W_D} = A\sqrt{x}(1 + Cx)^{2.5}(1 + Cx - B) \qquad (10\text{-}6)$$

where $A = \dfrac{1}{t_D^{2.5}(t_D - t_T)}$; $C = \eta_c\phi_D$ and $B = t_T = \dfrac{a}{a - \dfrac{1}{\eta_T}}$. Again, it may be seen that if $1 + Cx = B = t_T$ then $\dfrac{W}{W_D} = 0$.

Points to note are that (i) $\sqrt{x}$ is a measure of rotational speed N and (ii) since torque x N is proportional to W the torque-speed variation is very rapid for gas turbine engines so that though reduction gears may be (and often are) necessary there should rarely be need for change speed gears.

Example 10-3

Given the following data (a) $\phi_D = \dfrac{\theta_C}{T_C} = 1.5$; (b) $a = \dfrac{T_N}{T_O}\dfrac{1}{\phi_D} = 4.0$; (c) $\eta_C = 0.86$; (d) $\eta_T = 0.88$ and assuming that $\dfrac{N}{N_D} = \sqrt{x}$, what is the value of $\dfrac{N}{N_D}$ for $\dfrac{W}{W_D} = 0.75$?

$t_D = 1 + 0.86 \times 1.5 = 2.290$: $t_T = \dfrac{4}{4 - \dfrac{1}{0.88}} = 1.397$ ∴ the constants for Equation

(4) are: $A = \dfrac{1}{2.29^{2.5}}(2.290 - 1.397) = 0.141$ $C = 0.86 \times 1.5 = 1.29$

$B = t_T = 1.397$ ∴ from (6) $0.75 = \sqrt{x} \times 0.141 (1 + 1.29x)^{2.5}(1 + 1.29x - 1.397)$

or $\dfrac{0.75}{0.141} = \sqrt{x}(1 + 1.29x)^{2.5}(1.29x - 0.397) = 5.319$: from this x is found to be 0.917

hence $\dfrac{N}{N_D} = \sqrt{0.917}$, i.e., N is just under 96% of N_D.

SECTION 11
Gas Turbines With Recuperators

If the exhaust temperature of a gas turbine unit is higher than the temperature at compressor delivery, it should be possible in theory to transfer some of the waste heat of exhaust to the compressed air by means of a heat exchanger or recuperator (sometimes 'regenerator' is used) and thereby to reduce the amount of heat addition in combustion. The effect on overall efficiency can be very large when the pressure ratio is modest and T_M is high as may be seen from the diagrams of Figures 1a and 1b which have the same values of θ_C, T_m, η_c (= 0.87), η_T (= 0.87) and η_E (= 0.90). In Figure 1b it is assumed that a recuperator of 80% transfer efficiency (of counter flow type) is used to transfer exhaust heat to the compressor delivery air thereby raising its pre-combustion temperature to 793.2°K, thereby cutting θ_H from 862°C to 606.8°C. The effect on efficiency is striking. For the cycle of Figure 1a $\eta_o = \frac{293}{862} = 0.34$ or 34%. With the recuperator $\eta_o = \frac{293}{606.8} = 0.483$ or 48.3%.

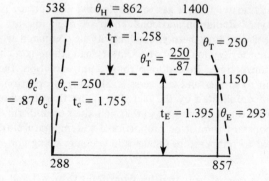
Fig. 11-1a

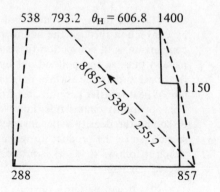
Fig. 11-1b

This example was a more or less random selection and is not an illustration of the whole possibilities of the use of a recuperator. Further, it is not a true comparison in that recuperator pressure loss would reduce both the effective power turbine efficiency and the effective delivery pressure, i.e., the effective value of η_c, but even if one reduces η_C to 0.85, and η_E to 0.88 to allow for recuperator losses, η_O is still 47%.

Unfortunately, recuperators of a capacity sufficient to achieve heat transfers of the order of 80% present severe technical problems in addition to pressure loss. Those problems include great bulk and weight, manufacturing costs, installation, thermal expansion, cleaning difficulties, and so on.

Heat exchangers are familiar features of everyday life, e.g., steam boilers, car radiators and cylinder jackets in liquid cooled piston engines, air conditioning units, refrigerators, etc., but, except for massive steam boilers, the rate of heat transfer is minute compared with that necessary in gas turbines. Thus, for example, if the cycles of Figure 1 were for 1000 HP the mass flow would be $\frac{1000}{293 \times 0.611}$ = 5.586 lb/sec, hence the rate of heat transfer would need to be 5.586 x 255.2 x 0.24 = 342 CHU/sec, equal to 871 HP or 87.1% of the power output. (A reader might say "yes!, but what of it?" auto engines have to get rid of about 25% of the heat value of the fuel consumed via the radiator and this is roughly the same as the power produced." The reply to this is that the transfer is from liquid (water plus anti freeze) to air and not from air to air as in the case of gas turbines.)

There are several types of heat exchanger, but it is not proposed to deal with the various alternatives – parallel flow, cross flow, rotating matrices (in which a cylindrical 'honeycomb' rotates with one portion in the low pressure hot stream, becoming heated up, and with the other portion in the high pressure 'cool' stream where it parts with the heat picked up in the hot stream. So far as is known to the writer, none of the many attempts to solve the sealing problem has succeeded.), etc., because only in the case of the counter flow type can there by any hope of transfers of the order of 80%. Any other type, even if the transfer were 100%, could not raise the pre-combustion temperature by more than 50% of the difference between exhaust temperature and compressor delivery temperature.

The theory of heat transfer in recuperators is very complex and is beyond the scope of this present work. A few major factors will, however, be pointed out, as follows:

(1) The main 'mechanism' of heat transfer from a gas to a metal dividing surface is by convection through the boundary layer. Radiation and conduction are negligible by comparison with convection (gases have a very low thermal conductivity as compared with metals) thus, as was pointed out by Osborne Reynolds about a century ago, there is a close relationship between surface drag and convection. Indeed, as a rough approximation, the rate of heat transfer per unit area from air to surface and vice versa is, like drag, proportional to This apparently simple rule, however, is complicated by the fact that as the fluid is heated or cooled the density varies inversely with the temperature, so that, in a duct of uniform section, since the product uρ remains constant u must be proportional to temperature, from which it follows that ρu^2 is proportional to temperature (changes in pressure being considered negligible).

(2) It will be fairly obvious that the rate of heat transfer from fluid to metal and vice versa will be proportional to the surface area and the temperature difference.

(3) The conductivity of metals is so high compared with gases that for thin tubes or plates separating gas streams, the temperature drop in the metal may be regarded as negligible.

(4) From (1) and (2) above it follows that the rate of heat transfer per unit area is proportional to velocity, since, if area is increased (by, say, increasing tube diameter) u is decreased.

Since pressure loss is proportional to surface drag it follows that a recuperator for large heat exchange and low pressure loss must be very bulky if of tubular construction. Even so, the potential fuel saving is so great that it is worth taking a more detailed look at gas turbine cycles with recuperators assuming that a transfer coefficient of 0.8 can be achieved. This may be conservative since the writer has heard that 0.85 has been obtained in at least one case though at what expense in pressure loss he was not told.

Figure 2 is a repeat of Figure 1b in non-dimensional form (T_E, T_C, H, etc., are multiples of T_0) in which it is assumed that 80% of the difference between T_E and T_C is added to T_C by a recuperator, to give a pre-combustion temperature of T_R.

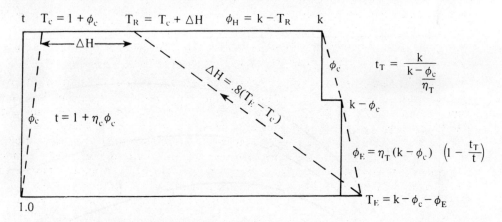

Fig. 11-2

Specific power for given values of ϕ_C and k is the same as for a power unit without a recuperator but, as long as T_E exceeds T_C, ϕ_H is reduced by $\Delta H = 0.8(T_E - T_C)$ and the overall efficiency correspondingly increased.

With a recuperator ϕ_H is found to be $\phi_H = 0.2k + 0.6\phi_C + 0.8\phi_E - 0.2$. Without a recuperator $\phi_H = k - \phi_C - 1$, hence the recuperator increases η_o in the ratio

$$\frac{0.2k + 0.6\phi + 0.8\phi_E - 0.2}{k - \phi_C - 1}$$

if one neglects recuperator pressure losses (an admittedly dubious assumption).

The curves of Figures 3a and 3b show the effect of an 80% recuperator as compared with non regenerated units for values of k = 4.0, 4.5 and 5.0 and over a range of ϕ_C = 0.3 to 1.1. The curve for air standard efficiency is also shown.

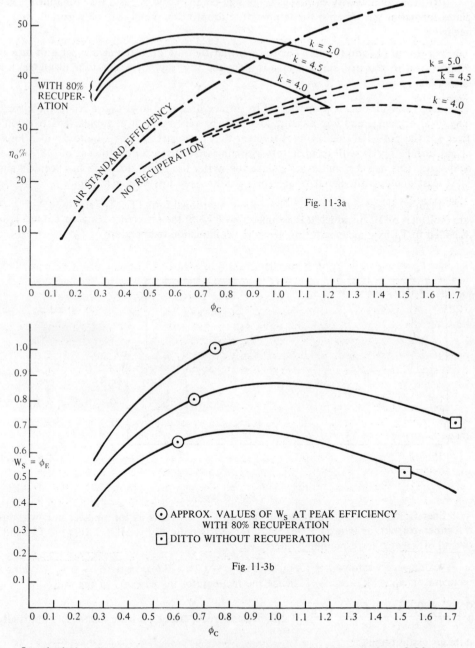

Fig. 11-3a

Fig. 11-3b

In calculating these curves it was assumed that η_C decreased linearly from 0.88 at $\phi_C = 0.3$ to 0.85 at $\phi_C = 1.1$ and that η_T decreased from 0.9 at k = 4.0 to 0.87 at k = 5.0 (to allow for blade cooling) but η_E was left unaltered at 0.88 which is probably rather 'unfair' for the unregenerated edition.

If the curves were prolonged beyond $\phi_C = 1.1$, then, for a given value of k they would intercept when $T_E = T_C$.

The most striking thing to note is, that with a recuperator, peak efficiencies occur at much lower pressure ratios. Indeed they exceed air standard efficiency very substantially. This means, however, that though specific power is the same at given values of ϕ_C and k and, for a given value of k, would not differ greatly at corresponding peak efficiencies (see Figure 3b), the much lower pressure ratio of the 'recuperated' plant means that for a given total power, the diameter of compressors and turbines would have to be substantially larger— which can be a very helpful fact for power units of low design power. Even so, a 100 HP unit with axial velocity at compressor entry of, say, $300'/sec$ and a tip hub ratio of 1.3 would still be under 4.5" diameter with first stage blades about 0.5" long. Nevertheless, this seems to be in the range of practicable possibility and well worth development effort to achieve it, even if efficiencies of less than 30% have to be accepted. Indeed, if a single stage centrifugal compressor could achieve 80% efficiency at the modest pressure ratio of 3.25, overall efficiencies of well over 30% should be possible, and with a centrifugal compressor there might be a reasonable hope that production cost could be competitive with the reciprocating engine.

Using some crude and rather hasty methods, the writer estimates that a counter flow heat exchanger of 80% transfer capacity in which the compressed air flows through a number of unfinned tubes in parallel would require a volume of about 21 cubic feet for a 100 HP unit at the expense of about 5 HP in pressure loss. The effective HP loss would, however, be less than this because some of the friction heating in the high pressure ducting (i.e., between compressor delivery and combustion) would be partially recovered in expansion (as in the case of compressor losses) and the dynamic pressure of the inevitable exhaust velocity from the power turbine might well be sufficient to discount the pressure loss in the hot low pressure ducting.

By using longitudinal fins on the tubing and other devices, the above quoted volume could probably be reduced by 50% or more, at the expense of a considerable increase in manufacturing cost.

Figure 4 is a diagrammatic representation of a counter flow heat exchanger in which the low pressure hot exhaust flows from right to left through the annulus of section $S_h - S_C$, and the cooler compressed air from the compressor delivery flows left to right through a tube of diameter d and section S_C. It is assumed that there is no heat loss through the outer wall of the insulated annulus.

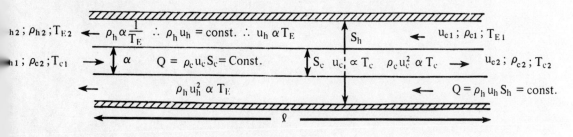

Fig. 11-4

It may be seen that the area available for heat transfer from annulus to tube per unit length is proportional to d. It will be obvious that if the single tube shown is subdivided into n tubes in parallel, each of section $\frac{S_c}{n}$ then the surface per unit length will be increased n times, so that for a given total transfer surface, ℓ may be reduced to $\frac{\ell}{n}$, hence the volume $\ell \times S_h$ tends to vary inversely as n (the word 'tends' is used because there will be scale effects on the heat transfer coefficients). For a given pressure difference, the thickness of the tube walls may also be decreased by $\frac{1}{n}$ but there is a practical limit to this.

If ΔT_1 is the temperature difference between the annulus and the surface of the tube (or tubes) and ΔT_2 is the temperature difference between the tube wall and the flow in the tube, though $\Delta T_1 + \Delta T_2$ will tend to remain constant, the ratio $\frac{\Delta T_1}{\Delta T_2}$ will be affected by the fact that ρu^2 increases in the direction of flow within the tube (or tubes) and therefore the rate of heat transfer per °C, while the opposite happens external to the tubes, therefore $\frac{\Delta T_1}{\Delta T_2}$ will increase from right to left. Again, how this will happen will depend on scale effect, the effect of temperature on viscosity, etc.

In short, in order to keep down the volume and weight of a tubular type counter flow heat exchanger, it is desirable to have the largest practicable number of tubes in parallel for the desired rate of flow and to mount them in such a manner as to avoid thermal expansion and other problems.

It is not intended to imply in the foregoing that only tubular types are feasible, but the writer does not wish to become involved in a long dissertation on ways of preventing distortion by pressure differences across non-tubular surfaces.

SECTION 12
Turbo-Compressors and Turbines

The wording of the heading is chosen to exclude discussion of many types of compressors which are not used as major components of gas turbines, e.g., piston type compressors, rotary compressors such as the 'Rootes Blower,' sliding vane types, etc.

The words 'continuous flow' are also implied in the heading, since it is not intended to cover intermittent flow 'explosion' turbines such as proposed by Holzwarth and others.

The compression process in a gas turbine is performed in a centrifugal compressor or an axial flow compressor or a combination of both. If both types are used, the axial flow compressor invariably precedes the centrifugal type.

In the case of air breathing aircraft engines, the deceleration of the intake air can contribute substantially to the total compression ('ram compression'). Indeed in ramjets, ram compression is the only compression.

The expansion process in gas turbines takes place in turbines having 'full peripheral admission.' They may be either of axial flow type or radial flow type, the former being by far the most common.

In the case of aircraft gas turbines, the latter part of the expansion is used to produce a 'propelling jet,' the expansion in the turbine stage or stages being just sufficient to drive the compressor or compressors (and low power auxiliaries such as fuel pumps, generators, etc.)

Torque and Angular Momentum

Newton's second law of motion states that the rate of change of momentum is proportional to the applied force and takes place in the line of action of the force. ('Momentum' is the product of velocity and mass.)

For many purposes, this law may be re-stated as 'the rate of change of angular momentum is proportional to the applied torque.' This may be regarded as the 'key law' for turbo machinery, the torque and angular momentum being measured relative to the axis of rotation. It should be pointed out, however, that the law is not particular to rotating components; it is absolutely general as may be seen by reference to Figure 1.

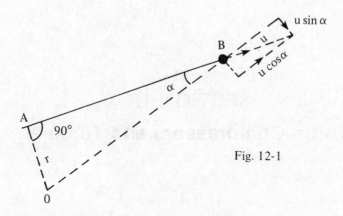

Fig. 12-1

If a particle of mass m is at B and is travelling at uniform velocity u in the direction AB then, if O is *any* point in space, the component of u normal to OB is $u\sin\alpha$, and the component in the direction OB is $u\cos\alpha$. This latter, passing through O, has no component of angular momentum about O ∴ the angular momentum about O is $u\sin\alpha \times$ OB, i.e., since OB$\sin\alpha$ = AO ≡ r ∴ u × OB $\sin\alpha$ = ur. Thus, if the linear momentum mu is constant so is the angular momentum about *any* point in space. (The 'rotational version' of Newton's first law.)

If the particle is being acted upon by a force F, also in the direction AB, then Fr = $mr\frac{du}{dt}$ or Fr = $\frac{d}{dt}$(mru), i.e., torque about O equals the rate of change of angular momentum about O.

This elementary piece has been included because, in dealing with rotary machinery one may be apt to forget that the relevant reference point is not necessarily the axis. For example, suppose Figure 2 represents a small disc A mounted near the rim of a large disc B on a frictionless bearing at O', then however great the angular acceleration of B there is no torque acting on A about O' so A will have reverse rotation relative to B, i.e., one revolution clockwise for every revolution of B, hence relative to a stationary observer the tangental velocity about O at a point on A at a radius greater than OO' will be less than if A were fixed to B and conversely for any point on A less than OO' from O. (This example is relevant to the concept of the 'relative eddy' which, inter alia, explains why the whirl velocity of the fluid leaving the tips of centrifugal compressor blading, is somewhat less than the tip velocity of the blades, i.e., there is a 'slip factor'.)

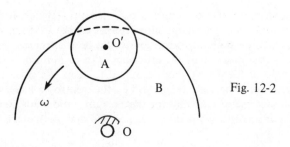

Fig. 12-2

Nevertheless, in the case of turbo machinery, the axis of rotation is the reference for torque and angular momentum. Thus *the torque generated by a turbine is equal to the rate of reduction of angular momentum in the driving fluid*. Similarly *the torque required to drive a compressor or pump is equal to the increase induced in the rate of angular momentum*. (These statements require minor modifications in that bearing friction and windage losses detract slightly from the effective torque generated by a turbine and add to the torque necessary to drive a compressor.) It follows that when a turbine is directly driving a compressor at uniform speed, except for the minor qualifications mentioned above, the rate of change of angular momentum in turbine and compressor must be equal and opposite. An interesting example of this principle is furnished by Nernst's proposal illustrated diagrammatically in Figure 3. Nernst proposed a 'U tube' arrangement rotating at high speed around axis AB, with combustion at C. Air in the ascending leg was compressed by centrifugal force, heated at C and then expanded down the descending leg to exhaust with a greater pressure and temperature than at entry thus having had its energy augmented. There being no net change of angular momentum between entry and exit no driving torque was necessary except for bearing friction and air resistance (which would be high unless there were an enclosing drum). Unfortunately this simple gas generator was impracticable because of the low compression obtainable at C with available materials. Moreover, the potential for increased compression by utilizing the kinetic energy at C could not be used. Nevertheless the device contains the germ of the concept of combining a radial flow compressor with a radial flow turbine as has been done, notably by Von Ohain in the first German turbo-jet engine.

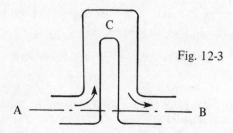

Fig. 12-3

Definition of Efficiencies

An outstanding characteristic of turbo machinery is the enormous volumetric capacity in proportion to size as compared with all other types of power generating machinery, e.g., reciprocating piston engines. So much so that heat loss through casings may be taken as negligible in comparison with the energy changes involved. In other words, the processes of compression and expansion may be regarded as 'adiabatic.' Hence, if there were no losses, i.e., no conversion of kinetic energy into internal energy by friction and turbulence, etc., the processes of compression and expansion would be isentropic. These losses, however, are unavoidable, hence the measure of the efficiency of a compressor — the 'adiabatic efficiency' — is the ratio of the energy input required to achieve a given pressure ratio isentropically to the energy input actually required, i.e., the isentropic increase of enthalpy divided by the actual increase of enthalpy for the pressure ratio. For constant specific heat, enthalpy is proportional to temperature, hence one may define adiabatic efficiency of compression as $\eta_c = \frac{\theta'_c}{\theta_c}$ where θ'_c is the isentropic total temperature increase and θ_c is the actual total temperature increase for the pressure ratio (total to total).

It could be argued that this definition of compression efficiency leaves something to be desired in that a part of the losses are recoverable on expansion and to that extent they are not a 'dead loss,' but this fact can be readily allowed for in cycle calculations.

In expansion through a turbine the adiabatic efficiency η_T is given by $\eta_T = \frac{\theta_T}{\theta'_T}$ if one assumes constant specific heat, where θ_T is the actual temperature drop (total to total) and θ'_T is the isentropic temperature drop (total to total) corresponding to the pressure ratio (total to total).

Occasionally, when the axial velocity at exhaust is high, it may be necessary to think in terms of shaft efficiency η_{TS} defined as $\eta_{TS} = \frac{\theta_T - \theta_a}{\theta'_T}$ where θ_a is the temperature equivalent of the kinetic energy of the exhaust axial velocity, i.e., treating θ_a as a loss so far as shaft power is concerned, though normally it is not a real loss because it forms part of the energy available for further expansion through another turbine or propelling jet.

Example 12-1

A compressor takes in air from the atmosphere of temperature = 288°K and pressure = 2116 lb/sq ft and compresses it to a total pressure of 14000 lb/sq ft; the total temperature at delivery is found to be 531°K. What is the adiabatic efficiency?

The pressure ratio is $\frac{14000}{2116} = 6.616$ ∴ the isentropic temperature ratio is $6.616^{0.286} = 1.717$ ∴ the isentropic temperature rise = $0.717 \times 288 = 206.5$, i.e., $\theta' = 206.5$. The actual temperature rise $\theta = 531 - 288 = 243$ ∴ $\eta_c = \frac{206.5}{243} = 0.85$ or 85%.

Example 12-2

Expansion in a turbine takes place from a total pressure of 1650 p.s.f. and total temperature of 1150°K to a total pressure at exhaust of 778 p.s.f. The total temperature at exhaust is found to be 950°K. Assuming the driving fluid is air and that specific heat is constant at $C_p = 0.24$, what is the adiabatic efficiency?

The pressure ratio $p_T = \frac{1650}{778} = 2.121$ ∴ the temperature ratio (isentropic) t_T is given by $p_T^{0.286} = t_T$ ∴ $t_T = 2.121^{0.286} = 1.24$ ∴ the isentropic temperature drop $\theta'_T =$

$1150 \left(1 - \frac{1}{1.24}\right) = 222.6°C$; θ_T, the actual temperature drop, $= 1150 - 950 = 200°C$

$\therefore \eta_T = \frac{200}{222.6} = 0.899$ or 89.9%.

Example 12-3

If the turbine of Example 2 above has an exhaust axial velocity of 800'/sec, what is the shaft efficiency and the static pressure at exhaust?

The temperature equivalent of 800'/sec $= \left(\frac{800}{147.1}\right)^2 = 29.6°C$ hence $\eta_{TS} = \frac{200 - 29.6}{222.6} = 0.766$ or 76.6%.

The total temperature at exhaust being 950°K, the static temperature at a velocity of 800'/sec $= 950 - 29.6 = 920.4°K$ $\therefore$ the total to static temperature ratio $= \frac{950}{920.4} = 1.032$. The corresponding total to static pressure ratio is $1.032^{3.5} = 1.117$ $\therefore$ the static pressure is $\frac{778}{1.117} = 696.4$ p.s.f.

This example illustrates the large difference there may be between adiabatic efficiency and shaft efficiency if exhaust axial velocity is high.

CENTRIFUGAL COMPRESSORS

Centrifugal (or 'radial flow') compressors and pumps have a longer history in engineering because of their fundamental simplicity but have been largely overtaken in gas turbine technology by axial flow types because intensive development of the latter has led to the achievement of substantially higher adiabatic efficiencies at the high pressure ratios of modern gas turbines. Centrifugal types are still extensively used, however, for many engineering purposes. Also there are a few gas turbine engines in which a centrifugal compressor is used for the final stage of compression where efficiency is rather less important than in the initial stages of compression. Moreover, there are still many Rolls Royce 'Dart' turbo-prop engines in service having two stage centrifugal compressors. Further, in small gas turbines for auxiliary purposes where efficiency is not all important, centrifugal compressors still have their uses, largely again because of their simplicity and the fact that they 'scale down' more easily.

(The terms 'radial flow' and 'axial flow' are somewhat unsatisfactory descriptions because the flow in compressors is never purely radial or axial except at entry or discharge. 'Radial flow' means that the component of velocity which carries the fluid through the compressor is radial. 'Axial flow' means that the component of velocity which carries the fluid through is parallel to the compressor axis. 'Radial flux' and 'axial flux' might be better descriptions.)

The large diameter of the centrifugal compressor is another big disadvantage as compared with the axial flow type. This is of special importance in aircraft gas turbines and more than compensates for the much greater length of the axial flow type. However, the use of a centrifugal stage at the high pressure end of a compound compressor is much less of an embarrassment from the size point of view because the volumetric capacity necessary is much reduced by the preceding compression.

There are several types of centrifugal compressors, but by far the most common is the unshrouded radial vane type as used in aero engine superchargers for many years before the advent of the jet engine, and as used by the writer in his early jet engine designs.

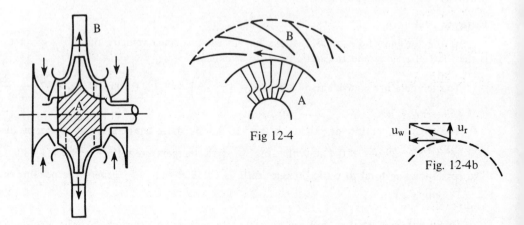

Fig 12-4

Fig. 12-4b

Figure 4 is a rough representation of the double sided radial vaned centrifugal compressor used in the first jet engine, the left hand sketch being a section through the axis and the right hand sketch being a partial view of the impellor A and diffuser blade ring B. The volute into which the diffuser channels discharged is not shown.

The velocity vector diagram for the air leaving the impellor periphery is shown in Figure 4b.

The choice of a double sided compressor was based on the necessity to get the maximum possible flow in proportion to size and thereby reduce the surface friction and other losses as compared with a single sided compressor of the same volumetric capacity. Moreover, this arrangement resulted in a much better match with its directly coupled driving turbine.

The design tip speed was 1470'/sec hence, with the materials available, stress considerations precluded the use of other than radial vanes and the fitting of shrouds. (Shrouds are 'cover plates' attached to the edges of the radial vanes to prevent leakage from the high pressure to the low pressure sides of the blading through the clearance between the high speed impellor and the stationary casing. Shrouded impellors are generally considered to be more efficient where rotational speeds are low enough to permit their use, despite the inevitable inward leakage from tip to intake through the clearance between shroud and casing.)

At the annular entrance to the impellor, axial extensions of the radial blades were curved to conform with the relative velocity of the air entering the impellor. Later, it became more usual to use rings of blades separately mounted to perform the same function, namely to 'steer' the inflowing air into the impellor with minimum turbulance. They came to be known as 'inducers.'

As may be seen from the left hand sketch of Figure 4, the initial direction of flow of the intake air was radially inwards. This was a necessity imposed by the general arrangement of the whole engine. But, at the entry, the air had no angular momentum, and therefore the change of angular momentum in the impellor was from nil to $u_w r_o$ where u_w was the tangential (or 'whirl') component of velocity at the tip radius r_o. (The radial velocity component u_r contributes nothing to angular momentum.) Thus the specific torque Ω_s is given by $\Omega_s = \dfrac{u_w r_o}{g}$ from which, since energy input per unit mass flow W_s equals $\dfrac{\omega \Omega}{g}$, where ω is the angular velocity of the impellor, $W_s = \dfrac{u_w r_o \omega}{g}$, or since $r_o \omega = u_b$, where u_b is the tip speed of the blading, $W_s = \dfrac{u_w u_b}{g}$. If the number of blades were infinite, u_w would equal u_b and so W_s would equal $\dfrac{u_b^2}{g}$. The corresponding temperature rise would thus be $\theta_c = \dfrac{u_b^2}{gK_p}$, i.e., $2\left(\dfrac{u_b^2}{2gK_p}\right)$ or $2\theta_b$. Thus, if $u_w = u_b$, the temperature rise θ_c is twice the temperature equivalent θ_b of the tip speed u_b. For 1470'/sec the temperature equivalent is 99.9°C and twice this is 199.8°. The actual temperature rise at design speed in the first jet engine was 190° approximately, which was in accordance with expectations; the pressure ratio, however, was well below design initially because the adiabatic efficiency hoped for, namely 80%, could not be obtained by a considerable margin. This efficiency was ultimately achieved after numerous modifications to the intake assembly and the diffuser system.

The fact that θ_c is always somewhat less than $2\theta_b$ in a radial bladed compressor is due to the 'slip factor' which in turn is a consequence of the 'relative eddy.' The relative eddy is explained by reference to Figure 5 which represents a 'cell' between two adjacent blades.

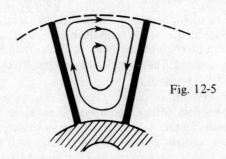

Fig. 12-5

If there were no through flow, then, if the impellor were started from rest and rotated anticlockwise, the body of fluid in the cell would also be accelerated anticlockwise as a whole, but, *relative to the cell itself*, except for skin friction there is no torque and therefore no change in angular momentum, i.e., there is a clockwise circulation within and relative to the cell as shown in Figure 5. This is the relative eddy which, when there is through flow, is superimposed on it, thus reducing u_w at the tip below u_b. Clearly, the 'slip' due to the relative eddy depends on the number of blades and would be zero for an infinite number. Hence, the slip factor is usually taken to be $1 - \frac{1}{n}$ where n is the number of blades, i.e., $u_w = \left(1 - \frac{1}{n}\right) u_b$ at the tip $\therefore \theta_c = \left(1 - \frac{1}{n}\right) \frac{u_b^2}{gK_p} = 2 \left(1 - \frac{1}{n}\right) \theta_b$.

Though the total temperature rise is the result of the specific energy input to the impellor, the rise in static pressure is partly due to centrifugal force within the impellor and partly due to the conversion of kinetic energy into pressure by the effect of the divergent channels in the diffuser blade ring; i.e., the total compression is shared between the impellor and the diffusers, though the latter, being stationary, do no work on the fluid.

If the radial velocity were small compared with the tangential velocity within the impellor then the pressure increase (assuming a very large number of blades) in the impellor would correspond to that due to the 'solid' rotation of a 'forced vortex' in which tangential velocity $u_w = r\omega$. Hence we would have $\frac{\delta P}{\delta r} = \frac{\rho}{g} \frac{u_w^2}{r}$ or $\frac{dP}{\rho} = \frac{r\omega^2 dr}{g}$ $\therefore$ from the relationship $\frac{dP}{\rho} = K_p dT$

$$dT = \frac{\omega^2}{gK_p} rdr \therefore \Delta T = \frac{\omega^2}{2gK_p} (r_o^2 - r_1^2) \equiv \frac{u_{wo}^2 - u_{wi}^2}{2gK_p} \equiv \theta_b - \theta_1$$

So, for $r_i = 0$, $\Delta T = \theta_b$. (But, in this, T is static temperature, so that total temperature at the tip radius would be increased by the temperature equivalent of the tip tangential velocity, i.e., tip blade speed, to give a total temperature increase of $2\theta_b$ as before.) Assuming an isentropic relationship between P and T, $1 + \frac{\Delta P}{P_1} = \left(1 + \frac{\Delta T}{T_i}\right)^{3.5}$ or $\frac{\Delta P}{P_1} = \left(1 + \frac{\theta_b}{T_1}\right)^{3.5} - 1$ where ΔP is the increase of static pressure within the impellor due to centrifugal force. (Note that the assumption that $r_1 = 0$ is equivalent to assuming that the inducers 'impose' a tangential velocity and pressure on the fluid 'as if' the fluid had in fact entered at $r_1 = 0$ which of course, is impossible.) The foregoing ignores the affect of the relative eddy, i.e., an infinite number of blades is assumed.

In practice what happens within the impellor is far more complex than the foregoing suggests. The mechanical simplicity is not matched by the aerodynamic simplicity. At a given radius the pressure on the 'driving face' of a blade must be higher than on the rear face otherwise there would be no torque. This pressure difference must increase with mass flow rate so, in every channel there has to be a substantial tangential pressure gradient superimposed on the radial pressure gradient with a corresponding variation of radial velocity, so that the vector diagram of Figure 4b represents an 'average' state of affairs. Moreover, the implication in Figure 5, that the channels run 'full' is almost certainly never true. Fig. 6 is more likely to represent what actually happens, i.e., a detachment of the flow from the rear face of each blade so that it issues from the impellor periphery in a series of 'jets' having a much higher radial velocity than one would calculate from the axial width at the tip.

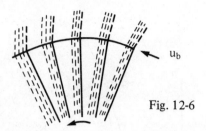

Fig. 12-6

In pre-jet engine days, superchargers and the like usually had about 16 blades, ostensibly to keep down skin friction over the blade surfaces; but, from the first, the writer contended that there ought to be as many blades as was practically possible. With the manufacturing techniques then available, 30 was the largest number which could be milled out of a solid forging. This was later reduced to 29 because tip failures were thought to be a consequence of some kind of resonant interaction between the 30 impellor blades and the 10 diffusers.

It is hard to say what happened in the divergent diffuser passages; a series of waves in all probability, i.e., a far from steady flow. At the design speed of 17,750 r.p.m. and with 29 blades, the flow at entry to each diffuser must have been fluctuating at the rate of 8,580 times/second.

As previously indicated, the torque necessary to drive the impellor is equal to the rate of change of angular momentum induced plus some torque resistance due to casing friction. At a given speed this latter would remain more or less constant irrespective of the mass flow rate and therefore becomes of decreasing importance with increase of flow rate. (The skin friction due to the radial component of flow cannot affect the torque.)

Once the fluid leaves the impellor its angular momentum remains constant (except for the small retarding effect of casing drag) until it engages and enters the diffusor blade ring, hence the tangential velocity decreases with radius, i.e., $u_w r$ = const., and there is a corresponding rise in pressure. It follows that it is possible to obtain a conversion of kinetic energy into pressure energy in a bladeless diffuser space, but unfortunately this greatly increases diameter. Thus, to reduce the tangential velocity to half that at the impellor tip would require that the diameter be twice that of the impellor, and even then 25% of the tip tangential kinetic energy would still remain. Nevertheless, in some commercial compressors where size is not of great importance, bladeless diffuser spaces are used. (On one occasion, the writer persuaded the designer of one such compressor – a diesel engine supercharger – to add a ring of diffuser blades at about twice the impellor diameter. There was an appreciable improvement in delivery pressure and efficiency.) Another big disadvantage of a bladeless diffuser space is the structural problem of bracing the casings against the pressure acting to force them apart.

The performance of a compressor is represented as in Figure 7 in the form of 'characteristic curves' in which pressure ratio is plotted against a mass flow rate parameter for a series of speeds, usually represented non dimensionally.

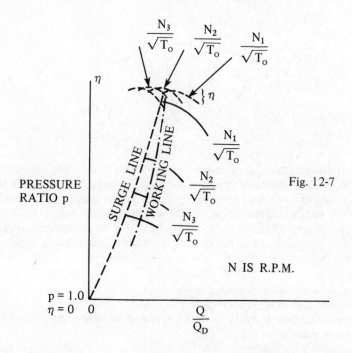

Fig. 12-7

In Figure 7, the speed parameter $\frac{N}{\sqrt{T_o}}$ is not truly non dimensional, but is frequently regarded as such for a given compressor because the quantities needed to make it so, namely diameter D, g and K_p, are constants. It would be more correct to use $\frac{ND}{\sqrt{2gK_pT_o}}$ or $\frac{u_b}{u_c}$ where u_b is tip speed and u_c the acoustic speed for temperature T_o (since $u_c \propto \sqrt{T_o}$), in which case the same curves would apply to a family of geometrically similar compressors.

The 'surge line' defines a region of instability. To the left of it, stable running cannot be obtained. There is a rapid series of flow reversals often quite violent depending to a large extent on what follows compressor delivery. Surging often makes its presence known by explosive bangs.

There is another kind of flow instability in addition to total flow reversal, namely a rotating stall where there is a 'domino effect' caused by a breakdown of flow in one channel 'triggering off' a breakdown in the next, and that in the next, and so on. This was first noted by the writer and his co-workers when a perspex casing was fitted to a centrifugal type jet engine compressor in an effort to detect the flow pattern with a stroboscope and showers of sparks. The main object of the experiment was not achieved but the rotating stall which developed, when the flow was reduced to a certain point, was very obvious. It was quite slow compared to the speed of rotation and in the opposite direction.

Stationary intake guide vanes are often used in both centrifugal and axial flow compressors, so shaped and disposed as to generate 'pre-whirl' in the fluid entering the compressor. Though this reduces the work capacity for a given tip speed, in that it reduces the rate of change of angular momentum induced by the impellor, it enables mass flow to be increased relative to intake size up to a point without exceeding the maximum permissible Mach number at the outer diameter of the inducers. It also reduces the amount of deflection necessary in the inducers. When intake guide vanes were introduced in early jet engines the improvement in overall performance was quite striking. (They had been in use for some time before the Rolls Royce Nene was designed, nevertheless, they were not used in the Nene as first designed. They were fitted into the engine later and, as a result, the static thrust jumped from 4000 to 5000 lbs.)

The 'working line' shown in Figure 7 represents the locus of the points at which the compressor operates and is determined by what happens after compressor discharge. It represents the points at which the compressor is 'in balance' with, say a combination of combustion chambers and driving turbine. It might also appropriately be called the 'matching line.' As may be seen it tends to approach the surge line as $\frac{N}{\sqrt{T_o}}$ is increased. It has often happened that a jet engine, completely surge free at ground level, has run into surging trouble at heights where the reduction of atmospheric temperature T_o has caused $\frac{N}{\sqrt{T_o}}$ to increase to the point where the working line intercepted the surge line.

Another point which has to be taken into account is that, during acceleration, extra driving power is needed to overcome rotor inertia. This means extra fuel and increased temperature at entry to a driving turbine, this in turn causes a temporary reduction in mass flow, i.e., the working line for acceleration is to the left (Figure 7) of that for normal operation thus increasing the tendency to surge. This must be allowed for in design, i.e., the steady speed working line must be sufficiently far to the right of the surge line to prevent the risk of surging during acceleration, or devices such as 'blow off valves' must be provided to allow for a temporary increase of mass flow through the compressor during acceleration. (This is particularly important in aircraft gas turbines where a rapid increase of power is frequently essential.)

Other forms of centrifugal compressors are depicted in Figures 8 and 9, having blades swept forward and backward respectively.

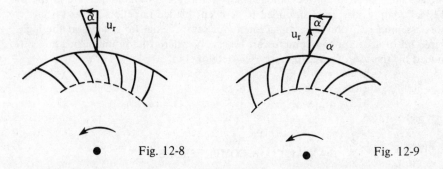

Fig. 12-8   Fig. 12-9

In the Figure 8 arrangement the tip whirl induced is increased by $u_r \tan\alpha$ where α is the angle with which the flow relative to the impellor leaves relative to the radius, so, as compared with the radial vaned type, the rate of change of angular momentum is greater, i.e., the specific work capacity is greater for a given tip speed. Moreover, since the radial velocity u_r increases with mass flow rate the specific work capacity also increases with mass flow rate. The opposite happens in the case of the back swept blades of Figure 9. The latter arrangement, however, is the only arrangement used in practice so far as is known to the writer.

Referring to Figure 7, the characteristic curves would, if efficiency did not vary with mass flow, tend to be horizontal with a radial vaned impellor (in the absence of stationary intake guide vanes), but would tend to have a positive slope with forward swept blades and a negative slope with back swept blades. The former is a very 'surge prone' state of affairs. The back swept blading has the further advantage that the curvature of the impellor blade channels helps to prevent the type of breakaway of flow depicted in Figure 6, so that improvement in efficiency may well offset the reduction in specific work capacity, especially if the blades are shrouded (as they usually are with this type of impellor).

There seems little doubt that compressors with impellors having back swept radial blading are intrinsically more efficient than any other kind of centrifugal compressor; indeed, when the writer visited the Swiss firm of Oerlicken in 1948, he saw a gas turbine with a compressor consisting of three stages of centrifugal compression; each stage having shrouded and back swept impellor blades, and was astonished to learn that the compressor efficiency was as high as 86% despite the fact that, in his opinion, the size of the compressor was far larger than it need have been for the design mass flow rate. Another unusual feature of this machine was that two intercoolers were used between stages. Seemingly the reduction of the negative work of the cycle by this means more than compensated for the increase in combustion heat addition for a given T_M due to the reduced compressor delivery temperature. And, of course, the pressure ratio of the second and third stages was increased by reduced inlet temperature. Further, since the machine also had a heat exchanger (or recuperator), the effectiveness of heat recovery was increased by the increased difference between turbine exhaust and compressor delivery temperatures.

Material developments and improvements in manufacturing techniques, especially the big advances in titanium technology, may well cause a notable increase in the use of centrifugal compressors using shrouded back swept bladed impellors in gas turbine compressors assemblies, if only as a replacement for several axial flow stages at the high pressure end of a compound compressor, especially where the optimum pressure ratio is reduced by the use of a recuperator (see Section 11).

AXIAL FLOW COMPRESSORS

Figure 10 is a rough representation of a seven stage axial flow compressor. The upper diagram represents a section through the axis and below is a partial view of the first stage rotor blades viewed from the front.

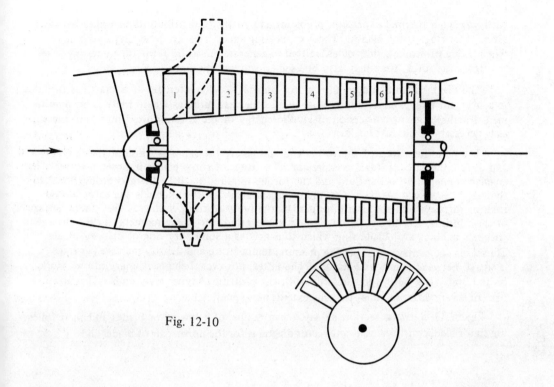

Fig. 12-10

Superimposed on the upper half of the longitudinal section is the 'ghost' of a single sided centrifugal compressor impellor of about the same volumetric and work capacity. The 'ghost' of the equivalent double sided impellor is superimposed on the lower half, (the speed of which would be about 41% greater than that of the single sided one for the same tip speed, i.e., same work capacity). This comparison gives some indication of the extent to which, in the axial flow compressor, efficiency and small diameter is 'traded' for mechanical complexity, length, etc. The design time alone required for the axial flow type is measured in months as compared with a matter of a day or two for the centrifugal type. The development time and cost is also vastly greater. Nevertheless, after many years of intensive development and accumulated experience, the axial flow type has virtually completely displaced the centrifugal type in modern gas turbines (research and development of the centrifugal type in the past thirty years or so has been sadly neglected).

In the sectional view of Figure 10, the rotor blade rings are numbered and the construction of the discs or drum on which they are mounted is not shown. The stationary stator blades (unnumbered) are 'rooted' in the casing which is normally made in two halves having longitudinal flanges by means of which they are bolted together. This is necessary for assembly purposes.

The vanes shown ahead of the first stage support the hub which carries the front bearing. They are usually three or four in number often offset from the radial direction, partly to avoid thermal expansion problems and partly to distribute their wakes between several first stage rotor blades. These supporting vanes have no deflecting function. When a ring of intake guide vanes is fitted to generate pre-whirl in the fluid entering the first stage rotor, they may also be used to support the hub.

The divergent intake duct shown in Figure 10 is characteristic of aircraft gas turbines. A convergent inlet is more appropriate in all other gas turbines where there is no requirement for the efficient conversion of kinetic energy of the entering fluid into 'ram pressure' as is necessary in aircraft engines.

The basic functions of the rotors and stators are the same as those of the impellor and diffusers of a centrifugal compressor. The rings of rotor blades impose an increase in angular momentum on the fluid and the torque required to drive them is proportional to the rate of increase. Also, due to the divergent shape of the channels the rotors act as rotating diffusers and so contribute part of the pressure rise per stage. The kinetic energy of the whirling flow at exit from a rotor is converted into pressure energy in the divergent channels of the stator blade ring which thus acts as a stationary ring of diffuser blades. Thus, as in the centrifugal type, the pressure, temperature and density increase per stage is shared between rotors and stators. The latter, however, being stationary, do no work on the fluid. But here, the similarity with the centrifugal type really ends. The flow pattern is very different, and, to some extent, more predictable.

Figure 11 shows a section at, say, mean radius, of a few rotor blades and their following stator blades with the relevant vector diagrams for the upper pair of blades shown.

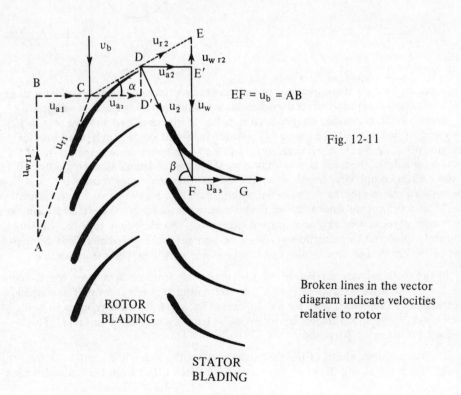

Fig. 12-11

Broken lines in the vector diagram indicate velocities relative to rotor

There are other ways of drawing the vector diagrams. The one shown in Figure 11 is the preferred one because it makes it easy to define the 'skeleton line' of the blade profiles.

In the diagram, the degree of deflection with its accompanying large increase in channel width, is exaggerated. It must, however, be remembered that the diagram is two dimensional and for one particular radius. The actual area change depends also upon what happens at other radii and on whether or not blade height (i.e., the length of the blade normal to the plane of the paper) varies from front to back of a blade row; thus if the trailing edges of the rotor blades were shorter than the leading edges as in Figure 12, the effective increase in channel section would be greatly reduced. Such a change in blade height would mean a substantial increase in the axial component of velocity at blade exit and this, in turn, would decrease the degree of deflection in the stator blading and increase it in the rotor blading, i.e., an increase of u_{a2} in Figure 11 would reduce the exit angle α of the rotor blading and the entry angle β of the stator blades.

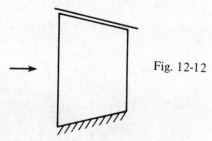

Fig. 12-12

The vector diagram of Figure 11 is drawn for a case where the flow into the rotor is purely axial ahead of the rotor (i.e., no intake guide vanes) and also purely axial flow at discharge from the stator blading.

If the blade speed is u_b at radius r then the vector AB is the whirl at entry relative to the rotor, i.e., $u_{wr1} = -u_b$. The vector BC is the entry axial velocity u_{a1}, so the resultant relative velocity u_{r1} is the vector AC. It is obvious that $u_{r1}^2 = u_{a1}^2 + u_b^2$ and that the temperature equivalents of these relative velocities are related by the equation $\theta_u = \theta_a + \theta_b$. If the total temperature of the entering air is T_o then the static temperature is $T_o - \theta_a$ so that the total temperature relative to the moving blade T_{TR} is ∴

$$T_{TR} = T_o - \theta_a + \theta_u = T_o + \theta_b \qquad (12\text{-}1)$$

Since there can be no change of total temperature in the blade channel, $T_o + \theta_b$ is also the total temperature relative to the rotor at exit. Total pressure (relative to the blade) would be reduced by blade friction and possible detachment of the flow on the convex face with resulting turbulence, especially if the adverse pressure gradient due to divergence is excessive.

Vector DE represents the relative velocity at exit from the rotor u_{r2} which is the resultant of the relative whirl u_{wr2} (= E'E) and the axial velocity u_{a2} (= DE'). Since $u_{r2}^2 = u_{wr2}^2 + u_{a2}^2$ then $\theta_{ur2} = \theta_{ua2} + \theta_{wr2}$, the static temperature in the interblade space; (which will be the same for both rotor and stator blades) T_s is thus $T_{Tr} - \theta_{ur2}$

or

$$T_s = T_o + \theta_b - \theta_{ua} - \theta_{wr2} \qquad (12\text{-}2)$$

The whirl relative to the stator blade u_w (i.e., the 'absolute whirl' induced by the rotor) is seen (from Figure 11) to be given by E'F

i.e., $$u_w = u_b - u_{wr2} \qquad (12\text{-}3)$$

It follows that the temperature equivalent of u_2 (= DF) is given by $\theta_{u2} = \theta_{ua2} + \theta_{uw}$.

From (3), $$\theta_{uw} = \frac{(u_b - u_{wr2})^2}{2gK_p} = \theta_b + \theta_{wr2} - \frac{u_b u_{wr2}}{gK_p} \qquad (12\text{-}4)$$

hence $$\theta_{u2} = \theta_{ua2} + \theta_b + \theta_{wr2} - \frac{u_b u_{wr2}}{gK_p} \qquad (12\text{-}5)$$

∴ the total temperature at stator entry (and exit). T_T is given by

$$T_T = T_S + \theta_{u2} \qquad (12\text{-}6)$$

Substituting for T_S from (2) and for θ_{u2} from (5)

$$T_T = T_o + \theta_b - \theta_{ua2} - \theta_{wr2} + \theta_{ua2} + \theta_b + \theta_{wr2} - \frac{u_b u_{wr2}}{gK_p}$$

which reduces to $$T_T = T_o + 2\theta_b - \frac{u_b u_{wr2}}{gK_p} \qquad (12\text{-}7)$$

Writing $\Delta T = T_T - T_o$ (the total temperature increase across the stage)

$$\Delta T = 2\theta_b - \frac{u_b u_{wr2}}{gK_p} \qquad (12\text{-}8)$$

ΔT is, of course, a measure of the specific work done by the stage.

(Note that if $u_{wr2} = 0$ the total temperature rise for the stage would be twice the temperature equivalent of the blade speed as in the case of a radial vaned centrifugal impellor with a very large number of blades having a negligible slip factor).

As may be seen from Figure 11, the whirl leaving (and relative to) the rotor u_{wr2} depends on the axial velocity u_{a2} and the angle α, i.e., $u_{wr2} = u_{a2} \tan\alpha$. It follows that, since u_{a2} depends upon the mass flow rate Q and the area of the annulus, from (8) ΔT decreases with increase of Q (as in the case of the centrifugal impellor with back swept blades).

At this point it is appropriate to point out that exit angles of both rotor and stator blades govern the performance of the compressor to a far greater extent than the entry angles. These latter, however, are important in that if they are not reasonably consistent with the flow direction then flow breakdown is likely due to stalling. Axial flow compressors are much more sensitive in this respect than axial flow turbines because of the adverse pressure gradient within the blade channels. Clearly a reduction of Q below the 'correct' value would be much more harmful than an increase. (See Figure 11. If a reduction of u_{a1} occurs as a result of a reduction of Q the 'angle of attack' is increased.)

Example 12-4

Given the following data: (1) T_o = 300°K; (2) u_{a2} = 600'/sec; (3) angle α = 45°; (4) u_b at radius r (unspecified) is 900'/sec; and (5) u_{a1} = 550'/sec $\therefore$ what is (a) the specific work capacity measured as ΔT, (b) the total temperature relative to the rotor, and (c) the static temperature between rotor and stator?

Figure 13 is the vector diagram differing slightly in arrangement from that of Fig. 11 (yet another variant would be to have joined C and F as the resultant velocity at entry to the stator) with the component and resultant velocities being shown with their temperature equivalents.

The total temperature relative to the rotor is seen to be 300 + 37.4 = 337.4°K (i.e., 300 + 37.4 + 14 − 14) $\therefore$ the static temperature after the rotor is 337.4 − 33.27 = 304.13°K which is also the static temperature ahead of the stator $\therefore$ the total temperature before, through, and after the stator is 304.13 + 20.8 = 324.93°K, thus ΔT for the stage is just below 25°. The same result is obtained by using Equation (8).

The static temperature before the rotor is 286° for which u_c = 1112.5'/sec $\therefore$ the Mach No. at rotor entry = $\frac{1055}{1112.5}$ = 0.948. This is sufficiently high to indicate that great care would be necessary to avoid a choking throat in the rotor blade channels.

This last point emphasizes the importance of designing the channels rather than the blade profiles. (In the early days of axial flow compressor design the tendency to regard the blades as a 'cascade' of aerofoils was so strong that designers tended to design the blade profiles on aerofoil principles and to overlook the greater importance of the channels.)

The fact that u_{a2} is greater than u_{a1} in the above example implies that the rotor blade annulus decreases from entry to exit as in Figure 12.

Other points to notice from the example are: (1) though the specific work done by the rotor blading is about that which was usual in the early days of axial flow compressor development, the rotor blade curvature is quite small at the radius at which u_b = 900'/sec; (2) the deceleration in the rotor blading is from 1055'/sec to 848.5'/sec; (3) the axial velocity at exit from the stator blading (unspecified) would be determined by the annulus, density etc.; It *could* be made to equal the entry velocity of 671'/sec by reducing the annulus (as in Figure 12) in which case there would be no deceleration in the stator blading at the radius concerned, and therefore no pressure rise. This point is mentioned to show that the way the stage pressure rise is distributed between rotors and stators is very much under the control of the designer; and (4) at the radius concerned the whirl velocity on leaving the rotor is 300'/sec hence the centrifugal field is $\frac{300^2}{rg}$, thus if r equalled, say, 1.0', the centrifugal acceleration would be nearly 3000g. Thus, from entry to exit of the rotor, the centrifugal force field increases from zero to 3000g approximately, while the reverse happens in the stator blading. Such powerful force fields would have a big effect on the retarded flow in the boundary layers and cause it to have a radial component of flow, i.e., the boundary layer is 'scavenged' away from the inner portions of the blading

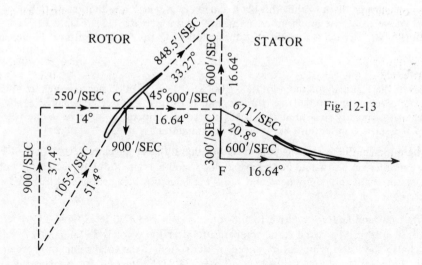

Fig. 12-13

where the blade deflection is greater. The writer believes this to be a major factor contributing to the high efficiency obtainable with axial flow compressors despite the normal poor efficiency of diffuser channels in general.

Figure 14 illustrates the effect of reducing u_{a2} as compared with the example shown in Figure 13. By increasing the exit annulus of the rotor blading u_{a2} is shown reduced from 600 to 500'/sec without change of rotor blade profile.

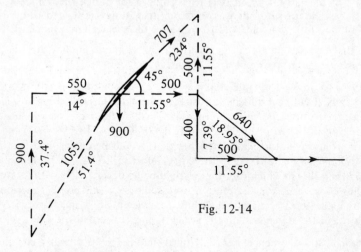

Fig. 12-14

From Equation (8) $\Delta T = 2 \times 37.4 - \dfrac{900 \times 500}{32.2 \times 336} = 33.2°$, i.e., a very substantial increase of work capacity. As may be seen, the velocity reduction in the rotor is much greater while that in the stator is reduced unless the leaving axial velocity is substantially below that of Figure 13 (not given). The increase in ΔT is, of course, due to the 33%

increase of whirl imparted by the rotor, which latter also increases the centrifugal acceleration in the ratio $\left(\frac{400}{300}\right)^2$ by approximately 78%.

The stator blades of Figure 13 would no longer be appropriate since the relative flow angle at entry of the Figure 13 example is $\tan^{-1} 0.5 = 26.6°$ while for Figure 14 it is $\tan^{-1} 0.8 = 38.7°$ approximately, so that if the stators of Figure 13 were used in the circumstances of Figure 14, the angle of attack would be 12.1° which would almost certainly cause them to stall.

There are some interesting variants of Equation (8), i.e., $\Delta T = 2\theta_b - \frac{u_b u_{wr2}}{gK_p}$.

Multiplying the second term on the right by $\frac{2u_b}{2u_b}$ we obtain $\Delta T = 2\theta_b - \frac{2u_b^2}{2gK_p} \frac{u_{wr2}}{u_b}$ and, since $\frac{u_b^2}{2gK_p} = \theta_b$, this becomes

$$\Delta T = 2\theta_b \left(1 - \frac{u_{wr2}}{u_b}\right) \tag{12-8a}$$

or, since u_{wr2}, the relative whirl from the rotor, $= u_{a_2} \tan \alpha$

$$\Delta T = 2\theta_b \left(1 - \frac{u_{a_2}}{u_b} \tan \alpha\right) \tag{12-8b}$$

or, since $u_{wr2} = u_b - u_w$, from (8a)

$$\Delta T = 2\theta_b \left(1 - \frac{u_b - u_w}{u_b}\right) = 2\theta_b \left(\frac{u_w}{u_b}\right) \tag{12-8c}$$

The foregoing may be described as the 'temperature equivalent method' of dealing with the subject, and illustrates the point that moving objects such as rotor blading may be regarded as having a temperature equivalent of their velocity. One simply has to multiply temperatures and temperature equivalents by K_p to obtain the energy values in mechanical units. It must be remembered, however, that, in the foregoing, ΔT is a measure of the *specific work*, and must be multiplied by Q to obtain the power input. Thus, if an axial flow compressor has a temperature rise of ΔT_1 in the first stage, ΔT_2 in the second and so on, then the power required W in mechanical units is given by $W = QK_p(\Delta T_1 + \Delta T_2 + \ldots \Delta T_n)$ for n stages.

An alternative method might be described as the 'linear momentum method.' If the change of whirl is Δu_w in passing through a rotor, then the component of tangential force F_s per unit mass in the plane of the rotor at the radius where Δu_w occurs then $F_s = \frac{\Delta u_w}{g}$ and the specific rate of doing work $W_s = F_s \times u_b = \frac{u_b \Delta u_w}{g}$.

Since $u_b = r\omega$ where ω is the angular velocity of the rotor $F_s \times r\omega = \frac{r\omega \Delta u_w}{g}$ or $F_s = \frac{r\Delta u_w}{g}$ and since $F_s r$ is the specific torque and $\frac{r\Delta u_w}{g}$ is the specific change of angular momentum at radius r, we are back to torque = rate of change of angular momentum. However, in each of these methods Δu_w and flux density may vary with radius so that the relationship between Δu_w and r and between ψ and r must be known for torque, etc., to be obtainable by integration. Thus the two dimensional analysis so far discussed is by no means the whole story, though, as will be seen hereafter, there are certain circumstances in which the two dimensional analysis may substitute for a three dimensional one for some of the performance characteristics, but a three dimensional study is essential for blade design.

Referring to Equation (8c), i.e., $\Delta T = 2\theta_b \left(\dfrac{u_w}{u_b}\right)$, this was arrived at on the assumption that there was no whirl in the fluid entering the rotor. More generally,

$$\Delta T = 2\theta_b \left(\frac{\Delta u_w}{u_b}\right) \tag{12-8d}$$

and since $\theta_b = \dfrac{u_b^2}{2gK_p}$ then $\Delta T = \dfrac{u_b^2}{gK_p}\left(\dfrac{\Delta u_w}{u_b}\right)$ or

$$gK_p \Delta T = u_b (\Delta u_w) \tag{12-8e}$$

Hence, since u_b is proportional to radius, for ΔT to be the same at all radii Δu_w must be inversely proportional to radius.

As shown in Section 3 on radial equilibrium, this condition is best satisfied if the whirl velocity conforms to that of a free vortex of uniform axial velocity, i.e., $u_w r$ = const. and u_a = const. both before and after blade rings.

In the following three dimensional analysis of the effect of radial equilibrium on blade design, a specific example is used on the basis of the following assumptions

(1) ΔT (i.e., the total temperature rise for the stage) is 35°C at all radii

(2) The air has axial velocity only at entry to the rotor and at exit from the stator which, with (1), means that the rotor induces a free vortex flow in the interblade space, i.e., $u_w r$ = const. = $u_{wm} r_m$ where the suffix m implies conditions at mean radius r_m

(3) The tip/hub ratio at rotor entry is 2:1, i.e., $\dfrac{r_o}{r_i} = 2$ where r_o and r_i are outer and inner radii respectively

(4) r_m is constant from entry to exit of the stage

(5) The blade annulus section varies to conform with the condition that axial velocity is 550'/sec at entry to and exit from the rotor

(6) The rotor blade speed at r_m is 1050'/sec

(7) Total temperature before the rotor is 288°K

(8) Total pressure before the rotor is 2116 p.s.f.

From Equation (8e), at r_m, $u_{wm} = \dfrac{gK_p \Delta T}{u_{bm}} = \dfrac{32.2 \times 366 \times 35}{1050} = 360.6'/\text{sec}.$

u_b and u_w (at rotor exit) at r, r_m and r_o are then as follows

	u_b	u_w
r_o	1400	270.5
r_m	1050	360.6
r_i	700	540.9

The resulting vector diagrams for root mean and tip are shown in Figure 15 with resultant and component velocities and their temperature equivalents shown for each vector.

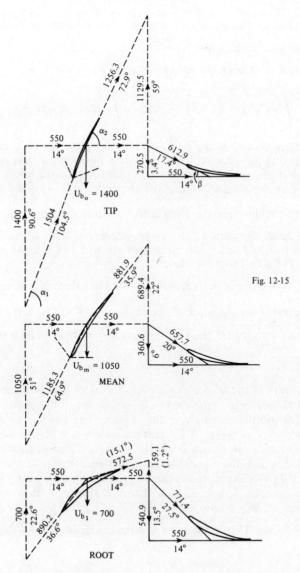

Fig. 12-15

The table below lists some of the salient features obtained from these diagrams.

	T_o	T_{S1}	θ_{r1}	T_R	θ_{r2}	T_{S2}	σ	$r\sigma$	α_1	α_2	$\beta_{\beta2}$	T_{S2}	θ_{rs}	T_2	ΔT	M_1	M'_1
r_o	288	274	104.5	378.5	72.9	305.6	1.314	2.628	68.6°	64.0°	26.2°	305.6	17.4	323.0	35	1.38	1.29
r_m	288	274	64.9	338.9	35.9	303.0	1.286	1.930	62.4°	51.4°	33.2°	303.0	20.0	323.0	35	1.09	1.01
r_i	288	274	36.6	310.6	15.1	295.5	1.208	1.208	51.8°	16.4°	44.5°	295.5	27.5	323.0	35	0.82	0.76

In the above T_{S1} and T_{S2} are the static temperatures before and after the rotor. $\sigma \equiv \dfrac{\rho_{S2}}{\rho_{S1}}$ where ρ_{S1} and ρ_{S2} are the densities before and after the rotor, and σ is calculated from $\sigma = \left(\dfrac{T_{S2}}{T_{S1}}\right)^{2.5}$, i.e., on the assumption that the flow through the blading is isentropic i.e., making no allowance for blade losses.

The mass flow rate Q through the annulus is given by

$$Q = 2\pi u_a \int_{r_1}^{r_o} \rho r \, dr \quad \text{or} \quad Q = \frac{2\pi u_a}{\rho_{S1}} \int_{r_1}^{r_o} \sigma r \, dr \quad \text{at rotor blade exit.}$$

Hence the significance of $r\sigma$ in the table. The mean of the values of $r\sigma$ at r_1 and r_o is 1.923 as compared with the value of 1.930, namely a difference of less than 0.4%, thus the error is very small if one assumes that $Q = \rho_m u_a S$ where S is the area of the annulus, and ρ_m is the density at mean radius. As, in the above example, u_a is the same before and after the rotor, the annulus decreases from entry to exit in the ratio $\dfrac{1}{1.286}$, and, since r_m remains the same, the width of the annulus decreases from 1.0 to 0.778, i.e., r_o decreases from 2.0 to 1.889 and r_i increases from 1.0 to 1.111 to maintain the same axial velocity before and after the rotor blading.

Columns M and M' require some explanation. M is the Mach No. at rotor entry based on the acoustic speed for the static temperature of 274°K. M' is the value of $\dfrac{u_r}{u_c}$ where u'_c is the choking velocity for the total temperature relative to the rotor blade (T_R), i.e., $u'_c = 147.1 \sqrt{\dfrac{T_R}{6}}$.

As may be seen, both M and M' are above unity for more than half the blade length. At one time this would have been considered unthinkable but is now quite usual. As may also be seen, blade profiles become very thin near r_m and beyond, so the shock waves to which they give rise are very weak. Their pattern, however, is complicated by the fact that the leading edge of each rotor blade lies behind the shock waves emanating from one or more preceding blades, and, for other than the first stage, their reflections from the preceding stator ring. To this extent the vector diagrams are a little misleading at high Mach Nos. since they do not take into account the small changes in flow direction which occur on passage through a shock wave.

The diagrams of Figure 15 and the tabulated figures illustrate a number of interesting points as follows:

(1) Though the deceleration is from 1504 to 1256'/sec at the rotor tip the deflection is less than 5° but increasing to nearly 36° at the root. Fortunately, at the root the pitch is half that at the tip, and, if necessary, channel divergence can be reduced by increasing the axial width of the blades.

(2) In the stators also, deflection and channel width increase from tip to root, but Mach numbers are well below unity.

(3) The centrifugal field in the interblade space is far more intense at the root than at the tip. In fact, in a free vortex, it varies inversely as the cube of the radius (since $u_w \propto \frac{1}{r}$, $u_w^2 \propto \frac{1}{r^2}$ $\therefore \frac{u_w^2}{r} \propto \frac{1}{r^3}$). This suggests that, to prevent breakaway due to the effect of divergence on the boundary layer, it would be desirable to design the rotor channels to have increasing rate of divergence from entry to exit, and vice versa for the stator channels.

(4) It is obvious that increased axial velocity would increase the deflection in the rotor channels and reduce it in the stators.

(5) The increase of blade section from tip to root is very satisfactory from the point of view of blade stress due to centrifugal force.

The relative eddy, briefly explained in the discussion of the centrifugal compressor, must also occur between each pair of rotor blades of an axial flow compressor but owing to the much greater number of blades its influence would be much reduced. It's effect would differ in that while still tending to cause slip at the blade tips it would tend to cause 'negative slip' at the root radius. It's effect could be countered by slight adjustments to the blade exit angles.

A Note on Cascade Tests

It has long been established that what happens in the boundary layer is crucial to the whole aerodynamic behavior of a body moving through a fluid, yet many researchers have spent much time on experiments with stationary finite 'cascades' of blades in the belief that the information obtained is applicable to the blading of axial flow compressors and turbines. In the opinion of the writer there can be little resemblance between the flow through such cascades and what happens in the blading of high speed rotors or even rings of stator blades. Any results from tests with cascades which may seem to correlate must be presumed to be purely coincidental for the following reasons:

(1) The intense centrifugal fields which are present in both rotors and stators cannot be reproduced in a cascade, yet these forces must have a profound effect on boundary layers.

(2) A ring of rotor or stator blades is, in effect, an infinite cascade varying from blade root to blade tip in relative velocity, blade angles and profiles, blade pitch and chord, etc.

(3) Rings of blades discharge into an annulus of fixed section area which largely determines the axial velocity which, in turn, in conjunction with exit blade angles, boundary layer effects and the (relatively minor) effect of the relative eddy, determines the angle and velocity of flow at blade exit.

(4) The effects of compressibility can rarely be reproduced in a stationary cascade.

There have been efforts to deal with (3) above by using extensions of the end blades of a cascade to simulate the controlling effect of the exit area on the axial velocity component in the hope that this will cause a finite cascade to be representative of an infinite cascade.

Of the above objections, (1) is almost certainly the most cogent. Even if means of meeting the others could be devised, the differences in boundary layer phenomena would alone be so great as to destroy any resemblance in the flow behavior. One might think that a cascade could be least be made to be reasonably representative of a ring of stator blades, but, as seen above, the whirling flow entering a compressor stator has a centrifugal field of force of great magnitude at entry diminishing to zero at exit if the exit flow is purely axial (or having negligible residual whirl). And vice versa for a turbine nozzle ring.

In the case of the rotor blading the state of affairs in the boundary layer is further complicated by the fact that it is subject to much larger centrifugal forces than the main flow, especially those portions of it in contact or near contact with the blade surface. This is best visualized by imagining that a very small diameter tube is 'buried' radially in the blade and open to the flow at its inner end. Since it would be rotating with the angular velocity of the blade the centrifugal pressure gradient within it would be defined by $\frac{dp}{dr} = \frac{pr\omega^2}{g}$ from which $\frac{dp}{\rho} = \frac{\omega^2 r dr}{g}$ or, since $\frac{dp}{\rho} = K_p dT$, $K_p dT = \frac{\omega^2 r dr}{g}$ whence ΔT between r_1 and r_o is given by $\Delta T = \frac{\omega^2 (r_o^2 - r_1^2)}{2gK_p}$ or $\Delta T = \frac{u_{bo}^2}{2gK_p} - \frac{u_{bi}^2}{2gK_p}$ where u_{bo} is the blade tip speed and u_{bi} is the root speed; or, using temperature equivalents, $\Delta T = \theta_{bo} - \theta_{bi}$.

Thus, assuming an isentropic relationship between temperature and pressure, i.e., $1 + \frac{\Delta P}{P_i} = \left(1 + \frac{\Delta T}{T_1}\right)^{3.5}$ (i.e., for air), $\frac{\Delta P}{P_i} = \left(1 + \frac{\Delta T}{T_i}\right)^{3.5} - 1$.

Taking the example used for Figure 15 as an illustration, $T_i = 288°$ and $P_1 = 2116$ p.s.f. (total temperature and pressure): $u_{bi} = 700'/\text{sec}$ ∴ $\theta_{bi} = 22.64°C$; $u_{bo} = 1400'/\text{sec}$ ∴ $\theta_{bo} = 90.58°C$ ∴ $\Delta T = 90.58 - 22.64 = 67.94°C$, hence $\frac{\Delta P}{2116} = \left(1 + \frac{67.94}{288}\right)^{3.5} - 1 = 1.1$ ∴ $\Delta P = 2325$ p.s.f.

If one now imagines that our radial tube has a series of holes along its length, fluid would flow through them into the 'outer' parts of the boundary layer and the main flow, especially at entry to the rotor blading where the main flow has not yet 'acquired' a centrifugal pressure gradient.

Our imaginary tube would be replenished by inflow at its inner end.

The outflow from a similar tube near the exit of a rotor blade would be diminished by the fact that the main flow has a centrifugal pressure gradient 'imposed' upon it in its passage through the rotor blade channels corresponding to that of a free vortex, but which would still be less than that of the forced vortex within the 'tube.'

There is, thus, a strong tendency for boundary layer fluid to be 'pumped' into the main stream. But it must be replaced. To guess how this happens it is necessary to take into account the pressure gradient across the blade channels at a given radius due to the curvature of the flow through them and which causes the pressure on the concave side of a blade to be substantially higher than on the convex side.

Matters are further complicated by the pressure rise along the flow path which varies from root to tip to conform with the free vortex generated by the blading. The three dimensional pressure distribution which exists in the channels is so complex that one can do no more than guess at the kind of secondary flows which are induced by the effect of the pressure field on the boundary layers along the blade surfaces.

Two of the many possibilities for the innermost part of the boundary layer are suggested in Figure 16. The type of secondary flow indicated in the left hand sketch is possibly representative of what happens at rotor entrance while at and near the exit the 'cross channel' pressure increase added to whirl pressure increase might be sufficient to cause an inward flow on the concave side of a blade, i.e., to overcome the centrifugal pumping effect. In any event the scavenging effect will be strongest at the exit on the convex side of a blade at the root where it is most needed to prevent breakaway.

In the stators, the secondary flows are likely to be quite different and less favorable to the prevention of breakaway owing to the absence of the centrifugal pumping effect.

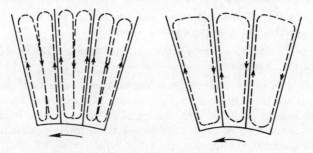

Fig. 12-16

In the case of the rotor blades, matters are yet further complicated by the 'spill' which must occur from the high pressure to the low pressure sides (i.e., from concave to convex) through the clearance between the blade tips and the casing. This could be prevented by mounting a shroud ring round the blade tips, but experience seems to indicate that this reduces compressor efficiency; presumably because it has some kind of adverse effect on the secondary flows within the channels. The use of shrouding would also provide a 'leakage path' from the exit to entry at the rotor tips through the clearance between shroud and casing which, without shrouding, would be obstructed to some extent by the 'tangential' tip leakage mentioned above.

Multi-Stage Axial Flow Compressors

Up to this point, the discussion of axial flow compressors has been limited to a single stage combination of rotor and stator of a kind which might well be representative of the first stage of a multi stage compressor as illustrated in Figure 10.

Multi stage units are essential to obtain the degree of compression required in gas turbines, because, in the present state of the art, about 40° rise per stage is the most that can be obtained with good efficiency. This, at 90% adiabatic efficiency and from an initial total temperature of 288°K, would yield a temperature ratio of 1.125 and a pressure ratio of

1.51. For a 400° compressor, i.e., one capable of producing a temperature rise of 400°, 10 stages would be necessary. It is, however, somewhat doubtful if this degree of compression can be achieved in a single compressor without special measures to prevent blade stalling and surging at low speeds. The need for these special measures is based on the fact that a design suitable for full design speed is mismatched from entry to exit at speeds less than design, primarily due to the difference in the increase of density as between low speed and full speed. The density increase calls for a progressive reduction of blade annulus from entry to exit as shown in Figure 10. If it were physically possible to have variable blade length so as to increase annulus area at the high pressure end for low speed operation in such a manner as to keep the axial velocity proportional to blade speed, the problem would not exist. But this, of course, is not practical and so other methods have to be employed to prevent the excessive changes in flow direction at entry to blading which would otherwise occur at 'off design' speeds, when the axial velocity at entry tends to be too low and that at exit too high. The former is more harmful than the latter because a reduction of axial velocity relative to blade speed increases the 'angle of attack' and therefore the tendency to blade stalling.

The usual preventive measures comprise one or more of the following:

(1) The provision of blow off valves part way along the compressor to increase the mass flow through the first few stages at low speeds.

(2) The use of variable stators for the first few stages. (This is characteristic of many General Electric Co. designs.)

(3) The division of the total compression between two or more compressors in series independently driven by turbines in series. Thus, if 14 stages were necessary then, say, two seven stage compressors might be used running at different speeds. In such cases (which are now common) the shaft of the L.P. turbine–L.P. compressor combination is 'threaded' through the HP assembly, the speed of the latter being the higher. Such arrangements are known as 'two spool' engines.

(4) Making possible a reduction of the effective back pressure of the expansion system so as to increase mass flow in proportion to speed. As, for example, by the opening up of a variable jet nozzle in a jet engine. (This, however, is not very effective with an axial flow compressor owing to its pressure ratio, mass flow characteristics − see below.)

As already mentioned, the temporary increase of turbine inlet temperature during acceleration aggravates the surging problem.

Some of these measures are partly self defeating; for example, the provision of blow off valves increases the power required to drive the compressor. Thus, if driven by a turbine, the turbine inlet temperature is increased which, since Q for the turbine is proportional to $\frac{P}{\sqrt{T}}$, in turn reduces the mass flow.

The nature of the problem of finding what happens at off design speeds is best explained (up to a point) by taking a specific case.

Figure 17 is a diagram for a design cycle based on the following assumptions: (1) T_o = 288°K and P_o = 2116 p.s.f.; (2) T_{max} = 1150°K; (3) η_T = 0.87; (4) η_r = 0.9; (5) θ_c = 245; (6) the mean radius for all stages is the same; (7) u_a at entry and exit is

550'/sec; (8) we are not concerned with what happens after the compressor turbine; and (9) the compressor has 7 stages, each having $\Delta T = 35°C$

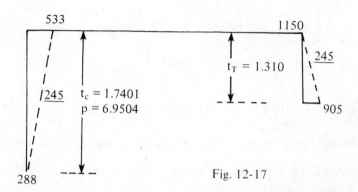

Fig. 12-17

The temperature and pressure ratios corresponding to the above assumptions are shown on the diagram. The stagnation density before entry ρ_o is $\dfrac{P_o}{RT_o}$ and is found to be 0.766. The temperature equivalent of 550'/sec is 14° ∴ the density ρ_i at rotor entry is $\rho_1 = \rho_o \left(\dfrac{288-14}{288}\right)^{2.5}$ from which $p_i = 0.0675$.

At compressor rotor exit, with compression total pressure of $1.7401 P_o$ and total temperature of 533°K, the stagnation density ρ_E is $\dfrac{1.7401 P_o}{533 R}$ which is found to be 0.2874 hence at 550'/sec exit axial velocity, the static density ρ_{ES} is given by $\rho_{ES} = \rho_E \left(\dfrac{533-14}{533}\right)^2$ and is found to be 0.269.

If S_i is the annulus area at entry and S_E is the annulus area at exit then $Q = \rho_1 \times 550 \times S_E$ from which $\dfrac{S_E}{S_i} = \dfrac{\rho_1}{\rho_{ES}}$. Since mean radius r_m is assumed the same, $S_i = 2\pi r_m \ell_i$ where ℓ_i is the annulus width (i.e., blade length) at entry, and $S_E = 2\pi r_m \ell_E$ where ℓ_E is the annulus width at exit. Therefore $\dfrac{S_E}{S_i} = \dfrac{\ell_E}{\ell_i} = \dfrac{\rho_1}{\rho_{ES}}$ i.e., $\dfrac{\ell_E}{\ell_i} = \dfrac{0.0675}{0.269} = 0.251$. Thus the blade length at exit is approximately 1/4 of that at entry.

Fig. 12-18 represents the cycle of the same assembly with θ_c reduced to 61.2°C and assumes that the off design conditions drop the compressor efficiency to 0.8 and the compressor turbine efficiency also to 0.8. It is further assumed that the assembly is just self-driving, i.e., that whatever follows the compressor turbine is 'absent,' i.e., the temperature and pressure ratios for the compressor and its driving turbine are the same and as shown in the diagram.

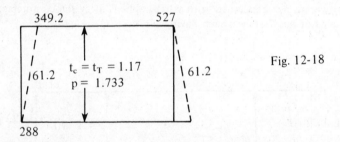

Fig. 12-18

It is readily found that to give an actual temperature drop of 61.2°C at 80% efficiency with a temperature ratio of 1.17 then $T_{max} = 527°K$ as shown. The temperature ratio of 1.17 is sufficiently near the critical ratio 1.2 to assume that $Q \propto \dfrac{P}{\sqrt{T_M}}$. If now Q_D is the design mass flow of the cycle shown in Figure 17, then Q for the cycle of Figure 18 is given by $\dfrac{Q}{Q_D} = \dfrac{1.733 P_o}{6.95 P_o} \sqrt{\dfrac{1150}{527}} = 0.368$

A θ_c of 61.2°C being 0.25 x design $\theta_c (= 245)$ would correspond to 50% design speed in a radial bladed centrifugal compressor but this would not be so in a multi stage axial flow machine, nevertheless, it is evident that Q decreases more rapidly than blade speed.

It now remains to find out what happens (so far as is possible) to axial velocities. This is not difficult if one makes the rather sweeping assumption that, in the off design condition, one may take the axial velocity at mean radius as representing an average for the annular section concerned. (Which would be true at entry to the first rotor.)

We have the following data from above:

(1) $Q = 0.368 Q_D$

(2) $Q_D = S_i \times 550 \times 0.0675$ at entry

(3) $Q_D = 0.25 S_i \times 550 \times 0.269$ at exit

For the off design condition of Figure 18, at entry we have: $Q = S_i u_{ai} \rho_i$, but ρ_i depends upon u_{ai} since $\rho_i = \rho_o \left(\dfrac{288 - \theta_{ai}}{288} \right)^{2.5}$. However, the approximations involved do not justify the successive approximation necessary to obtain 'exact' values of u_{ai} and ρ_i so it will be assumed that $\theta_{ai} \cong 2°C$, from which $\rho_1 = 0.0766 \left(\dfrac{286}{288} \right)^{2.5} = 0.0753$. S_i being unchanged, it follows from (1) and (2) that u_{ai} is reduced in the ratio $\dfrac{0.0675}{0.0753} \times 0.368 =$ 0.33, i.e., $u_{ai} = 0.33 \times 550 = 181'/\text{sec}$ (for which $\theta_{ai} \cong 1.5°$, hence the error in assuming 2° is negligible).

Thus it may be seen that a reduction of blade speed by something of the order of 50% means a reduction of axial velocity of the order of 67% at entry, which, of course, means quite a large increase in angle of attack and therefore in the tendency to stall.

At compressor exit, the opposite happens owing to the lower exit density and small annulus.

Using $Q = 0.368 Q_D$ and (3) we have

$$0.368 \times 0.251 S_1 \times 550 \times 0.269 = 0.251 S_1 \times u_{aE} \times \rho_{ES}$$

and given that the total pressure and temperature are $1.733 P_o$ and $349.2°K$ (see Figure 18) ρ_{ES} is found to be 0.108 and $u_{aE} = 505'/\text{sec}$, i.e., 92% of design value though again the blade speed will probably be in the 50–60% range.

But this is by no means the whole picture. For the design cycle of Figure 17 it was assumed that the total temperature rise of 245°C was equally divided between the stages, but, as seen from the vector diagram of Figure 19 where AB is the relative whirl at rotor entry ($= -u_b$), DE is the relative whirl u_{wr2} at rotor exit and EF is the absolute whirl u_w, (i.e., relative to the stator). The length of DF is equal to AB, i.e., to u_b. BC, CE and FG are the axial velocity vectors as indicated.

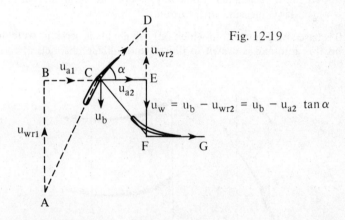

Fig. 12-19

As we have seen (Equation (8e)), with no whirl prior to rotor entry, $gK_p \Delta T = u_b u_w$. As seen in Figure 19, $u_w = u_b - u_{a2} \tan \alpha$

$$\therefore gK_p \Delta T = u_b(u_b - u_{a2} \tan \alpha) \tag{12-8f}$$

Tan α, of course, varies with radius and the relationship between tan α and r is determined by the design condition where, for free vortex whirl at rotor exit, $u_w r = u_{wm} r_m$ or $u_w = u_{wm} \dfrac{r_m}{r}$ and $u_{bD} = r\omega_D$ (suffix D implying the design value). It follows that $\tan \alpha =$

$$\dfrac{u_{bD} - u_{wD}}{u_{a2D}} \quad \text{or} \quad \tan \alpha = \dfrac{r\omega_D - u_{wmD} \dfrac{r_m}{r}}{u_{a2D}} \tag{12-9}$$

Writing $a = \dfrac{\omega_D}{u_{a2D}}$ and $b = \dfrac{u_{wmD}}{u_{a2D}}$, Equation (9) may be written

$$\tan \alpha = ar - \dfrac{b}{r} \tag{12-9a}$$

whence Equation (8f) becomes

$$gK_p \Delta T = u_b^2 \left[1 - u_{a2} \left(ar - \dfrac{b}{r} \right) \right] \tag{12-8g}$$

ΔT, being the increase of total temperature across a stage, is thus seen to be a function of radius when u_b is different from the design value. It also depends on the axial velocity after each rotor, decreasing as u_{a2} increases from entry to exit of the assembly. Indeed ΔT for the first stage may well be two or three times greater than for the last at, say, 70% design speed assuming that the flow angle at rotor exit differs little from the blade angle.

Using variable stators to create pre-whirl in the direction of rotation at rotor entry both reduces ΔT and the angle of attack of the entering flow and therefore the risk of stalling and flow breakdown.

No attempt will be made herein to deal with the complex calculations necessary to determine how ΔT varies across the radius and from stage to stage in off design conditions. Enough has been said in the foregoing to indicate the nature of the complexities involved and to indicate that multi stage axial flow compressors call for great skill in design and development — plus a large measure of intuition.

Figure 20 is based on Figure 15 in which only single blade sections were drawn for root, mean, and tip radii, and is drawn to illustrate the blade channels at root, mean, and

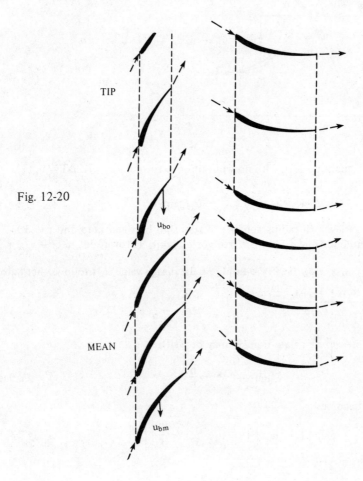

Fig. 12-20

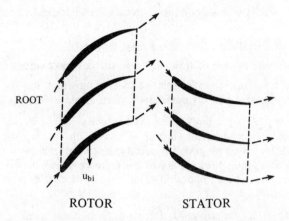

ROTOR STATOR

tip. For the rotor blades the pitch/chord ratio has been chosen to be approximately 0.5 at the root, and somewhat less for the stator blades. The pitch is, of course, proportional to radius, so the pitch/chord ratio increases from root to tip, especially in the case of the rotor where, as shown, the chord decreases from root to tip.

It will be recalled that Figure 15 was based on an example for the first stage of a multi stage compressor. Subsequent stages have progressively shorter blades so that if, say, all stages have the same mean radius, the root blade speed increases and the tip speed decreases. This latter fact coupled with the progressive rise in temperature means a decrease in Mach No. at the blade tips.

It will be noted from Figure 20 that there is considerable 'overlap' at the rotor blade roots but virtually no overlap at mean radius, while at the tips the trailing edges are pitched well ahead of the leading edges of the following blades. This is probably not a 'good thing' for supersonic relative velocity at entry in that the leading edge shock waves on the forward face side will 'miss' the trailing edge of the blade ahead, and no deflection can occur before passage through the shock wave. It seems desirable that the leading edge shock wave should at least 'strike' the trailing edge of the blade ahead.

The way the pitch/chord ratio will vary in successive stages will depend on the variation in the number of blades. As the blades get shorter it is usual to reduce the axial width and progressively increase the number of blades especially in aero engines where weight is all important.

The stator blades of Figure 20 are seen to have no overlap throughout their length.

The axial clearances shown in Figure 20 are purely arbitrary. From the aerodynamic and noise points of view, the larger the axial clearance the better to minimize the effect of the pressure field ahead of leading edges on the flow leaving trailing edges and to minimize the 'siren effect.' Large axial clearances, however, increase the overall length and therefore the weight, and increase sundry structural problems, which are themselves matters of considerable complexity in axial flow compressor design.

A Note on Axial Velocity

The choice of axial velocities for the design condition is a matter of judgement for the designer, but on balance, it seems desirable to design for high axial velocity for the following reasons:

(1) reduction of overall size for a given mass flow rate

(2) reduction of the change of flow angles in off design conditions

(3) increase of resultant velocities relative to blading which has the effect of reducing the load coefficient (analogous to lift coefficient of an airplane wing).

There is, however, an upper limit. The axial velocity cannot exceed the local speed of sound otherwise choking would occur. Indeed it is desirable to keep the Mach No. of the axial velocity well below unity to allow for aerodynamic disturbances which well might provoke choking. Moreover, though in (3) above the increase of resultant velocity is suggested as an advantage there could be an undesirable increase of Mach No. relative to rotor blading at entry at outer radii.

In any event the effect on size reduction tends to become small when at high axial velocity Mach Nos., the density reduction increasingly offsets the velocity increase (and offsets it completely at Mach 1.0).

Another disadvantage which partially offsets the above advantages is that reduction of blade length is not accompanied by a proportionate reduction of tip clearance, the necessity for which contributes to losses.

Characteristic Curves

The form of the characteristic curves for a centrifugal compressor are shown in Fig. 7 of this section. The characteristics of an axial flow compressor may be similarly plotted. There are substantial differences however. The negative slope of the $\frac{N}{\sqrt{T_o}}$ curves is far greater, tending almost to the vertical especially at high values of $\frac{N}{\sqrt{T_o}}$. This is because, as has been shown, ΔT decreases sharply with increase of Q and so does the efficiency, hence, at a given value of $\frac{N}{\sqrt{T_o}}$, very small changes in Q result in large changes of pressure ratio. To some extent this is a stabilizing factor in a gas turbine since anything which tends to cause a reduction of Q is offset by the increase of delivery pressure which results.

The surge line may well start from a point on the ordinate for a pressure ratio well above p = 1.0 and may often show a pronounced 'kink' at an intermediate value of $\frac{N}{\sqrt{T_o}}$

Losses

At the design condition, if the design is good, the main source of loss is blade skin friction and turbulence associated with the boundary layer. There is also possible flow breakaway if the rate of channel divergence is excessive, due to the adverse pressure gradient. But as already mentioned, the centrifugal scavenging effect on the boundary layer must be a big saving factor in spite of the secondary flows caused. These latter, however, inevitably have a disturbing effect on the flow leaving a blade row and probably induce increased losses as the flow progresses from entry to exit of a multi stage compressor.

Tip leakage and its effect on secondary flows has already been discussed above. It is obviously desirable to keep the tip clearance as small as possible. The limitations to this will be discussed briefly below.

The axial velocity at exit would be a loss if it were not utilized, but in gas turbines it invariably is used. It may well more than compensate for combustion chamber pressure loss.

A Note on Tip Clearance

The problem of tip clearance increases from entry to exit of an axial flow compressor because of decrease in blade length and increase of temperature. The designer must allow for possible mechanical distortion, thermal distortion, the effect of coefficients of thermal expansion, the thermal capacity of the components, changes in dimensions due to stresses, etc. Thermal capacities of components in particular are of special importance in gas turbines and is possibly one of the main reasons why aircraft gas turbines adapted for industrial purposes have, surprisingly, proved more successful than machines designed specifically for such purposes by steam turbine manufacturers who tend to cling to the heavy scantling usual in steam turbines but which are quite inappropriate with gas turbines.

As we have seen, rotor blade tips are very thin so a tip rub can be disastrous. Hence the quest for casing linings which will render a tip rub relatively harmless.

The temperature changes which can occur as operating conditions are varied can be quite large, thus, for example, a rapid shut down or speed reduction may, and probably will, reduce tip clearances because the casing cools, and therefore contracts, more rapidly than rotor blades and discs. On the other hand, the light scantlings of aircraft gas turbines allow them to adjust to rapid changes in operating conditions almost as rapidly as the changes themselves. (The writer knows of one rather extreme case, where a marine gas turbine, built by a steam turbine manufacturer, had to be 'barred over' for 24 hours after shut down to counter damaging thermal distortions.)

TURBINES (Discussion):

General

The function of a turbine is exactly the opposite of that of a compressor. Whereas the function of the latter is to convert mechanical work into enthalpy as near isentropically as possible, the objective in a turbine is to convert enthalpy into mechanical work, also as near isentropically as possible. To that extent the basics of turbine theory have been covered 'in reverse' in the preceding discussion of compressor theory; thus, in the depiction of the centrifugal compressor of Figure 4, if one reversed the flow directions the result would represent a radial flow turbine (though these are always single sided so far as is known to the writer). The diffuser then becomes a nozzle ring, etc.

The same basic law applies, namely 'Torque equals the Rate of Change of Angular Momentum.' Also the flow must conform to the requirements of radial equilibrium wherever it is whirling.

Turbines are, however, altogether easier design propositions than compressors except for the limitations imposed by the ability of the blading, etc. materials to withstand the combination of high stress and high temperatures.

One great advantage of a turbine as compared with a compressor is that the heat (i.e., enthalpy) change per stage can be far higher than in an axial flow compressor. A single stage turbine for example can be adequate to drive a seven stage axial flow compressor.

Another big advantage is that the flow is in the direction of the pressure gradient (except in occasional designs, e.g., the first jet engine – which called for a small degree of recompression in the rotor channels at and near the root radius) and so the tendency of an adverse pressure gradient to cause flow breakdown is absent.

Yet another advantage is that the conversion of losses into heat at the early part of the expansion is partially recovered in the later part of the expansion, thus the heating effect of blade friction in a nozzle (or stator) ring partly compensates for the loss of total pressure. Similarly, in a multi stage turbine, the heating effect of losses in one stage adds to the energy available in subsequent stages.

A further advantage is that the high temperatures, because of their effect on raising the local acoustic velocity, permit the use of high axial velocities. (Though the energy value of high axial velocity at exhaust is lost as far as shaft power is concerned, it usually contributes to the energy available for further expansion – especially in the case of a jet engine, to be discussed later.) As will be seen below, high axial velocities are desirable for several reasons.

Turbine Types

It is beyond the scope of this treatise to attempt to cover the vast range of turbine types – from windmills, water wheels, Hero of Alexandria's 'aelopile' (perhaps the first known heat engine) through Pelton wheels, DeLaval's partial admission 'impulse' turbine to Parson's full peripheral admission 'reaction' turbine and Ljungström's contra rotating, multi stage radial flow type, and many others. Even radial flow turbines, though used in the first German jet engine designed by Von Ohain, and still a component of small gas turbine units, will not be further discussed. The reader must be satisfied with the statement in the foregoing, that they are centrifugal compressors acting in reverse, plus the additional statement that flow conditions in a radial flow turbine are much more favorable than those in a centrifugal compressor because of the absence of adverse pressure gradients. The following discussion is, therefore, limited to the common forms of full peripheral admission axial flow turbines used in nearly all modern gas turbines. (There are one or two small gas turbine designs having radial flow turbines.)

The terms 'impulse' and 'reaction' call for some explanation. They are really obsolete and are 'left overs' from days before the importance of radial equilibrium was recognized, i.e., when turbine blading was designed on the assumption that the tangential and axial components of the velocity of the steam or other fluid issuing from the nozzle ring was constant across the annulus. In the case of blades which were short in proportion to the mean radius they usually had the same profile section from root to tip. In the case of longer blades there would be a certain amount of 'twist' to compensate for the variation of blade speed, but no attempt to compensate for the variation in whirl speed since the presence of this variation was not recognized. (When the writer's first jet engine was built he left the design of the turbine blades to the blade design specialists of the manufacturing firm – The British Thompson-Houston Co., otherwise known as the B.T-H Company, assuming that they were fully aware of the variation in whirl velocity which must occur to satisfy radial equilibrium or 'vortex flow' as the writer then called it.

He was then quite unaware that there was any novelty in the concept, and did not find out that there was a major difference in thinking between himself and the B.T-H designers until the latter proposed a design change which, they said, would cause a large increase in end thrust to a figure well beyond the capacity of the then existing thrust bearings. This completely mystified the writer and he started to inquire into the matter. He casually asked one of the designers most concerned, "What pressure difference do you calculate between root and tip in the annular gap between nozzle discharge and rotor entry?" and was astonished to receive the reply, "What pressure difference?" This started a somewhat acrimonious controversy, not helped by the fact that the writer, realizing that his views on turbine design were novel to the turbine industry, filed and obtained a patent for 'vortex design.' The B.T-H designers were, however, under the terms of the contract, obliged to conform to the writer's insistence that the blades be made to be compatible with a free vortex flow from the nozzle ring. So, so far as is known to the writer, the first jet engine had also the first turbine to incorporate design for radial equilibrium. This meant that the blades had about double the twist compared with those in which only variation in blade speed with radius was allowed for. It also meant that the calculated end thrust was negligible – a fact subsequently verified experimentally.

In fairness it should be mentioned that other designers were beginning to be suspicious of the then conventional design assumptions. In one of their bulletins, Escher Wyss reported their finding of a substantial pressure difference between root and tip ahead of a turbine rotor but did not explain it. Also the small gas turbine group at the Royal Aircraft Establishment, at that time in the early stages of the design of a turbo-prop engine for aircraft, recognized the need for designing for radial equilibrium and supported the writer in his controversy with the B.T-H.)

In order to explain the terms 'impulse' and 'reaction' the reader is asked to visualize blades so short in proportion to mean radius that the blade profiles are not affected by variation of whirl velocity and blade speed with radius.

An impulse turbine is then one in which the entire pressure drop (neglecting blade losses) takes place in the nozzles and the rotor blades merely reverse the direction of relative whirl. This means that, if the direction of the flow at rotor exit is purely axial, the absolute whirl before the rotor is exactly twice the blade speed so that the whirl relative to the rotor is equal to blade speed both at entry and exit. The appropriate vectors and blade profiles are shown in Figure 21.

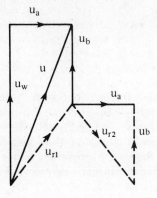
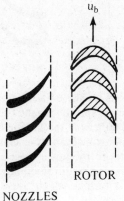

Fig. 12-21
Impulse Blading

In the left hand (vector) diagram the full lines represent absolute velocities and the broken lines represent velocities relative to the rotor. The axial component u_a is, of course, common to both nozzles and rotor blades, and has been chosen to be the same before and after the rotor blades.

The few blades sketched in the right hand diagram are drawn to conform with the vectors shown. As may be seen, the channel width of the rotor blading is substantially constant from entry to exit. In practice the rotor blades have leading edges 'rounded off' to a greater extent than shown to render them less sensitive to changes in flow direction at off design speeds.

Since, at design speed, $u_w = 2u_b$ the whirl velocity relative to the rotor equals u_b both at entry and exit $\therefore \Delta u_w = 2u_b$ and the change in angular momentum $\Delta r u_w = 2u_b r$ $\therefore$ the specific torque is proportional to $2u_b r$ and the specific power $W_S \propto 2u_b r \omega$, so, since $r\omega = u_b$, $W_S \propto 2u_b^2$, or $W_S \propto 2\theta_b$. More exactly, $W_S = \dfrac{2u_b^2}{g} = \dfrac{2u_b^2}{g}\left(\dfrac{2gK_p}{2gK_p}\right) = 4K_p\left(\dfrac{u_b^2}{2gK_p}\right)$, so writing $\dfrac{u_b^2}{2gK_p} = \theta_b$, $\quad W_S = 4K_p\theta_b$

In a 'pure reaction' turbine, the whirl component of the flow issuing from the nozzles is equal to the blade speed and, for axial exit flow so is the whirl relative to the rotor at exit. The fact that the absolute whirl before the rotor equals blade speed means that there is no relative whirl at rotor entry, i.e., the resultant velocity relative to the rotor is purely axial and so expansion takes place in the rotor blade to generate the relative whirl at exit. The appropriate vectors and blade profiles are shown in Figure 22, for the same blade speed and axial velocities as in Figure 21. This arrangement is usually referred to as '50% reaction' in that half the conversion of heat drop into kinetic energy occurs in the nozzles and the other half in the rotor blades, which are, in effect, rotating nozzles.

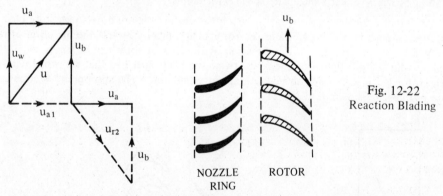

Fig. 12-22
Reaction Blading

There are many variants on these basic arrangements. One may design for 'residual whirl' and thereby increase the specific torque of rotor blading. This residual whirl would, of course, be a loss unless either (a) the kinetic energy it has is efficiently reconverted into heat energy by a ring of stator blades in the same manner as in an axial flow compressor, or (b) its energy contributes to the power generated in subsequent expansion as, for example, by driving a contra-rotating turbine. (It should be noted that residual whirl is unavoidable and unusable in a turbine which consists of a single rotor only, e.g., most windmills and so called 'wind turbines.')

As may be seen from Figure 22, $\Delta u_w = u_b$, i.e., half that of Figure 21 $\therefore$ the specific torque is proportional to $u_b r$ and the specific work W_S is proportional to u_b^2, i.e., half that of impulse type blading for a given blade speed, i.e., $W_S = 2K_p \theta_b$.

Figure 23 shows the vector diagram and blade arrangement of a 'two row Curtis Wheel' as often used at the high pressure end of a steam turbine and as proposed to be used to drive a compressor in the writer's first jet engine patent. As shown, the initial whirl u_{w1} is four times u_b and the 'residual whirl' is twice u_b on leaving the first rotor (of impulse sections). The intermediate stator acts like a stationary set of impulse blading and merely reverses the direction but not the magnitude of the absolute velocity from the first rotor, i.e., $u_3 = u_2$. Thus the whirl at entry to the second rotor is $2u_b$ (as in Figure 21).

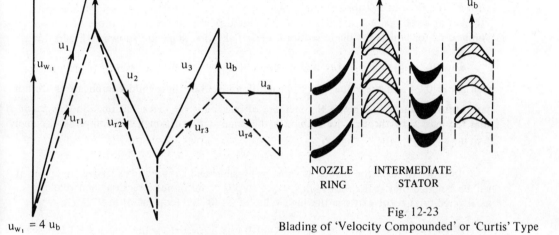

Fig. 12-23
Blading of 'Velocity Compounded' or 'Curtis' Type

The change of whirl in the first rotor is seen to be $6u_b$ and to be $2u_b$ in the second rotor to give a total of $\Delta u_w = 8u_b$. Thus the specific torque and power for a given blade speed is four times that of impulse blading (as in Figure 21). Conversely, the blade speed could be reduced by half for a given specific power.

There are many variants on the arrangements shown in Figures 21, 22, and 23. For example the one of Figure 23 might be converted to a contra rotating two stage turbine by removing the intermediate stator and reversing the direction of rotation of the second rotor. The total specific power would remain unchanged.

The foregoing discussion is, however, somewhat unrealistic in the light of the need to design for radial equilibrium which requires that the 'degree of reaction' must increase from root to tip.

In the writer's opinion, the terms 'impulse' and 'reaction' should be abandoned. One might define a turbine type by the ratio of blade speed to change of whirl at the mean radius, i.e., by $\dfrac{u_{bm}}{\Delta u_{wm}}$. Alternatively one might use $\dfrac{\theta_{bm}}{\Delta T}$ where θ_{bm} is the temperature equivalent of blade speed at mean radius. Thus, for blading which, at mean radius, had profiles corresponding to 50% reaction (as in Figure 22), $\frac{u_{bm}}{\Delta u_w} = 1.0$, or since $\Delta T = \frac{u_b \Delta u_w}{gK_p}$ (see Equation (8e) of this section) and for '50% reaction' $\Delta u_w = u_b$ and $\theta_{bm} = \frac{u_{bm}^2}{2gK_p}$ then $\frac{\theta_{bm}}{\Delta T} = 0.5$.

It is now necessary to take a look at the effects of radial equilibrium on blade design, and for this purpose a specific example will be chosen, it being assumed that the whirl distribution before the rotor is that of a free vortex (which is the condition for constant ΔT at all radii, as has been shown earlier), i.e., $u_w r = u_{wm} r_m$, and that the exhaust flow is purely axial.

The assumptions are based on a single stage turbine to drive directly a 245° compressor with a mass flow of 350 lb/sec and which, at entry to the first stage, has a blade speed of 700'/sec at a root radius of 1.0 ft (note that these figures are basically those of a seven stage compressor for which the blade sections, etc. of the first stage are shown in Figure 15).

Though, in practice, at the high temperatures concerned, the value of C_p would be well above 0.24 and the value of $\frac{\gamma}{\gamma-1}$ slightly above 4.0, these mild complicatons are not taken into account. Neither is the increase of mass flow by fuel addition. So it is assumed that the mass flow through the turbine is the same as that through the compressor, namely 350 lb/sec, and that $K_p = 336$ ft. lb/lb°C and $\frac{\gamma}{\gamma-1} = 3.5$.

In order to keep the blades as short as possible and thus avoid excessive tip speed, it will be assumed that at the root radius, $u_w = 2u_b$ (i.e., an impulse section) and that the axial velocity at rotor exit has the high value of 1450'/sec (equivalent to 97.2°C).

Other assumptions are: (1) Total temperature T_{max} before the nozzles = 1200°K; (2) Total pressure P_{max} before the nozzles is 14,700 p.s.f.; and (3) Adiabatic efficiency $\eta_T = 0.9$.

The actual temperature drop θ_T is thus $245 + 97.2 = 342.2$°C (since the energy remaining in the axial velocity contributes nothing to the shaft power) ∴ the static temperature of exhaust is $1200 - 342.2 = 857.8$°K (for which the acoustic velocity is 1927'/sec. hence the Mach No. of exhaust is $\frac{1450}{1927} = 0.75$); θ'_T, the isentropic temperature drop (total to static) is $\frac{342.2}{0.9} = 380.2$ to give a temperature ratio $t_T = \frac{1200}{1200-380.2} = 1.464$ for which the pressure ratio p_T is $p_T = 1.464^{3.5} = 3.795$ to give a static pressure at exhaust $P_E = \frac{14,700}{3.795} = 3,874$ p.s.f. which, with a static temperature of 857.8, gives the density ρ_E at exhaust, i.e., $\rho_E = \frac{3874}{96 \times 857.8} = 0.047$. It now remains to find the annulus area S from $S = \frac{Q}{\rho u_a} = \frac{350}{0.047 \times 1450} = 5.13$ ft.^{2}

From $gK_p\Delta T = 2u_{b_1}^2$ for the root radius r_i, $u_{bi} = \sqrt{\dfrac{gK_p\Delta T}{2}} = \sqrt{\dfrac{32.2 \times 336 \times 245}{2}} = 1151'/\text{sec}$. It follows that, since the blade speed of the compressor was taken to be $700'/\text{sec}$ at a radius of 1 ft., r_i for the turbine must be $\dfrac{1151}{700} = 1.645'$. The tip radius may now be found from $S = \pi(r_o^2 - r_i^2)$ and with $S = 5.13$ (from above) $r_o = 2.083$ ft. to give a tip speed $u_o = \dfrac{2.083}{1.645} \times 1151 = 1457'/\text{sec}$. The mean radius $r_m = \dfrac{2.083 + 1.645}{2} = 1.864$ ft.

It now remains to find the axial velocity before the rotor on the assumption that the annulus area does not change in the rotor, i.e., S is also 5.13 sq. ft. at rotor entry.

As we have seen in the case of the axial flow compressor with free vortex whirl distribution the density at mean radius may be used to calculate axial velocity from mass flow and vice versa. Having assumed that root whirl is twice blade speed at the root, i.e., $2 \times 1151 = 2302'/\text{sec}$, the whirl velocity u_{wm} at mean radius is given by $\dfrac{r_1}{r_m} \times 2302 = \dfrac{1.645}{1.864} \times 2302 = 2031'/\text{sec}$ the temperature equivalent of which is $190.7°C$ ∴ the static temperature at mean radius $T_m = 1200 - 190.7 - \theta_a = 1009.3 - \theta_a$ where θ_a is the temperature equivalent of the (uniform) axial velocity before the rotor.

The total density before the nozzles $\rho_{max} = \dfrac{P_{max}}{RT_{max}} = \dfrac{14,700}{96 \times 1200} = 0.128$ ∴ assuming substantially isentropic expansion through the nozzles, the density at mean radius ρ_m after the nozzles is given by $\rho_m = \rho_{max}\left(\dfrac{T_m}{T_{max}}\right)^{2.5} = 0.128\left(\dfrac{1009.3 - \theta_a}{1200}\right)^{2.5}$. Also, $\rho_m = \dfrac{Q}{Su_a} = \dfrac{350}{5.13 u_a} = \dfrac{68.2}{u_a}$. Using $u_a = 147.1\sqrt{\theta_a}$, $\rho_m = \dfrac{68.2}{147.1\sqrt{\theta_a}} = \dfrac{0.464}{\sqrt{\theta_a}}$ ∴

It follows that $\dfrac{0.464}{\sqrt{\theta_a}} = 0.128\left(\dfrac{1009.3 - \theta_a}{1200}\right)^{2.5}$ or $\sqrt{\theta_a}\left(\dfrac{1009.3 - \theta_a}{1200}\right)^{2.5} = 3.623$ from which (by trial) $\theta_a = 38°C$ and $u_a = 907'/\text{sec}$. Thus there is a very considerable increase of axial velocity in passage through the rotor blades if the annulus area remains constant. This means acceleration in the moving blades even at the root which is, therefore, not a true impulse section.

All the information is now available to draw the vector diagrams and blade sections and this is done in Figure 24 with the aid of the following table:

	r	u_w	u_{a1}	u_b	u_{a2}
r_i =	1.645	2302	907	1151	1450
r_m =	1.864	2032	907	1304	1450
r_o =	2.083	1818	907	1457	1450

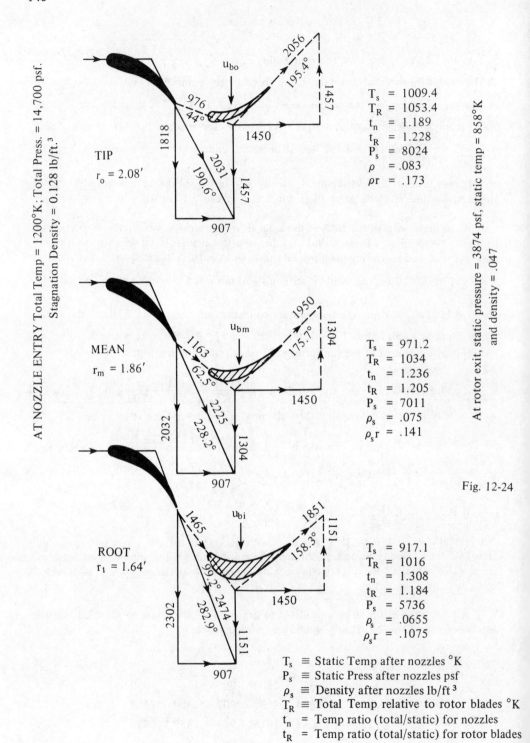

Fig. 12-24

In Figure 24, full lines indicate absolute component and resultant velocities and broken lines whirl (at exit) and resultant velocities relative to the rotor blades. The velocities are shown on all vectors and their temperature equivalents on resultant velocity vectors. To the right of each diagram the values of several important variables are listed. From these a number of important points should be noted, namely:

(1) The total temperature relative to the rotor blades decreases from tip to root which is a good thing from the temperature-stress point of view.

(2) The temperature ratio (total to static) across the nozzles is above the critical ratio 1.2 from root to mean radius and below at greater radii, i.e., the nozzles are choking, but 'only just.'

(3) The reverse is the case for the rotor blading, i.e., the greatest temperature ratio is at the tips.

(4) The static pressure upstream of the rotor blades increases from 5736 p.s.f. at the roots to 8024 p.s.f. at the tips.

(5) The increase of flux density is also considerable from root to tip in the interblade space, which means that there has to be an outward radial component of flow in the nozzles and vice versa in the rotor blading. The effect of these radial velocity components on radial equilibrium would, however, be insignificant, i.e., $\dfrac{d\theta r}{\theta_w}$ would be negligible.

(6) The value of $\rho_s r$ at mean radius is within 1% of the mean of its values at root and tip which, since $Q \propto \int_{r_i}^{r_o} \rho r dr$, justifies the use of density at mean radius for calculating the axial velocity ahead of the rotor blades.

The tip/root radius ratio is $\dfrac{2.08}{1.64} = 1.27$ approximately, i.e., the blades are relatively short, yet, as may be seen, the 'degree of reaction' increases considerably from root to tip. The effect of this on blade profile section is very useful from the stress point of view.

It is, of course, possible to deal with vector diagrams, etc., entirely non dimensionally by making all velocities multiples of a reference velocity (which might conveniently be the blade speed at mean radius); all dimensions multiples of, say, mean radius; all temperatures and temperature equivalents multiples of, say, T_{max}, and using pressure ratios, etc., rather than absolute pressures as was done for the calculations relating to Figure 24. The writer, however, deemed it preferable to present a picture of actual magnitudes, especially in relation to the temperatures to which the rotor blades are subjected.

If non-dimensional methods had been used then the diagrams in Figure 24 would be appropriate for a range of geometrically similar turbines.

The high axial velocity at rotor exhaust would be a serious loss if not utilized, since its energy represents a considerable proportion of the heat drop. In fact though the adiabatic efficiency in the case on which Figure 24 is based was taken to be 0.9 the shaft efficiency η_s would be given by $\eta_s = \dfrac{245 \times 0.9}{245 + 97.2} = 0.644$ (since θ_a at exhaust = 97.2°C). But normally the energy of the exhaust contributes to the energy available for further expansion in, say, a succeeding turbine or propelling jet. If exhaust axial velocity energy cannot be so utilized, then it would be desirable to design for a far lower value or reduce it in a diffusing exhaust annulus. In the latter case, the static pressure at rotor exit would be substantially lower than that of final exhaust, i.e., there would be recompression in the exhaust ducting.

It will be noted that the tip radius of the turbine of Figure 24 is only 4% greater than that of the first stage rotor of the compressor. It was designed to drive directly so that compressor and turbine are well matched from the point of view of size. They could in fact have been made equal by reducing the turbine root diameter if there was good reason to do so.

With the temperatures used, blade speed, and the relatively short rotor blades, blade cooling should not be necessary with present day materials, but for aircraft gas turbines at least, the trend is for much higher turbine inlet temperatures, greater tip speeds, etc., for which blade cooling becomes essential.

Calculation of End Thrust

At root radius of $1.64'$ the static pressure before the rotor is 5736 p.s.f. and after is 3874 p.s.f. $\therefore$, neglecting the shaft section, the end thrust due to differential pressure across the rotor disc is $(5736 - 3874) \pi \times 1.64^2 = 15.730$ lbs. To this must be added the pressure differential across the blading which increases from root to tip and is given by

$$2\pi \int_{r_i}^{r_o} (P_s - 3874) r \, dr,$$ and is found by plotting, to be 15,800 lbs. approximately, to give

a total 'downstream' end thrust of 31,500 lbs. approximately. If coupled directly to the compressor rotor by shafting able to transmit the tension this goes some way to countering the opposite end thrust of the compressor rotor (not calculated).

Losses

These are generally of the same type as those already covered in the discussion of losses in compressors earlier in this section. As with compressors, the boundary layers in the blading will be profoundly affected by the intense centrifugal field, but with turbine blading, losses due to fluid friction are much reduced by the fact that the flow is in the direction of 'descending' pressure gradient so that the likelihood of flow breakaway is much reduced. For the same reason turbine blades are far more tolerant of changes of flow direction at rotor entry at off design conditions. With sufficiently rounded leading edges a variation of angle of attack of the order of $\pm 15°$ would not cause serious flow breakdown unless the pitch/chord ratio were too large. (In the writer's experience the number of rotor blades should be as large as structural and stress considerations will permit.)

Tip leakage is probably the largest source of loss with uncooled blading and is a bigger problem than in a compressor. For stress and temperature reasons the turbine rotor is the most massive component in a gas turbine and, therefore, the one with the largest thermal capacity, hence there is a considerable time lag before contraction or expansion is complete after temperature changes, and this must be allowed for by relatively large tip clearance. This coupled with the large difference in static pressure from entry to exit means a very undesirable 'short circuit' of fluid. If stresses will permit, shrouding the tips (each blade carrying a shroud segment) is very beneficial.

Blade cooling adds to blade losses and reduces turbine efficiency but reduced turbine efficiency is more than offset by the gain in overall cycle efficiency resulting from the increase in allowable turbine entry (i.e., maximum cycle) temperatures.

SECTION 13
Aircraft Propulsion General

Introductory

Up to this point this work has been concerned with air breathing gas turbines in general. From here on we shall be studying their application to aircraft propulsion in the forms of turbo jets, turbo fans, etc., plus a brief discussion of the ram jet.

In its use for aircraft propulsion the gas turbine has several advantages as compared with its use for power generation at sea level which are the reasons for the fact, in general, that its application in such fields as marine propulsion, power generation, oil pipe line pumping, etc., has lagged badly behind its aircraft applications.

Except for light aircraft the gas turbine in various forms has completely displaced the piston engine-propeller combination which, until World War II, was the sole form of aircraft power plant.

The main advantages referred to above are:

(1) Part of the total compression is provided at high efficiency by ram compression due to forward speed (though at the expense of momentum drag).

(2) The low atmospheric temperatures at height substantially increase the cycle efficiency for a given maximum thermal cycle temperature.

(3) Power plant weight and bulk are greatly reduced in proportion to power.

(4) Because of (1) and (2), overall efficiency increases with height and speed.

(5) Only part of the expansion is subject to turbine losses. The part which provides the energy of a propelling jet takes place at very high efficiency.

There are certain disadvantages, however, e.g., though the aircraft gas turbine can operate with a much wider range of fuels than the piston engine it is less amenable in this respect than sea level gas turbines which can operate with almost all the range of distillate fuels whereas the aircraft gas turbine is restricted to fuels with a waxing point not above

−40°C approximately. (It is possible that this restriction may be eased for supersonic aircraft or if fuel tank heating should prove to be practicable for subsonic aircraft.) This means that the fuel must be in the kerosene–gasolene range.

Another disadvantage is the greater complexity of the combustion and fuel system due to the far greater range of fuel consumption requirements as between that for take off and initial climb on the one hand and idling on descent from great heights on the other.

As compared with the piston engine–propeller combination, the advantages are immense— far higher power/weight ratio, lack of vibration, an astonishing increase in reliability, etc.

International Standard Atmosphere (ISA)

The physical properties of the atmosphere vary so much from day to day, from season to season, with latitude, height, etc., that comparisons of performance of aircraft and engines would be very poorly based unless corrected to some generally accepted 'standard atmosphere.' This difficulty was recognized some fifty years ago and led to the international acceptance of such a standard atmosphere on which design could be based and to which observations of performance could be corrected. The result was the now generally accepted "International Standard Atmosphere."

The properties of I.S.A. are summarized as follows:

Temperature: taken as 288.3°K at sea level and the rate of fall with height (the 'lapse rate') as 6.5°C per km. which is almost exactly 2°C per thousand feet. The drop off of temperature at this lapse rate is assumed to extend to the 'tropopause' at 36,090' (11 km) above which (the 'stratosphere') temperature is assumed constant at 216.8°K, i.e., the stratosphere is isothermal and is deemed to extend to heights far above the foreseeable operational limits of aircraft. (At very great heights – the mesosphere – the absorption of short wave solar radiation causes temperature to increase with height.)

Pressure: taken as 2116.2 p.s.f. (= 760mm of mercury at 0°C) at sea level and to decrease with height in the troposphere (below the tropopause) according to the law $\frac{P_h}{P_o} = \left(\frac{T_h}{T_o}\right)^{5.256}$ (which may be deduced from $P = \rho RT$ and $dP = -\rho dh$). In the stratosphere, the pressure at 36,090' is 472.7 p.s.f. and above that is given by $20{,}813 \, \text{Ln.} \frac{472.7}{P_h} = h - 36{,}090$ (which may be deduced from the isothermal relationship $P = \rho R \times 216.8$ and $dP = -\rho dh$).

Density: taken as 0.07646 lb/ft³ at sea level and to decrease with height in the troposphere according to the law $\frac{\rho_h}{\rho_o} = \left(\frac{T_h}{T_o}\right)^{4.256}$ to the value 0.02272 lb/ft³ at the tropopause. In the stratosphere, density is proportional to pressure.

Viscosity: the coefficient of viscosity μ decreases with temperature up to the tropopause and remains constant in the stratosphere. According to Sutherland's formula $\mu = A \left(\frac{T^{3/2}}{T + B}\right)$ in slugs/ft sec where $A = 3.059 \times 10^{-8}$ slug/ft sec and $B = 114°$.

The kinematic viscosity $\nu = \frac{\mu}{\rho}$ increases with height the rate of increase being greater in the stratosphere than in the troposphere.

In practice the 'real' atmosphere varies very widely from I.S.A. For example, according to season, the tropopause may be as high as about 70,000' in tropical latitudes and as low as 25,000' over polar regions. Moreover, in unstable atmospheric conditions such as those associated with thunderstorms or when winds are forced up or down by ground contours the lapse rate will tend to conform to the 'isentropic value' of 3°C per thousand ft. These facts coupled with the very large variations of sea level temperature with season, latitude and weather must mean that the occasions when the real atmosphere coincides with I.S.A. above any given point on the earth's surface must be very rare. Nevertheless the I.S.A. is essential for design purposes and for performance corrections. But the designer must make allowance for departures from I.S.A., e.g., the loss of power for take off which occurs with high sea level temperature as in the tropics.

In all that follows, the atmosphere is assumed to conform to I.S.A.

Aircraft Drag

For subsonic aircraft in steady horizontal flight there are two major components of drag, namely 'parasite' drag (or aerodynamic resistance drag) and 'induced' drag or 'lift drag' which is the penalty paid for lift.

Parasite drag may be broken down into 'skin friction' due to viscosity, form drag due to 'shape' and 'interference drag' arising from junctions of components, e.g., wings and fuselage. These components have one thing in common — that they are proportional to the dynamic pressure relative to the aircraft, namely $\frac{\rho u^2}{2g}$ (where u is the air speed) and to the surface area. They (especially skin friction) are also a function of Reynold's number R_e where R_e is defined as $R_e = \frac{u\ell\rho}{\mu}$ or $R_e = \frac{u\ell}{\nu}$ where ℓ is a characteristic dimension of length in the direction of u, e.g., wing chord or fuselage length and ν is kinematic viscosity. But, for a given airplane, the effect of $f(R_e)$ is normally small so one may assume that parasite drag $D_P = C_D \frac{\rho}{2g} S u^2$ where S is the 'wetted area' and C_D is the coefficient of parasite drag. The effect of R_e cannot, however, be ignored when relating the results of wind tunnel model tests to full scale aircraft.

Induced or lift drag is a consequence of obtaining lift by the downward displacement of air by the aircraft's wings. It is beyond the scope of this work to explain how this gives rise to induced drag. In brief, with a wing of finite span, the wing tip vortices resulting from wing tip 'spill' from the under (high pressure) side to the upper (low pressure) side are a continuation of the 'bound vortex' surrounding the wing which, superimposed on the general flow, constitutes 'circulation' which, ahead of the wing, causes an upward component of flow or 'upwash' which compliments the 'downwash' astern. The aircraft is, as it were, flying uphill. This effect is inversely proportional to wing span and ρu^2. With infinite wing span there would be no induced drag. So, for a finite span, induced drag $D_i \propto \frac{1}{\rho u^2}$.

Total drag D is therefore given by $D = A\rho u^2 + \frac{B}{\rho u^2}$ and the constants A and B are such that D is a minimum when parasite drag and induced drag are equal, i.e., Lift/Drag ratio is a maximum when $D_P = D_i$.

At transonic and supersonic speeds the aerodynamic situation changes dramatically due to the formation of shock waves, and the fact that the air ahead of the aircraft cannot sense its approach (since small pressure disturbances cannot propagate faster than the speed of sound). There is therefore an additional shock drag, and the lift drag arises from the downward deflection of the air passing through the shock wave. The net result is a large increase of drag and therefore reduction of lift drag ratio as the penalty for supersonic speeds. Fortunately, improvement in propulsion efficiency with speed goes quite a long way to offsetting the reduction of lift drag ratio.

In fact, aircraft drag begins to rise sharply at speeds a little below Mach 1.0 because air velocity over aircraft surfaces (outside the boundary layer) is greater than flight speed especially over the forward parts of wings, etc. The drag rise 'peaks' at values of Mach No somewhat greater than unity and then begins to drop quite rapidly at higher Mach Nos though always being well above low and moderate subsonic drag.

Aircraft Weight

Aircraft weight is, of course, greatest at take off and diminishes as fuel is consumed. Hence since total fuel weight may be as much as 50% of take off weight, there is a very considerable weight change with time and distance flown.

At modern and future heights and speeds, certain small factors affecting weight are beginning to become significant, e.g., at a height of 12 miles g is reduced by nearly 2/3 of 1%, but this is about half offset by the negative buoyancy of a pressurized cabin, which for a civil aircraft flying at 50,000' and pressurized to 5,000' is nearly 1/3 of 1%. This is equal to about 8 passengers in a 'wide bodied' airliner. Again as speeds rise to, say, 2,000 mph relative to the earth's surface, there is a significant reduction of effective weight due to centrifugal force relative to the earth's axis. At orbital speed, i.e., 26,000'/sec approximately, W = 0.

The reduction of the effective weight is proportional to the square of the speed about the earth's axis and this is affected by latitude and direction of flight, thus a supersonic aircraft travelling at 2,000 mph from west to east (i.e., in the direction of the earth's rotation) over the equator would have a speed of 3,000 mph or 4,400'/sec approximately relative to the axis so its effective weight would be reduced by $\left(\frac{4400}{26000}\right)^2 = 0.029$ approximately, i.e., a reduction of nearly 3% which could be worth about 116 miles of extra range for a 'normal' still air range of 4,000 miles. But if flying from east to west the speed relative to the axis would be only 1,000 mph and the effective reduction of weight only 0.003 approximately, i.e., less than 1/3rd%.

Functions of Aircraft Power Plant

Apart from important but minor functions such as cabin pressurization, powering electric and hydraulic systems, the primary function of aero engines is to provide thrust to overcome drag and for climb and acceleration.

The 'ceiling' is reached when the thrust is sufficient only for drag at or near the maximum value of lift drag ratio. This ceiling, of course, increases slowly as fuel consumption reduces weight.

If an aircraft is climbing at angle α to the horizontal then the component of weight W along the flight path is $W \sin \alpha$ and the component normal to the flight path is $W \cos \alpha$, i.e., Lift $L = W \cos \alpha$.

The thrust required for acceleration $F_{acc} = \dfrac{W}{g}\dfrac{du}{dt}$ or $\dfrac{W}{g}u\dfrac{du}{ds}$ where du and ds are along the flight path.

Hence to deal with drag D, climb $\dfrac{dh}{dt}$, and acceleration, total thrust F is given by

$$F = D + W \sin \alpha + \frac{W}{g} u \frac{du}{ds} \qquad (13\text{-}1)$$

Since $D = \dfrac{W \cos \alpha}{\ell}$ where ℓ is the lift/drag ratio, Equation (1) may be written

$$\frac{F}{W} = \frac{\cos \alpha}{\ell} + \sqrt{1 - \cos^2 \alpha} + \frac{1}{g}\frac{du}{dr} \qquad (13\text{-}1b)$$

or $$\frac{F}{W} - \frac{1}{g}\frac{du}{dt} = \frac{1}{\ell}\cos\alpha + \sqrt{1 - \cos^2 \alpha} \qquad (13\text{-}1c)$$

It should be noted that if the angle of climb α is large and the lift component normal to the flight path correspondingly reduced then induced drag is reduced and the weight/drag ratio correspondingly increased. In the extreme case of a straight vertical climb the lift is zero and so induced drag disappears.

For most aircraft α is significantly large only in the initial stages of climb. As the angle flattens out to the point where $\alpha \simeq \tan \alpha \simeq \sin \alpha$ equation (1) may be simplified to

$$\frac{F}{W} - \frac{1}{\ell} = \frac{1}{u}\frac{dh}{dt} + \frac{1}{g}\frac{du}{dt} \qquad (13\text{-}2)$$

F, W, and ℓ, however, vary with height and speed so Equation (2) can only be used for time intervals sufficiently short for reasonably accurate mean values of F, W, and ℓ to be used.

Equations (1) and (2) are mainly of use in rate of climb calculations. In such calculations where take off thrust is high in proportion to weight (e.g., S.S.T.s) the writer finds it is sufficient to take height intervals of 5,000' up to heights where acceleration has ceased and the rate of climb is approaching zero. Such calculations are necessary for calculating the weight at the end of climb and the horizontal distance covered during climb — usually of the order of 150–250 miles in still air.

The Breguet Range Formula

According to this well known formula the still air cruising range R from the end of climb to the beginning of descent is given by $R = \eta \ell C \, \text{Ln.} \, \dfrac{W_1}{W_2}$ where η is the overall efficiency of the power plant, ℓ is the lift/drag ratio, C is the calorific value of the fuel, in ft lb/lb W_1 is the weight at end of climb and W_2 is the weight at beginning of descent, i.e., $W_1 - W_2$ is the weight of fuel consumed during cruise. W_2, of course, includes fuel for descent and reserves for diversion and 'stand off.'

The formula assumes that as weight is decreased by fuel consumption the aircraft climbs slowly to maintain constant indicated air speed which is the condition for constant lift/drag ratio ℓ.

The Breguet range formula is easily obtained as follows: If the fuel consumed is $-\delta W$ in distance δs, the work done against drag is $\dfrac{W}{\ell}\delta s$. The power for this must come from the power plant, hence $-\eta C \delta W = \dfrac{W}{\ell} \delta s$ or $\delta s = -\eta \ell C \dfrac{\delta W}{W}$ from which $s_2 - s_1 \equiv R = \eta \ell C \operatorname{Ln} \dfrac{W_1}{W_2}$.

Example 13-1:

If the take of weight is W_O; climb fuel is $0.1 W_O$; reserve fuel is $0.1 W_O$; and fuel for descent is $0.01 W_O$, what is the still air range if 1) take off fuel = $0.5 W_O$, 2) $\eta = 41\%$, 3) $\ell = 7.5$, and 4) the calorific value of the fuel is 10,500 C.H.U./lb? (1 C.H.U. = 1400 ft lbs.)

The fuel available for cruise is $W_O (0.5 - 0.1 - 0.1 - 0.01) = 0.29 W_O$ ∴ $W_1 = 0.9 W_O$ and $W_2 = W_O (0.9 - 0.29) = 0.61 W_O$ ∴ $R = 0.41 \times 7.5 \times 10{,}500 \times 1{,}400 \operatorname{Ln} \dfrac{0.9}{0.61} = 17{,}581{,}000$ ft = 3,330 statute miles.

The distance covered on climb would be 200 miles and on descent about 150 miles to give a total still air range of 3,680 miles. (This example is roughly based on the Concorde.)

It should be mentioned that the slow climb required by the Breguet equation is not allowed by Air Traffic Control regulations except in the case of aircraft such as Concorde which cruises far above the usual traffic lanes.

Both η and ℓ vary with height and speed especially η in the case of jet engines where it increases almost linearly with speed up to very high speeds, so, until the ceiling is reached, the maximum value of the product $\eta \ell$ occurs at a higher speed than that for the maximum value of ℓ. Even at the ceiling, if the indicated air speed remains constant, the true speed is inversely proportional to $\sqrt{\sigma}$, where σ is the relative density, and therefore increases as the aircraft's ceiling increases (if permissible) with decrease of weight.

The importance of the product $\eta \ell$ is more relevant to climb and diversion than to continuous cruise at ceiling height were its greatest value is used anyway. On the climb one may not be able to operate at maximum $\eta \ell$ because of gust load limitations on indicated air speed (Concorde is limited to 400 knots I.A.S. for climb to about 40,000'). If conditions at the intended destination (weather, traffic congestion, etc.) require that diversion to an alternate airfield is necessary, this normally occurs at heights far below that for optimum cruise and again gust load limitations may prevent flying at optimum $\eta \ell$ especially as the power required for level flight at low height is much below maximum cruise power which means that η for low level cruise is considerably less than for full cruise power. (This situation could be alleviated by shutting down, say, two out of four engines, but this procedure is apparently not so far considered acceptable.) For stand off the fuel consumption per mile is irrelevant. The requirement here is that the time rate of fuel consumption should be as low as possible.

Propulsion in Fluids

In all forms of propulsion in fluids, i.e., in water or air, other than rocket devices, the 'drive' is obtained by generating sternwards momentum in the fluid. This is true of birds, insects, fish, etc., as well as ships, aircraft, etc.

The propulsive force is proportional to the rate of change of momentum induced and, if steady, is given by $dF = \dfrac{dQ(u - u_o)}{g}$ or, since $dQ = u \rho \delta S$

$$\delta F = \frac{\rho}{g} u(u - u_o) \delta S \qquad (13\text{-}3)$$

where u_o is the speed of motion of the main mass of fluid relative to the body and u is the increased velocity passing through an element of area δS in a plane behind and normal to the direction of motion. If the momentum increase is not steady as with a squid, the waving tail of a fish, the flapping wings of a bird, the oars of a rowing boat, etc., then u would be a function of time and the matter would be much more complicated and equation (3) would be quite inadequate. Herein, however, it will be assumed that we may limit ourselves to a continuous and steady rate of change of momentum.

$$\text{From (3)} \quad F = \frac{\rho}{g} \int_0^S u(u - u_o) \, \delta S \qquad (13\text{-}4)$$

where S is the total area outside which $u - u_o$ is negligibly small.

The expenditure of energy, i.e., the increase of kinetic energy corresponding to the increase of momentum, is given by $\delta E = \frac{\rho}{2g} u(u^2 - u_o^2) \, \delta S$ (13-5)

from which $E = \frac{\rho}{2g} \int_0^S u(u^2 - u_o^2) \, dS$ (13-6)

The 'useful' work done (or effective power) W_p by increasing momentum is Fu_o which, from (4) is

$$W_p = \frac{\rho u_o}{g} \int_0^S u(u - u_o) \, dS \qquad (13\text{-}7)$$

Propulsive efficiency η_p may thus be defined as $\eta_p = \frac{W_p}{E}$ This, from (6) and (7) is given by

$$\eta_p = \frac{\frac{\rho u_o}{g} \int_0^S u(u - u_o) dS}{\frac{\rho}{2g} \int_0^S u(u^2 - u_o^2) dS} \quad \text{or} \quad \frac{2u_o \int_0^S u(u - u_o) \, dS}{\int_0^S u(u^2 - u_o^2) \, dS} \qquad (13\text{-}8)$$

which by writing $u = u_o \frac{u}{u_o}$ converts to

$$\eta_p = \frac{2 \int_0^S \left[\left(\frac{u}{u_o}\right)^2 - \frac{u}{u_o} \right] dS}{\int_0^S \left[\left(\frac{u}{u_o}\right)^3 - \frac{u}{u_o} \right] dS} \qquad (13\text{-}8a)$$

Beyond this we cannot go unless the values of $\frac{u}{u_o}$ are known at all points within the area S. We shall, however, be concerned with cases in which $\frac{u}{u_o}$ may be presumed uniform over S. With this assumption (8a) reduces to

$$\eta_p = \frac{2}{\frac{u}{u_o} + 1} \tag{13-9}$$

Equation (9) is the well known 'Froude efficiency' originally derived by Froude for aircraft and ship propellors using the rather dubious assumption that $\frac{u}{u_o}$ was uniform immediately behind these devices. It is a safe assumption to make with jet engines if $\frac{u}{u_o}$ is calculated or measured for a plane sufficiently close to the jet nozzle for the interchange of momentum between the jet and the external air to be taken as negligible. This interchange would not affect the total momentum increase but would have a very marked influence on the variation of $\frac{u}{u_o}$ at some distance downstream from the jet nozzle.

For the purpose of calculating propulsive efficiency with jet propulsion, Equation 9 is sufficient and (3) to (8a) may be disregarded unless for some special purpose (e.g., noise reduction) devices are used to produce a non-uniform jet discharge. Neither can it be used directly for turbo fans with separate cold and hot jets.

As may be seen from (9), η_p approaches 100% as $\frac{u}{u_o}$ nears unity, but if $\frac{u}{u_o} = 1$, $u - u_o = 0$, i.e., no thrust. However, it is obvious from (9) that for high propulsive efficiency it is better to obtain a given thrust by a combination of high mass flow rate and low values of $\frac{u}{u_o}$ than vice versa. (In the early jet engines $\frac{u}{u_o}$ at full speed was approximately 3.0 to give a propulsive efficiency of 50%. With high bypass ratio turbo fans η_p can be in the neighborhood of 80%.)

Note: The foregoing neglects the modification to thrust if expansion in the propelling nozzle is incomplete, (see the early part of Section 5) but the effect is normally negligible.

SECTION 14
'Ram' Compression and Intake Design

With stationary (or slow-moving) gas turbines the appropriate kind of intake is a convergent or 'bell shaped' duct in which the air accelerates from zero to the axial velocity at the compressor face. Since the inspired air has no kinetic energy relative to the intake before entering, there is no contribution to the total compression ratio of the thermal cycle. Indeed there may be a slight loss, but this is negligible unless the intake is very badly designed. With aircraft gas turbines, however, at high speeds, the relative kinetic energy is utilized to contribute to the total pressure ratio in the form of 'ram compression,' and it is essential that the process of conversion should be as efficient as possible.

A useful 'rule of thumb' is that the temperature equivalent or 'kinetic temperature' of the entering air is proportional to the square of the speed in hundreds of miles per hour. (This arises from the fact that to convert from mph to ft/sec the multiplying factor is 1.467 and the value of $\sqrt{2gK_p}$ = 147.1 so $\frac{147.1}{1.467}$ = 100.27.) Thus if θ_u is the temperature equivalent of the flight speed u then for u = 100 mph θ_u = 1°C, for u = 500 mph θ_u = 25°, for u = 1000 mph θ_u = 100° and so on. Thus the kinetic heating problems of high supersonic speeds can be readily seen, e.g., at 4000 mph the 'rule' gives a kinetic temperature of 1600°C. At these and higher speeds, however, the effects of increased specific heat, dissociation, ionization, etc. would make the 'rule' very unreliable. But it is reasonably valid up to 2000 mph.

For other systems of units, one may use $\frac{\theta_u}{\theta_{u_1}} = \left(\frac{u}{u_1}\right)^2$ where θ_{u_1} is a known kinetic temperature rise for a flight speed u_1

Except at take off and low subsonic speeds, for reasons which will be given below, the efficiency of ram compression is very high at moderate and high subsonic speeds if the axis of the intake differs little from the line of flight, i.e., in the absence of yaw or other aerodynamic deflection of the air at intake entry, and there is no error of significance in assuming that the efficiency of ram compression is 100%. It may therefore be assumed that the temperature ratio of ram compression t_u is given by $t_u = \dfrac{T_o + \theta_u}{T_o}$ or $t_u = 1 + \dfrac{\theta_u}{T_o}$ (14-1)

and that the corresponding pressure ratio $p_u = \left(1 + \dfrac{\theta_u}{T_o}\right)^{3.5}$ (14-2)

and the density ratio $\sigma = \left(1 + \dfrac{\theta_u}{T_o}\right)^{2.5}$ (14-3)

(these ratios being total/static) where T_o is the atmospheric temperature.

Example 14-1

If the atmospheric temperature is 220°K and the flight speed 600 mph (= 880'/sec) what are the temperature and pressure ratios t_u and p_u of ram compression assuming 100% for adiabatic efficiency, i.e. $\eta_u = 100\%$?

$$\theta_u = 36° \therefore t_u = 1 + \frac{36}{220} = 1.164 \text{ and } p_u = 1.164^{3.5} = 1.70$$

Example 14-2

Using the same T_o and u as in Example 1 and assuming that the relative density σ of the atmosphere is 0.2463 (=I.S.A. at 40,000') what is the static temperature T_s, pressure P_s, and density ρ_s at compressor entry if the entry axial velocity u_a is 550 ft/sec.

Sea level density being 0.0766 the density (atmospheric) for $\sigma = 0.2463$ is $\rho_o = 0.2463 \times 0.0766 = 0.01887$. The temperature equivalent θ_a for 550'/sec is $\left(\dfrac{550}{147.1}\right)^2 = 14° \therefore$ the static temperature increase $\Delta T = 36 - 14 = 22°C$ to give $T_s = 242°K$.

The temperature ratio (static to static) $= 1 + \dfrac{22}{220} = 1.10 \therefore$ the density ratio (static to static) $= 1.1^{2.5} = 1.269 \therefore \rho_s = 1.269 \times 0.01887 = 0.02395$.

$$P_s = \rho_s R T_s = 0.02395 \times 96 \times 242 = 556.3 \text{ p.s.f.}$$

Subsonic Intakes

In these it is usual to allow, at design speed, a substantial proportion of the intake deceleration to take place ahead of the intake orifice as shown in Figure 1. Thus much of the ram compression is entirely loss free.

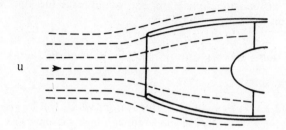

Fig. 14-1

It would be desirable to have the intake orifice as large as possible for take off and low speed, but the limit to this is set by the degree of divergence of the streamlines permissible at the rim of the orifice at design speed which, if too great, would cause flow breakaway external to the casing. This could be avoided by increasing the external diameter of the casing and 'rounding off' the lip of the entry orifice but at the price of increased nacelle drag.

(In the case of the first British jet plane (the Gloster–Whittle E28/39) the centrifugal type engine was mounted in a 'plenum chamber' in the fuselage behind the cockpit and fuel tank, i.e., the fuselage was also the engine nacelle. The air intake was at the nose of the fuselage, so, because of the size of the latter, the outer casing was well rounded at the intake orifice so it was possible to have most of the ram compression take place ahead of the orifice. In fact, there was deceleration from flight speed to 200'/sec at orifice entry. Behind the intake orifice two ducts led to the plenum chamber and were designed to reduce the velocity down to 100'/sec at plenum chamber entry.)

It is possible to increase the effective entry orifice for take off and low speeds by various means, e.g. by providing intake 'doors' to allow additional air to be drawn in through the wall of the intake duct between the entry orifice and compressor entry.

If intake ducts are alongside and close to a large adjacent surface such as a fuselage or wing it would be very undesirable for the boundary layer air of the said surface to enter the intake, so slots are used between the intake and the surface to allow the surface boundary layer to bypass the intake. (The first U.S. jet – the Bell XP59A – was the first jet plane to use this device on the writer's advice.)

Supersonic Intakes

The type of intake illustrated in Figure 1 would be quite unsuitable for supersonic speeds. There would be a shock wave ahead of the intake orifice somewhat as shown in Figure 2 much of which would comprise a 'normal shock' as indicated, across which there would be a very sharp pressure rise without change of direction with a corresponding big velocity reduction. The adiabatic efficiency of compression through a normal shock wave is very low as compared with that for oblique shocks (see Section 4). At Mach 2.0 in the stratosphere it would be about 80% or less, whereas one may expect about 95% or even more for an intake designed for oblique shocks as in Figures 3 and 4 (based on guesswork). Figure 3 illustrates the kind of intake used on Concorde and Figure 4 shows an alternative arrangement for breaking down shock compression into stages of oblique conical shocks ending in a weak normal shock.

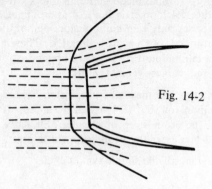

Fig. 14-2

Such intakes, unfortunately, unless of variable geometry, are suited to one particular speed and so variable geometry is virtually essential. Thus it is usually necessary for the forebody of Figure 4 to be capable of being moved fore and aft, and for the arrangement of Figure 3 to have adjustable 'ramps.'

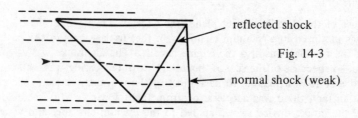

Fig. 14-3

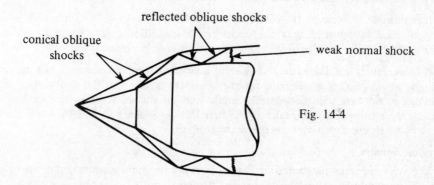

Fig. 14-4

Much development is still in progress on the forebody type of intake to reduce adverse boundary layer effects, sensitivity to yaw, etc. These disturbances apparently cause rapid fore and aft fluctuations in the position of the shocks.

Ramjets

At high supersonic speeds ram compression can be so high that additional compression becomes unnecessary, hence no compressor-compressor turbine combination is needed. It follows that the maximum temperature of the thermal cycle is no longer restricted by the stress-temperature properties of turbine blade materials and so, if desired, it is possible to use air/fuel ratios up to the stoichiometric limit of approximately 15.1. The ramjet thermal cycle therefore comprises only ram compression, combustion at substantially constant pressure, and expansion through a propelling jet. There are no major moving parts but there must be a certain amount of auxiliary apparatus for such purposes as fuel injection, for varying intake and exhaust geometry, etc.

The ramjet can never be the sole means of propulsion because of its dependence on high speed for adequate compression, but if 'boosted' to speeds of the order of Mach 3 by rockets it can 'take over' the propulsion for missiles. Similarly, if boosted to high speeds by turbo jets it could take over the propulsion for aircraft.

Figure 5 is the cycle diagram for the simple ramjet cycle.

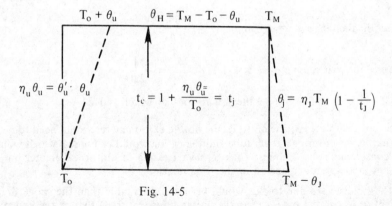

Fig. 14-5

It would be desirable to have a ram-jet body with a purely cylindrical casing and have complete expansion in the final nozzle, but this, however, is not possible at the speeds at which ramjets have good efficiency. This will be demonstrated by example. Figure 6 shows the cycle diagram for a ramjet based on the following assumptions: (1) $T_o = 220°K$, (2) $u = 3000$ mph $= 4400'$/sec ($\theta_u = 900°C$), (3) the efficiency of ram compression $\eta_u = 85\%$ (allowing for combustion pressure loss), (4) the efficiency of expansion is 100% to the nozzle throat and 98% thereafter, and (5) $T_{max} = 2000°K$.

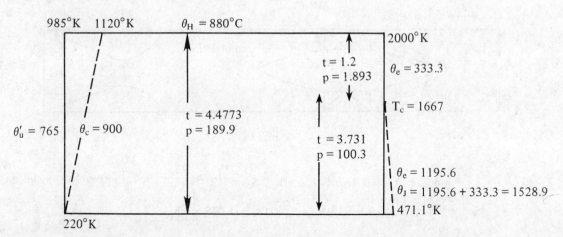

Fig. 14-6

Several of the relevant figures are shown on the diagram. Others are as follows:

Jet velocity (for complete expansion) $u_j = 147.1\sqrt{1528.9} = 5751.8'$/sec

Specific thrust $F_s = \dfrac{5751.8 - 4400}{32.2} = 41.98$ lbs/lb/sec

Overall efficiency $\eta_o = \dfrac{F_s u}{K_p \theta_H} = \dfrac{41.98 \times 4400}{336 \times 880} = 0.625$

If S_o is the cross section of the entering air ahead of the intake and S_e the nozzle exit section then $S_e \rho_e u_j = S_o u \rho_o$ $\therefore \dfrac{S_e}{S_o} = \dfrac{u \rho_o}{u_j \rho_e}$.

Since the exit temperature is 471.1°K, $\dfrac{\rho_o}{\rho_e} = \dfrac{471.1}{220} = 2.141$ and $\dfrac{u}{u_j} = \dfrac{4400}{5751.8} = 0.765$ $\dfrac{S_e}{S_o} = 2.141 \times 0.765 = 1.638$ the square root of which is 1.28.

So for complete expansion to P_o the nozzle exit diameter would need to be 28% greater than that of the entering stream tube and the casing would be conical which would, of course, cause considerable shock drag. Thus to have a casing of uniform diameter one would have to accept incomplete expansion in the final nozzle.

The small losses due to this would be more acceptable than the shock drag of a conical casing. For a 'parallel pipe' casing the writer finds (by trial) that at the exit plane the pressure $P_o = 2.049 P_o$, $u_j = 5561'$/sec and $T_e = 1096°K$ thus the further acceleration after discharge would only be $5751 - 5561 = 190'$/sec.

This external expansion would itself be 'bounded' by a shock wave as the issuing and expanding jet encounters the external air somewhat as shown in Figure 7.

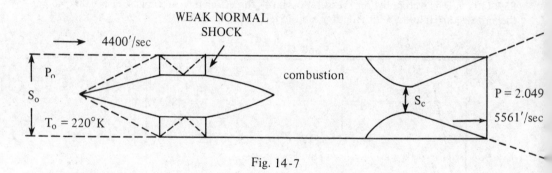

Fig. 14-7

For a flow of 1.0 lb/sec. $S_o \rho_o u = 1$ $\therefore S_o = \dfrac{1}{\rho_o u} = \dfrac{RT_o}{P_o u}$. So for incomplete expansion $F_s = \dfrac{5561 - 4400}{32.2} + 1.049 P_o S_o = 36.06 + 1.049 \dfrac{RT_o}{4400}$

$1.049 \dfrac{RT_o}{4400} = 1.049 \times \dfrac{96 \times 220}{4400} = 4.8$ lb hence total thrust (specific) $F_s = 36.06 + 4.8 = 40.86$ c.f. 41.98 for complete expansion. This reduces the overall efficiency η_o from 62.5% to $\dfrac{40.86}{41.98} \times 62.5 = 60.8\%$ which is still a very high figure.

In the foregoing example, the figure of 2000°K used for maximum temperature is far below that which would result from near stoichiometric combustion, but much higher values of T_{max} would increase the degree of under expansion and reduce propulsive efficiency. Specific thrust would, of course, be increased.

It should be mentioned that the choice of 85% for ram compression efficiency was purely arbitrary and has no basis on information available to the writer. To achieve it would probably require a much more elaborate intake design than that shown very diagrammatically in Figure 7. Obviously, however, ramjets would be acceptable propulsive devices with far lower efficiencies (overall) than 60%. If, in the above example, the ram efficiency had been taken as low as 70%, the overall efficiency would still be 40% approximately, but F_s would be greatly reduced (to 26.5 lb/lb/sec. c.f. 40.86).

SECTION 15
Combustion

Introductory

In the early development of the jet engine the greatest of the several technical problems was that of obtaining the very high intensity of combustion with reliability of combustion chamber components coupled with low pressure loss and high combustion efficiency. The need was to reduce the volume of combustion chambers to about 5% of the best then obtained for industrial applications in proportion to the rate of fuel consumption, e.g. as in steam boilers, etc. There seemed no chemical or physical reason why the target should not be achieved and it was in fact quite quickly achieved in combustion chamber test experiments but it took some three years before the promising results of these experiments could be obtained in seemingly identical combustion chambers installed in a complete engine. The main reason for this very frustrating situation was that the pattern of air flow at combustion chamber entry (i.e. from compressor delivery), in the engine was very different from that in the component testing. Eventual realisation of this underlined the fact that there were major aerodynamic factors to be taken into account in combustion chamber design. Unfortunately these are too complex for analysis so, for this and other reasons, combustion chamber design is still more an art than a science. There are, however, certain basic principles which must be observed and these will be discussed below.

The limitations imposed by size and weight in aircraft gas turbines are not, of course, so stringent in gas turbine power units for other than aircraft propulsion, but it is generally desirable for installation and other reasons that the bulk and weight of the combustion apparatus should not be disproportionate to the other major components of the power plant.

The Composition of the Atmosphere

The atmosphere for all practical purposes is almost entirely a mixture of nitrogen and oxygen in the proportions 78.1% N_2 to 20.95% O_2 by volume or 75.52% N_2 to 23.15% O_2

by weight. The balance is made up of the inert gases Argon, Neon, Helium, Krypton and Xenon; carbon dioxide, methane, oxides of nitrogen, hydrogen and ozone, and variable quantities of water vapour. With the exceptions of Argon (1.3% by weight), CO_2 (0.5%) and water vapour, the other constituents range from 0.4 to 12 p.p.m. For combustion purposes one may therefore safely assume that air is a mixture of nitrogen and oxygen.

Hydrocarbon Fuels

Virtually all fuels derived from the 'fossil fuel' petroleum are hydrocarbons, i.e. compounds of carbon and hydrogen and fall into two main groups namely the paraffinic series with the general formula $C_n H_{2n+2}$ and the aromatic series of the genral formula $C_n H_n$. Both series have hundreds of members, especially the paraffinic series ranging from CH_4 (methane) with boiling point of $-160°C$ approx. to bituminous compounds which are solids at room temperature.

Crude liquid petroleum is rarely used directly as a fuel (though some crudes are so 'light' that this is possible). It is refined in a number of processes of which the principle one is distillation whereby the crude is 'divided' into distillates ranging from heavy fuel oil to light and volatile gasolenes. The residue after distillation and many other refining processes comprises residual fuel oil, bituminous asphalts, etc. (And even coke after certain cracking processes) which contain most of the mineral impurities of the original crude.

Many attempts have been made to use 'residual' (and even powdered coal and peat) in industrial gas turbine power plant but without success to date because of the erosive effect of the ash on turbine blades. Also, since much of the ash is molten at combustion temperatures, it tends to solidify on the blading. In short, gas turbines are limited to distillate fuels whereas steam powered ships, power stations, etc. may, and usually do, operate on the much cheaper heavy residuals.

Even distillates are unusable if the content of sulphur, vanadium, etc. exceeds certain small limits.

For aircraft gas turbines the range of distillates which may be used is further limited by certain strict specification requirements the chief of which is that the waxing point must not exceed $-40°C$ owing to the low temperatures to which the tanks and fuel lines are subjected. Further the water content must be extremely low, and the proportion of aromatics is restricted: this last because aromatics are more prone to soot formation than the paraffin series hydrocarbons. All this adds up to the fact that distillate fuels for aircraft must be in the kerosene-gasolene range. (The degree to which low grade gasolene may be used has been the subject of much controversy because of the fire risk in a crash—kerosene is undoubtedly much safer in this respect). Vapour pressure with temperature, flash point, etc. are other factors to be considered (it has been claimed that as much as 18% of a low grade gasolene can be lost through boiling on a rapid climb). It is in any case desirable to replace air in fuel tanks with an inert gas as fuel is consumed to prevent the formation of explosive mixtures of fuel vapour and air which may form at certain temperatures and which may be ignited by static electricity.

The calorific value of hydrocarbon fuels varies little from 10,500 C.H.U./lb (1 C.H.U. = 1.8 B.Th.U = 1400 ft lbs) so the slightly denser kerosene occupies a somewhat lower volume than gasolene for a given fuel weight.

The Basic Chemistry of Combustion

Though kerosenes and gasolenes are mixtures of many hydrocarbons one may select any member of the family as representative of the mixture for the chemistry of the combustion process. So the writer elects to choose octane C_8H_{18} for the purpose. The chemical equation of combustion is then:—

$$C_8H_{18} + 12\tfrac{1}{2} O_2 = 8 CO_2 + 9 H_2O + \text{Heat}$$

The molecular weights are as follows:—

$$C_8H_{18} = 8 \times 12 + 18 \times 1 = 114$$
$$O_2 = 32$$
$$CO_2 = 12 + 32 = 44$$
$$H_2O = 2 \times 1 + 16 \times 1 = 18$$

∴ the 'mass balance' is:—

$$1 \times 114 + 12.5 \times 32 = 8 \times 44 + 9 \times 18 = 514$$

i.e. 114 lb. of C_8H_{18} combines with 400 lbs. of O_2 to produce 352 lbs. of CO_2 plus 162 lb. of H_2O or:—

1 lb. of C_8H_{18} combines with 3.509 lb of O_2 to produce 3.088 lb of CO_2 and 1.421 lb. of H_2O plus 10,500 C.H.U.

But, as we have seen above, for every lb. of O_2 in the air there is 3.262 lb. of N_2, so, assuming for the moment, that N_2 plays no part in the reaction, the chemical equation may be re-written as:—

$$1.0\, C_8H_{18} + 3.509\, O_2 + 11.446\, N_2 = 3.088\, CO_2 + 1.421\, H_2O + 11.446\, N_2 + 10{,}500 \text{ C.H.U.}$$

or 1 lb. C_8H_{18} + 14.96 lb. of air yields 3.088 lbs. CO_2 + 1.421 lb. H_2O + 11.446 lb. N_2 + 10,500 C.H.U.

Thus it may be seen that, for, complete combustion, the air/fuel ratio is 15:1 approx. This is the 'stoichiometric ratio.'

The average specific heat K_p of the N_2 plus combustion products at, say, 400°K is 0.268 C.H.U./lb./°K, so, if there were no change of K_p with temp. the temp. rise of combustion would be $\dfrac{10{,}500}{.268 \times 14.96} = 2619°C$ (this does not allow for the latent heat of vaporisation of the liquid fuel). In practice, however the temp. rise would be substantially lower due to 1) the increase of K_p with temp. 2) the dissociation of some of the CO_2, H_2O and N_2 into C, CO, the radicals OH, O, and N, 3) the formation of oxides of nitrogen ('NOX') and ozone (O_3).

In gas turbines the air/fuel ratio is far greater than stoichiometric, being of the order of 55-70:1, nevertheless the combustion temp. rise is appreciably lower than would be calculated on the assumption of constant specific heat.

On expansion to lower temps. the heat absorbed by dissociation, etc. is recovered by recombination, especially if the air/fuel ratio is well above stoichiometric so the exhaust of gas turbines has negligible carbon monoxide (CO) O_3, etc. in contrast to reciprocating engines operating with near stoichiometric mixtures the exhaust of which often contains unburnt hydrocarbons as well as CO, etc. When operating at high turbine entry temps. however, the exhaust of a gas turbine contains NOX; sometimes as much as 18 gms. per kgm of fuel, and, if combustion chamber design leaves something to be desired, carbon (visible as a trail of smoke) and unburnt hydrocarbons. These two latter, however, though creating an impression of poor combustion efficiency, represent a minute fraction of the fuel.

Attempts are in progress to cut down the emission of oxides of nitrogen by reducing peak combustion chamber temperatures. These efforts are based on the belief that NOX has a powerful catalytic action in breaking down the 'ozone layer' high in the stratosphere and so increases the amount of certain harmful wavelengths of ultra violet light from the sun which can reach the earth's surface and cause skin cancer and other biological damage. In the writer's opinion, the 'ozone scare' is based on several dubious assumptions which support the belief, and neglect others which counter it, chief of which is that the various constituents of NOX have a strong affinity for the water vapour present in jet exhausts to form nitrous and nitric acids which have none of the catalytic properties of NOX. However, a full discussion of this subject would be lengthy and beyond the scope of this work.

The combustibility of an air-fuel mixture depends on a number of factors, the chief of which is the air/fuel ratio. Clearly, if this is less than stoichiometric, complete combustion is not possible, and even in a completely homogenous mixture—not easy to achieve in very short time intervals—there are upper and lower limits to the air/fuel ratio necessary for satisfactory combustion. If the air/fuel ratio is less than about 12:1 the mixture is too 'rich'. If greater than about 20:1 it is too 'lean'. This fact dominates gas turbine combustion chamber design in that there has to be a primary combustion zone into which only about 20% of the total air is admitted, i.e. in the primary zone the mixture is near stoichiometric. The remaining secondary air is added in a secondary mixing zone in such a manner as to mix as rapidly as possible with the extremely hot output of the primary zone.

Ignition and Flame Propagation

Unless a catalyst is present, ignition of even a completely homogenous combustible mixture requires a certain minimum temp. to initiate combustion. Except in compression ignition engines (diesels) where the air is heated by compression to a sufficiently high temp. for injected fuel to ignite, the most common means of 'triggering' combustion is by an electric spark. Sometimes 'glow plugs' or pilot flames may be used, but in gasolene piston engines and gas turbines a high tension spark plug (or plugs) are the means of initiating combustion. With the continuous combustion of gas turbines, the spark is needed only for starting (unless there is a 'flame out').

Since initial ignition is inevitably very local to the spark (or other initiating means) the flame must propagate through the rest of the mixture in a wavelike manner. Thus the speed of propagation is very important considering that the residence time in the primary combustion zone is of the order of .001 secs. Unfortunately the rate of flame propagation in a homogenous combustible mixture at rest is very slow and it would be quite impossible to maintain combustion in a gas turbine combustion chamber primary zone if the air flow were steady—the flame would be swept downstream immediately in the high air speeds necessary to keep combustion chambers small and light. Means must therefore be provided to stabilise the flame in the primary zone. The writer first attempted to achieve this by 'upstream injection', i.e. by 'squirting' the fuel into the air stream so that the flame blew back on to the injected fuel and ignited it. This was partially successful but required about 6 ft. length of combustion chamber to achieve moderately complete combustion. Another defect of this system was that the fuel injector was located in the hottest part of the flame and suffered from overheating despite attempts to use the fuel as a cooling jacket before injection. Flame stabilisation is thus another major factor in combustion chamber design.

Fuel Injection

It is, of course, essential to achieve the speediest possible rate of mixing of fuel and primary air. With kerosene-gasolene type fuels it is possible to vaporise the fuel before injection through multiple jets, but the writer's several attempts (some thirty types of vaporisers were tried) to achieve this all failed. Though excellent primary zone combustion was often obtained, the vaporisers failed either through overheating or coke formation (due to cracking) within them. However, successful vapour injection was later achieved by others using vaporisers having air mixed with the fuel in the vaporising tubes. Nevertheless the most common form of fuel injectors are 'atomising burners' which inject a 'cone' of liquid droplets. In these the 'atomising' is achieved by admitting the fuel into the interior of the injector through tangential slots which impose an intense spin on the fuel before it emerges from the injector orifice. On leaving the orifice the rapidly spinning fuel bursts into tiny droplets under the action of centrifugal force. This procedure, of course, requires that the fuel pressure must be far higher than combustion chamber pressure to provide the necessary kinetic energy plus the surface tension energy contained in the droplets.

The behaviour of the burning droplets is a very complex matter. Their interaction with the primary air and rate of burning depends on the turbulence of the primary air the size and speed of the droplets, etc. etc. One would expect that the smaller the droplet the shorter the burning time but against this Stokes law operates to slow down relative movement and therefore the rate at which a droplet can 'engage' still unused oxygen. Further, each droplet has a 'cloak' of combustion products tending to interfere between the air and the still unburnt 'core'. Moreover, after a very short time the intense heat vaporises partially burnt droplets and this almost certainly slows down the rate of mixing for a given degree of turbulence.

Soot Formation

The carbon build up which can occur due to the cracking of the fuel in the intense heat and to locally over rich mixtures must be prevented. If it happens around the injector orifice it can distort the spray. If build up occurs on the nearby surfaces which control the motion of the primary air it alters their geometry, and therefore their intended effect on the primary air. The problem of coke formation bedevilled early combustion development. The main solution was to force some of the primary air over surfaces where coke formation tended to occur to cause it to burn off as rapidly as it was formed.

Mixing of Secondary Air

A complete combustion chamber normally has an outer casing and an inner flame tube or tubes. It is in the latter that both primary combustion and secondary air mixing takes place. As previously indicated about 20% of the total air is admitted directly to the primary combustion zone. The remainder is admitted in stages downstream of the primary zone in such a manner as to achieve as complete a mixing as possible without excessive pressure loss. This is not easily done and to obtain a satisfactory result much trial and error is usually necessary. One conflicting requirement is that the inner surface of the flame tube must be shielded from excessive temperatures by a protective 'skin' of comparatively cool air without causing a serious non-uniformity of temperature at entry to the turbine nozzle ring.

To obtain virtually complete mixing of 80% of comparatively cool air with 20% of extremely hot combustion products in a time interval of the order of .02 seconds is indeed a formidable problem and can only be done by inducing considerable turbulence over and above that created in the primary zone, so that there has to be a rapid momentum interchange only obtainable at the price of pressure loss.

It would take far too long to describe the sundry arrangements of holes, slots, perforated stub pipes, etc. by means of which the secondary air enters the flame tube.

Flame Stabilisation in the Primary Zone

This has been briefly touched upon above, but some amplification is desirable.

In brief, the most effective means in use is still that used in the first jet engine, namely to admit the primary air to the flame tube through a ring of swirl vanes surrounding the fuel injector. These vanes generate a vortex, preferably oppositely directed from that of the spin of the injected fuel spray. The success of this device was not understood until a transparent plastic model of a combustion chamber, equipped with wool tufts to indicate flow behaviour, demonstrated that there was flow reversal in the core of the vortex, i.e. the flow was towards the injector on and near its axis. Thus the initial fuel injection was upstream in effect.

At this point it should be mentioned that there was nothing novel in the use of swirl vanes surrounding a fuel injector—it had been common practice in steam boilers, etc. for some years where the volume available for combustion was far larger.

Swirl vanes are by no means the only method of flame stabilisation. Sundry devices which produce turbulence in the neighborhood of fuel injectors can be quite effective, e.g. perforated cones or plates surrounding the injector, annular V sectioned 'troughs' such as are used in after-burners of jet engines, etc.

Counter-flow Combustion Chambers

Figure 1 illustrates one of the ten counter flow type combustion chambers used in the first British jet engine. This arrangement was not a matter of choice initially; it was dictated by the need to make it conform to the rest of the engine because shortage of money proscribed the making of a completely new engine and dictated the continued use of the more expensive components—rotor assembly, compressor casing, etc. which had been designed for an entirely different (and unsuccessful) combustion chamber. Nevertheless the counter-flow arrangement had several advantages in the circumstances of the time. It was free from

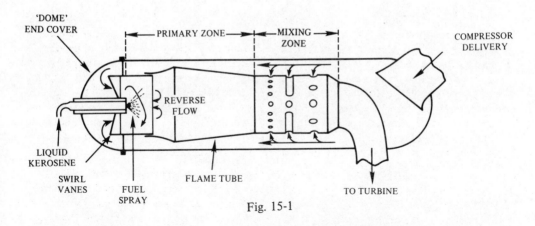

Fig. 15-1

problems of thermal expansion; it allowed the continued use of a very short shaft in the rotor assembly; the dome shaped end covers could be easily removed for inspection and modifications to the swirl vanes, flame tubes, etc. (of which there were many), and so it was retained in several later designs. It was very much a matter of "leave well alone" since experience had shown that any major change would entail another long period of development. It was, of course, eventually displaced by the 'straight through' type which, however, used the same basic principles of design in burner-flame tube arrangements.

SECTION 16
The 'Straight' Turbo Jet

The straight turbo jet is a gas turbine type gas generator in which the energy remaining at exhaust from the compressor turbine is used to provide a high velocity propelling jet by expansion through a nozzle at the rear of an aircraft fuselage or engine nacelle. It may be thought of as an 'energiser' within a duct.

Fig. 1 shows the general arrangement of the W1 engine (with combustion chamber details omitted) which powered the first successful British jet plane, the Gloster-Whittle E 28/39. Fig. 2 shows the general arrangement of the fuselage of the E. 28.

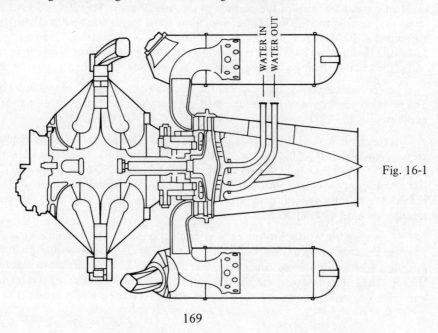

Fig. 16-1

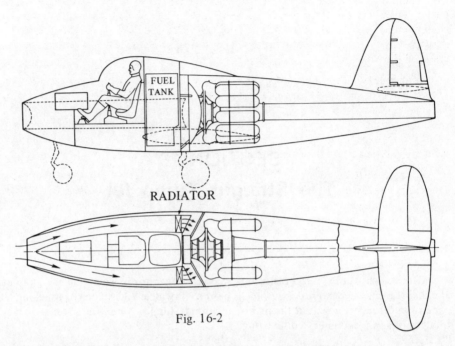

Fig. 16-2

Some facts of interest in brief are as follows:—

The W1 engine was designed for a static thrust of 1240 lbs. at 17,750 r.p.m. which it eventually produced, but for the purpose of the first flights in May 1941 it was, on Ministry insistence, derated to 16,500 r.p.m. at which the static thrust was 860 lbs. One flight was authorised at 17,000 r.p.m. and static thrust 1000 lbs. At this rating the aircraft reached a speed of 370 m.p.h. at 20,000 ft. (well below its ceiling), which was higher than expectations and slightly better than the Spitfire, then the best piston engined fighter in service.

The engine weight was 520 lbs. and the all up weight of the E. 28 was 3000 lbs. approx. The fuel tank capacity was only 80 imperial gallons which limited flight time to 50-60 minutes.

Throughout W.W. II the E. 28 was used to flight test a series of Whittle type engines of greater power—the W1A (1340 lbs. static thrust); the W2B (1400-1600 lbs. static thrust); the W2/500 (1650 lbs. static thrust) and the W2/700 (2000-2500 lbs.) The W2/700 was the last centrifugal type engine with counter flow combustion chambers. It proved too powerful to be fully opened up in level flight because the aircraft ran into compressibility troubles at about 470 m.p.h.

Meanwhile the W2B went into limited production as the Rolls-Royce Welland which powered the twin engined Gloster Meteor 1. The W2B was also the prototype of General Electric's I 14 which powered the twin engined Bell P59A, which in fact flew before the Meteor. The latter, however, was the only Allied jet fighter to go into operation during the war. (The German twin engined Me. 262 was operational a few weeks earlier than the Meteor).

There then followed a series of successful centrifugal types with 'straight through' combustion chambers—the Rolls-Royce Derwent 1; Derwent V, Nene, etc., The deHavilland Goblin and Ghost; the General Electric I.40, etc. Perhaps the most successful of these was the Derwent V (3,600 lbs. static thrust) which went into large scale production and, inter alia, powered the Meteor IV.

In parallel with the development of the centrifugal type, development of the axial flow type was in progress at the Royal Aircraft Establishment in cooperation with Metropolitan Vickers, in Germany and in the U.S.A. Eventually the early difficulties were overcome and the advantages of small frontal area increased in importance with increasing speeds to the point where, except for very small engines, the axial flow type completely displaced the centrifugal type despite the much higher cost and complication.

As previously mentioned, the jet engine as a gas generator differs from the types of gas generators already discussed in that part of the compression is provided by the ram compression in the intake at moderate and high speeds. As has been shown, the ram compression temp. rise is proportional to the square of the flight speed.

Fig. 3 is a very diagrammatic representation of an axial flow jet engine, and Fig. 4 is the corresponding Pressure–Specific Volume diagram.

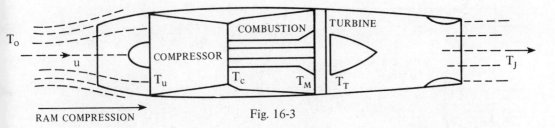

Fig. 16-3

The portion representing ram compression would, of course, be absent from the static sea level cycle.

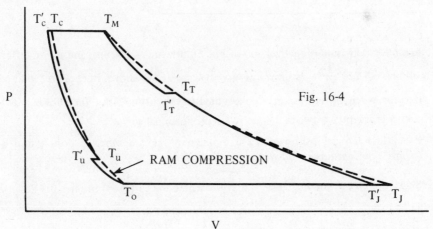

Fig. 16-4

The cycle diagrams, Figs. 5a and 5b, as used by the writer, represent sea level static and high speed flight.

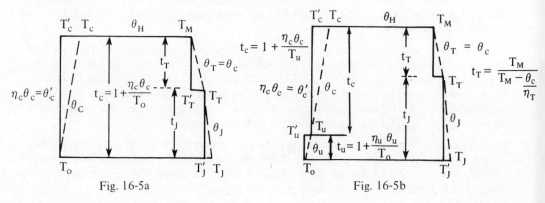

Fig. 16-5a   Fig. 16-5b

For sea level static, $t_J = \dfrac{t_c}{t_T}$; in high speed flight, $t_J = \dfrac{t_u t_c}{t_T}$

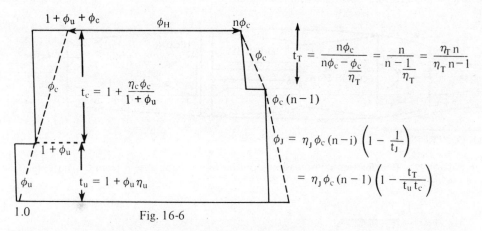

Fig. 16-6

Fig. 6 is a non-dimensional version of Fig. 5b in which all temps. and temp. differences are multiples of T_0. $\dfrac{T_M}{T_0}$ is shown as $n\phi_c$ because it was shown in Section 5 that over the range for which η_c and η_T may be assumed to vary little, with fixed choking nozzles in series with energy extraction between them, T_M is proportional to θ_c.

From the relationships shown in Fig. 6, one may find a number of 'doubly non-dimensional' relationships as follows:—

Since $u_j \propto \sqrt{\phi_J}$ then $\dfrac{u_J}{u_{JD}} = \sqrt{\dfrac{\phi_J}{\phi_{JD}}}$ where the suffix D implies the design

condition $\quad \therefore \quad \dfrac{u_J}{u_{JD}} = \sqrt{\dfrac{\phi_c}{\phi_{cD}}\left(1 - \dfrac{t_T}{t_u t_c}\middle/ 1 - \dfrac{t_{T0}}{t_{uD} t_{cD}}\right)}$ \hfill (16-1)

or
$$\left(\frac{u_J}{u_{JD}}\right)^2 = \frac{\phi_c}{\phi_{cD}} \left(\frac{t_{uD} t_{cD}}{t_u t_c}\right) \left(\frac{t_u t_c - t_T}{t_{uD} t_{cD} - t_{TD}}\right) \tag{16-1a}$$

Merging the design values into a constant A

$$u_J^2 = A\phi_c \left(\frac{t_u t_c - t_T}{t_u t_c}\right) \tag{16-1b}$$

Since spec. thrust $F_s \propto u_J - u$
$$\frac{F_s}{F_D} = \frac{u_J - u}{u_{JD} - u_D} = B\left(\sqrt{\phi_J} - \sqrt{\phi_u}\right) \tag{16-2}$$

where $B = \left(\sqrt{\phi_{JD}} - \sqrt{\phi_{uD}}\right)^{-1}$

Since mass flow rate $Q \propto P_o (t_u t_c)^{3.5} / \sqrt{\phi_c}$

$$\frac{Q}{Q_D} = \frac{P_o}{P_{oD}} \left(\frac{t_u t_c}{t_{uD} t_{cD}}\right)^{3.5} \sqrt{\frac{\phi_{cD} T_{oD}}{\phi_c T_o}} \tag{16-3}$$

Using (2) and (3)
$$\frac{F}{F_D} = \frac{F_s Q}{F_{sD} Q_D} = C \frac{P_o}{\sqrt{\phi_c}} (t_u t_c)^{3.5} \left(\sqrt{\phi_J} - \sqrt{\phi_u}\right) \tag{16-3}$$

where $C = \dfrac{B \sqrt{\phi_{cD}}}{P_{cD} (t_{uD} t_{cD})^{3.5}}$

In similar manner the ratios of other quantities—fuel consumption, efficiency, etc. to design values can be found.

From the foregoing it should be clear that, for a given design, the performance characteristics are functions of four variables, namely atmospheric temp. T_o, atmospheric pressure P_o, the temp. equivalent of flight speed θ_u, and the compressor temp. rise θ_c.

Many variants of several of the above equations may be obtained by substituting for t_c and t_u from $t_c = 1 + \dfrac{\eta_c \phi_c}{1 + \phi_u}$ and $t_u = 1 + \eta_u \phi_u$.

It should be noted that none of the quantities which can be derived from the cycle diagram involve atmospheric pressure and so depend on the temp. parameters only (for given efficiencies).

The foregoing methods will now be illustrated with examples.

Example 16-1.

Fig. 7 illustrates the design cycle diagram for a Mach 2 (1950'/sec.)

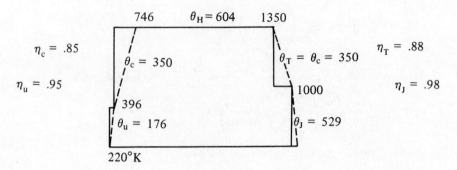

Fig. 16-7

aircraft at 40,000'. Assumptions not shown on the diagram are:— 1) the required thrust is 10,000 lbs. 2) the atmospheric pressure is 392 p.s.f.

Using the data on the diagram the following figures are obtained:—

$$t_{uD} = 1 + \frac{.95 \times 176}{220} = 1.76 \; ; \; t_{cD} = 1 + \frac{.85 \times 350}{396} = 1.751$$

$$t_{TD} = \frac{1350}{1350 - \frac{350}{.88}} = 1.4177 \; ; \; t_{jD} = \frac{t_{uD} \, t_{cD}}{t_{TD}} = 2.174$$

$$\theta_{JD} = .98 \times 1000 \left(\frac{t_j - 1}{t_j}\right) = 529.2°C$$

$$\therefore \quad u_{jD} = 3384'/\text{sec.} \quad (\text{i.e.} \quad 147.1\sqrt{529.2})$$

$$\therefore \quad F_{sD} = \frac{3384 - 1950}{32.2} = 44.54 \text{ lbs/lb/sec}$$

$$\therefore \quad Q_D = \frac{10,000}{44.54} = 224.5 \text{ lb/sec.}$$

$$f_s = \frac{\theta_H \times .24 \times 3600}{10,500 \times F_s} = 1.116 \text{ lb/hr/lb thrust}$$

$$\eta_o = \frac{F_s \times u}{604 \times 336} = \frac{44.54 \times 1950}{604 \times 336} = .428$$

$$\phi_{uD} = \frac{176}{220} = 0.8 \quad \phi_{cD} = \frac{350}{220} = 1.591 \therefore n = \frac{1350}{350} = 3.857$$

$$\phi_{JD} = \frac{529}{220} = 2.405$$

A for equation (1b) $= \dfrac{t_{uD}\, t_{cD}}{\phi_{cD}} \left(\dfrac{1}{t_{uD}\, t_{cD} - t_{TD}} \right) u_{JD}^2 =$

$$\frac{1.76 \times 1.751}{1.591} \left(\frac{1}{1.76 \times 1.751 - 1.418} \right) u_{JD}^2 = 1.164 \times 3384^2$$

B for equation (2) $= \dfrac{1}{\sqrt{2.405} - \sqrt{0.8}} = 1.523$

C for equation (4) $= \dfrac{1.523}{392\,(1.76 \times 1.751)^{3.5}} = 7.557 \times 10^{-5}$

From the above results it is now possible to calculate the performance of the engine at other heights, flight speeds and engine speeds provided that it is assumed that there is no appreciable change in component efficiencies and that θ_c is proportional to N^2, where N is engine r.p.m.

Example 16-2.

What would be the performance characteristics of the engine of Example 1 at 20,000' at a speed of 600 m.p.h. (880'/sec.) at the same engine speed (i.e. same θ_c and T_M) assuming $T_o = 250°K$ and $P_o = 970$ p.s.f.

For 880'/sec. $\theta_u = 36 \quad \phi_u = \dfrac{36}{250} = 0.144 \quad t_u = \dfrac{.95 \times 36}{250} + 1 = 1.137$. The total temp. of ram compression $= 250 + 36 = 286 \therefore t_c = 1 + \dfrac{.85 \times 350}{286} = 2.04$

$\phi_c = \dfrac{350}{250} = 1.4 \quad t_T$ is the same as for Example 1, i.e. $t_T = 1.4177$. At this point one may either use the above equations or draw the cycle diagram though equation (3) is necessary for finding Q.

Fig. 8 shows the cycle diagram with the lines omitted, and with temp. differences in boxes.

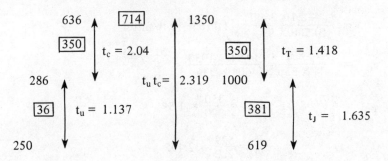

Fig. 16-8

Since θ_J is found to be $381°C$ $u_J = 147.1\sqrt{381} = 2871'/\text{sec}$.

$$F_s = \frac{2871 - 880}{32.2} = 61.84 \text{ lbs/lb/sec.} \quad (\text{or } 61.84 \text{ secs.})$$

$$\eta_o = \frac{F_s u}{714 \times 336} = \frac{61.84 \times 880}{714 \times 336} = .227$$

Since $\dfrac{C_p \times 3600}{10{,}500} = \dfrac{.24 \times 3600}{10{,}500} = .0823$

$$f_s = \frac{.0823 \times 714}{61.84} = .95 \text{ lbs/hr/lb thrust}$$

From equation 3, since $\varphi_{cD} T_{oD} = \phi_c T_o$, (i.e. $\theta_{cD} = \theta_c$),

$$\frac{Q}{Q_D} = \frac{P_o}{P_{oD}} \left(\frac{t_u t_c}{t_{uD} t_{cD}}\right)^{3.5} \sqrt{\frac{220}{250}}$$

From Example 1 $Q_D = 224.5$ lb/sec. $P_{oD} = 392$ p.s.f.

and $t_{uD} t_{cD} = 1.76 \times 1.751 = 3.082$

and P_o for this example is 970 p.s.f. From Fig. 8, $t_u t_c = 2.319$

$$\therefore \frac{Q}{Q_D} = \frac{970}{392} \left(\frac{2.319}{3.082}\right)^{3.5} \sqrt{\frac{220}{250}} = .858$$

$$\therefore Q = .858 \times 224.5 = 192.7 \text{ lb/sec.}$$

$$\therefore F = QF_s = 192.7 \times 61.84 = 11{,}916 \text{ lbs.}$$

Comparison of the results of Examples 1 and 2 illustrates the extent to which high speed and low atmospheric temp. T_o can largely compensate for a much reduced air density at height. Also the striking effect of low T_o and high speed on overall efficiency.

In the case of Example 2 the thrust would be well in excess of that necessary for cruise so the aircraft would be climbing or accelerating or both.

Example 16-3.

What would be the part load performance of the same engine at 400 m.p.h. (587'/sec., $\theta_u = 16°C$) at 10,000' if $T_o = 270$ and $P_o = 1455$ p.s.f. and $N = .9N_D$ assuming negligible change in component efficiencies and that $\theta_c \propto N^2$.? i.e. $\theta_c = .81 \times 350 = 283.5$.

Since, as we have seen, T_M is proportional to θ_c, it is reduced to $.81 \times 1350 = 1093.5$.

The cycle diagram now becomes as shown in Fig. 9.

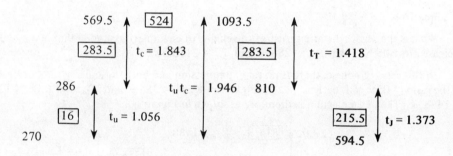

Fig. 16-9

Since $\theta_J = 215.5 \quad u_J = 2160.2'/\text{sec.}$

$$F_s = \frac{2160.2 - 587}{32.2} = 48.85 \text{ lbs/lb/sec}$$

$$\eta_o = \frac{48.85 \times 587}{524 \times 336} = .163$$

From equation (3) $\quad \dfrac{Q}{Q_D} = \dfrac{P_o}{P_{oD}} \left(\dfrac{t_u t_c}{t_{uD} t_{cD}} \right)^{3.5} \sqrt{\dfrac{1350}{1093.5}}$

As before $P_{oD} = 392$ p.s.f.; $t_{uD} t_{cD} = 3.082$ and P_o for this example is 1455 p.s.f.

From Figure 9, $\quad t_u t_c = 1.946$

$$\therefore \frac{Q}{Q_D} = \frac{1455}{392} \left(\frac{1.946}{3.082} \right)^{3.5} \sqrt{\frac{1350}{1093.5}} = .825 \quad \therefore \text{ since } Q_D = 224.5$$

$$Q = .825 \times 224.5 = 185.2 \text{ lbs/sec.}$$

$$\therefore \quad F = QF_s = 185.2 \times 48.85 = \underline{9,047} \text{ lbs}$$

On reviewing the reasoning leading to equation (3) it occurs to the writer that, since $t_u t_c$ is the overall temp. ratio of the cycle t_o, and since $\phi_c T_o = \theta_c$ and $\theta_c \propto N^2$, equation (3) can be simplified to

$$\frac{Q}{Q_D} = \frac{P_o}{P_{oD}} \left(\frac{t_o}{t_{oD}}\right)^{3.5} \frac{N_D}{N} \qquad (16\text{-}6)$$

Another query which may occur to the reader is why, in the above examples, the ram efficiency has been assumed to remain the same, namely 95%, when it has been previously stated that subsonic intakes are virtually 100% efficient for ram compression? There are two answers to this, namely: 1) An intake designed for supersonic flight is not equally appropriate for subsonic flight. 2) In subsonic flight, ram compression contributes far less to the total compression (as may be seen in Examples 2 and 3 above) so that (within limits) the cycle calculation is insensitive to ram efficiency assumptions.

Example 16-4.

What is the sea level static thrust of the engine of example 1 assuming that N may be increased to 1.05 N_D, that $T_o = 288°K$ and $P_o = 2116$ p.s.f.?

In this case, of course, there is no ram compression and both θ_c and T_M are increased in the ratio $(1.05)^2$ c.f. design; i.e. $\theta_c = (1.05)^2 \times 350 = 385.9$ and $T_M = (1.05)^2 \times 1350 = 1488°K$. The cycle diagram is therefore as shown in Figure 10.

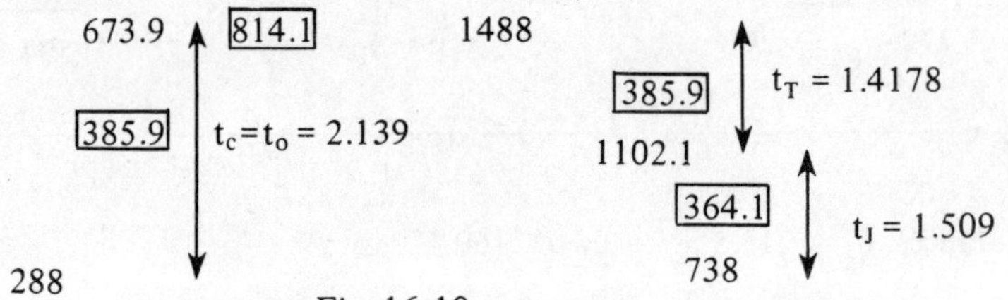

Fig. 16-10

Since $\theta_J = 364.1$ $u_j = 147.1\sqrt{364.1} = 2807'/\text{sec}.$

$$\therefore F_s = \frac{2807}{32.2} = 87.17 \text{ lbs/lb/sec}.$$

From equation (6) (and as before $Q_D = 224.5$, $t_{oD} = 3.082$, $P_{oD} = 392$)

$$\frac{Q}{Q_D} = \frac{2116}{392} \left(\frac{2.135}{3.092}\right)^{3.5} \times \frac{1}{1.05} = 1.406$$

$$\therefore Q = 1.406 \times 224.5 = 315.7 \text{ lb/sec}.$$

$$\therefore F = QF_s = 315.7 \times 87.17 = 27,522 \text{ lbs}.$$

$$f_s = \frac{.0823 \times \theta_H}{F_s} = \frac{.0823 \times 814.1}{87.17} = .769 \text{ lbs/hr/lb thrust}.$$

The overall efficiency is, of course, zero but the gas generator efficiency = $\frac{364.1}{814.1}$ = .447.

The results of this example are probably over-optimistic in that, with the higher pressure ratio and a T_M of 1488°K both compressor and turbine efficiencies would be lower than those assumed in Example 1 especially as extra air cooling for the turbine would be necessary.

Up to this point it has been assumed that $\theta_T = \theta_c$, but, as has previously been mentioned, if the axial velocity at turbine exhaust is high then its temp. equivalent should be debited to the turbine and credited to final expansion. i.e., $\theta_T = \theta_c + \theta_a$ where θ_a is the temp. equivalent of turbine exhaust axial velocity. This means that the static pressure at turbine exhaust is lower than if $\theta_T = \theta_c$, and this, in turn, means that the partial recovery of turbine loss in final expansion is reduced since the degree of recovery depends on the final expansion static to static temp. ratio. The error due to the assumption that $\theta_T = \theta_c$ will now be examined in Example 5 below.

Example 16-5.

Taking the engine of Example 1, what is the effect of assuming that $\theta_T = \theta_c + \theta_a$ if the turbine exhaust axial velocity is 1200'/sec. for which the temp. equivalent is $\left(\frac{1200}{147.1}\right)^2 = 66.5°$, but also assuming that turbine losses are the same as in Example 1, i.e. $\frac{\theta_c}{.88} - \theta_c$

The turbine losses θ_L (converted into internal energy) are given by

$$\theta_L = \frac{\theta_c}{.88} - \theta_c = .136\,\theta_c = .136 \times 350 = 47.7°C$$

Figures 11a and 11b below indicate a method of dealing with the problem using the values

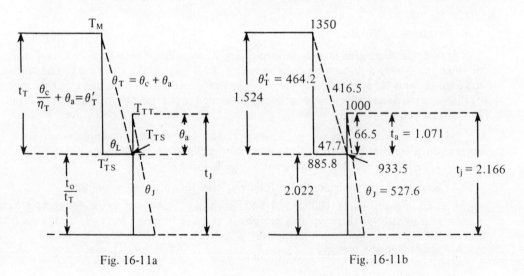

Fig. 16-11a   Fig. 16-11b

of Example 1 for T_M, θ_c etc., it being assumed that the isentropic heat drop in the turbine =
$\frac{\theta_c}{\eta_T} + \theta_a = \frac{\theta_c}{.88} + 66.5 = \frac{350}{.88} + 66.5 = 464.2$ to give a static exhaust temp. (isentropic)
of 885.8°K which is then increased by turbine losses to 933.5°K (since $.136\,\theta_c = 47.7°C$).
It is then assumed that the total temp. before final expansion equals $933.5 + \theta_a =$

$933.5 + 65.5 = 1000°K$.

$t_T = \frac{1350}{885.8} = 1.524$. From Example 1, $t_o = 1.76 \times 1.751\ (= t_u t_c) = 3.082$

∴ the static to static temp. ratio of the exhaust $= \frac{3.082}{1.524} = 2.022$. The total to static temp. ratio $t_J = 2.022 \times t_a$ (i.e. as though the exhaust axial velocity had been brought to rest isentropically to give a total temp. of 1000° and $t_a = \frac{1000}{933.5} = 1.071$)

Thus $t_j = 2.166$ (cf. 2.174 for Example 1) from which θ_J is found to be 527.6 (cf. 529.2 for Example 1).

$u_j = 3378.8'/\text{sec.}$ and $F_s = \frac{3378.8 - 1950}{32.2} = 44.37$ lbs/lb/sec. (cf. 44.54 for Example 1) i.e. the error is less than ½%. With the lower final expansion ratios of Examples 2, 3 and 4 the error would be somewhat larger. The assumption of an axial velocity as high as 1200 ft/sec. is, however, somewhat extreme.

It could be argued that Example 5 exaggerates the error, small though it is, in assuming that $\theta_T = \theta_c$, because tip leakage—the main item of the turbine losses—contributes directly to the exhaust kinetic energy and the axial component of leakage at least should be added to the energy available for final expansion.

Variable Propelling Nozzles

If complete expansion in a convergent-divergent propelling nozzle is desired—as it should be—then the divergent portion needs to be variable (See Section 5). There are several mechanical arrangements for achieving this which it is not proposed to describe herein, but the amount of the variation necessary will now be briefly examined with reference to Examples 1, 2 and 3 of this section.

In these examples it was assumed that the throat section of the final nozzles was fixed otherwise the assumption that $T_{max}/\theta_c =$ const. could not have been used. But this does not carry the implication that the exit nozzle area is also fixed.

In Example 1 (the design case) the following relevant quantities were found:—
$Q_D = 224.5$ lb/sec., $T_{JD} = 1000 - 529 = 471°K$, $u_{JD} = 3384'/\text{sec}$. Since, for complete expansion, the exhaust pressure equals atmospheric pressure P_o which was given as 392 p.s.f.

∴ the exhaust density $\rho_e = \frac{P_o}{RT_J}$.

If S_{eD} is the exit area, then $Q_D = S_e \rho_e u_J$

$$\therefore \; S_{eD} = \frac{Q_D}{\rho_e u_J} \quad \text{or, substituting for } \rho_e$$

$$S_{eD} = \frac{Q_D \, R \, T_{JD}}{P_{oD} \, u_J} \quad \therefore \; \text{from the figures from Example 1 above,}$$

$$S_{eD} = \frac{224.5 \times 96 \times 471}{392 \times 3384} = 7.652 \text{ ft.}^2$$

From Example 2, $Q = 192.7$ lb/sec.; $T_J = 619°K$; $u_J = 2871'$/sec. and $P_o = 970$ p.s.f.

$$\therefore \; S_e = \frac{192.7 \times 96 \times 619}{970 \times 2871} = 4.112 \text{ ft.}^2$$

From Example 3, $Q = 185.2$ lb/sec.; $T_J = 594.5$; $u_J = 2160'$/sec. and $P_o = 1455$ p.s.f.

$$\therefore \; S_e = \frac{185.2 \times 96 \times 594.5}{1455 \times 2160} = 3.363 \text{ ft.}^2$$

However, as may be seen, the section area of the final nozzle exit is somewhat larger than that of the entering stream (in the design condition, 224.5 lb/sec. at $T_o = 220$,

$P_o = 392$ and $u = 1950'$/sec., $S_o = \dfrac{224.5 \times 96 \times 220}{392 \times 1950} = 6.202$ ft.^{2} i.e. 0.81 S_{eD}).

From the point of view of minimising external shock drag, it is obviously desirable to have a nacelle of constant section from front to rear. For the design case of Example 1 this means that S_{eD} would have to be restricted to 6.202 ft.^{2} which in turn means incomplete expansion in the final nozzle, entailing some loss of thrust. It could therefore be argued that, since effective net thrust is actual thrust less nacelle drag, it might be worth 'trading' some degree of under expansion for reduction of nacelle drag to give the optimum net thrust. Alternatively the design cycle could be altered to give $S_{eD} = S_o$, or nearly so, but the writer has not, so far, explored these alternatives.

SECTION 17

Effect on Cycle Calculations of More Accurate Methods

Introductory

Up to this point the methods used have ignored such important factors as the increase of specific heat with temperature, the fuel mass addition in combustion and the effect on the gas constant R of combustion products.

With monatomic gases the energy of atomic motion is largely kinetic energy of linear motion but with diatomic gases such as O_2 and N_2, triatomic gases such as CO_2, H_2O, etc., part of the energy is in the form of molecular rotation. The increase in this with temperature is presumed to account for the increase of C_v (R is not affected by temperature in the absence of dissociation so that the increase of C_p is the same as the increase of C_v). At very high temperatures the effective specific heat is affected by dissociation and formation of radicals. Also by the formation of oxides of nitrogen ('NOX').

It is the purpose of this section to examine the degree to which the approximations are justified for comparative purposes using the rather extreme case of a straight jet engine designed for supersonic cruise as a basis for the analysis.

More accurate methods require that enthalpy h be used instead of temperature, enthalpy differences Δh instead of temperature differences θ, pressure ratios instead of temperature ratios, etc. Also it must always be borne in mind that the expansion mass flow is greater than the compression mass flow by the amount of fuel added in combustion (usually of the order of 1.5% to a little over 2%).

For present purposes, the approximate method (i.e. the assumption of constant specific heat and neglect of fuel mass addition) will be referred to as the 'F. W. Method' and the more accurate method as 'The h–p Method'.

Symbols and Suffixes Used

Symbols

h	enthalpy
$\bar{h}$	the mean enthalpy for a process
Δh	enthalpy difference
e	enthalpy ratio
P	absolute pressure
p	pressure ratio
T	absolute temp.
ρ	density
ψ	flux density
S	section area
Q	mass flow rate
$\bar{\alpha}$	the exponent in $e^{\bar{\alpha}} = p$
δ	the numerical difference between $T°K$ and h in kj/kg
m	fuel mass per unit compressor mass flow rate
u	velocity

Suffixes

M	the greatest value of a quantity
t	the critical value of a quantity in a choking throat (i.e. when $\psi = \psi_t$)
u	pertaining to ram compression
c	pertaining to the compressor
H	pertaining to combustion
T	pertaining to the turbine
j	pertaining to final expansion through the propelling jet
0 or 1	pertaining to an initial condition
0	is also used to denote 'overall' e.g. p_0 means overall pressure ratio
a	pertaining to the atmosphere
D	implies the design condition

As before a prime denotes an isentropic value.

Many of these have been used before but are included above for convenience.

The Cycle Diagram

The cycle diagram is now as shown in Fig. 1 in which several important relationships are indicated.

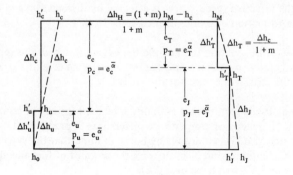

Fig. 17-1

It is now no longer possible to say that $e_u \, e_c = e_T \, e_j$ and so e_j, etc., must be calculated from the fact that $p_u p_c = p_T p_J = p_0$.

The Relationships between T, h, δ and $\bar{\alpha}$

For numerical convenience the units of h used are kJ/kg but mass flow rates remain in lb/sec. (The writer apologises for this mixture of unit systems, but it eases calculation.) To convert kJ/kgm to ft lb/lb multiply by 334.6 (cf. 336 for K_p). Using this, one finds that a calorific value of 10,500 C.H.U./lb is equivalent to 43,800 kJ/lb.

For the h–p method of dealing with problems it will frequently be necessary to refer to Fig. 2 for the value of δ and to Fig. 3 for the value of $\bar{\alpha}$ corresponding to $\bar{h}$. (It is a convenient fact that $e^{\bar{\alpha}} = p$. It is suggested that the reader verifies this by splitting an expansion from h = 1600 to h = 1000 into three or four parts and calculates the pressure ratio for each. He will find that the products of the series of pressure ratios will give the same total pressure ratio as would be obtained by taking e as 1.6 and $\bar{\alpha}$ corresponding to $\bar{h}$ = 1400.)

For the convenience of the reader who may wish to plot these curves on squared paper the necessary tabulations are given on Figs. 2 and 3.

The Calculation of Fuel Mass m

This is simple. The quantity of gas at h_M is $1 + m$, ∴ the heat added

$$\Delta h_H = (1+m)h_M - h_c, \therefore \Delta h_H = 43{,}800\,m = (1+m)h_M - h_c \qquad (17\text{-}1)$$

from which

$$m = \frac{h_M - h_c}{43{,}800 - h_M} \qquad (17\text{-}2)$$

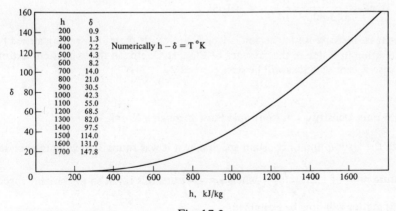

Fig. 17-2

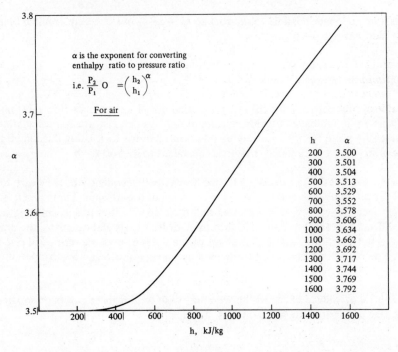

Fig. 17-3

Example 17-1

What is the value of m if $T_c = 746°K$ and $T_M = 1350°K$?

From Fig. 2, for T_c, $\delta = 18$, $\therefore h_c = 746 + 18 = 764$. For T_M, $\delta = 106$, $\therefore h_M = 1456$,

$$\therefore m = \frac{1456 - 764}{43,800 - 1456} = .01634$$

The mass increase for expansion is 1.63% approx. It therefore seems justified to ignore the slight alteration in R due to the presence of CO_2, H_2O, etc., in the expansion part of a cycle. (In any case R cancels out as will be seen.)

Maximum Flux Density ψ_t in Isentropic Flow through a Nozzle

With the assumption of constant specific heat it was found that ψ_t corresponded to a temperature ratio of $1.2 \left(= \frac{\gamma + 1}{2} \text{ when } \gamma = 1.4 \right)$ but this is not so for variable specific heat. This matter will now be examined.

No simple formula can be derived so it is necessary to explore the problem numerically with the aid of Figs. 2 and 3.

In order to determine how e_t varies with h_M it is necessary to take a series of values of e and find the value of ψ for each and then to plot ψ against e to find the values of e_t and ψ_t and to repeat the procedure for a series of values of h_M. The process is best explained by an example as follows:

Example 17-2

What is the value of ψ corresponding to $h_M = 1600 \, \text{kJ/kg}$ if $e = 1.17$?

The value of h after expansion is $\dfrac{1600}{1.17} = 1367.5$

$\therefore \bar{h} = \dfrac{1600 + 1367.5}{2} = 1483.8.$

From Fig. 3 the corresponding value of $\bar{\alpha}$ is 3.766

$\therefore$ the pressure ratio $(= e^{\bar{\alpha}})$ is $1.17^{3.766} = 1.8063$

$\therefore$ the pressure P after expansion from P_M is $\dfrac{P_M}{1.8063}$.

From Fig. 2, for $h = 1367.5$, δ is found to be 92, $\therefore T = 1367.5 - 92 = 1275.5°K$.
We now have sufficient information to find ρ and u and therefore ψ.

$\rho = \dfrac{P}{RT}$ and since $P = \dfrac{P_M}{1,8063}$ and $T = 1275.5, \rho = \dfrac{P_M}{1.8063 \times 1275.5R}$,

$\Delta h = h_M \left(1 - \dfrac{1}{1.17}\right) = 1600 - 1367.5 = 232.5$

for which

$u = \sqrt{2g \times 334.6} \sqrt{232.5} = 146.8 \sqrt{232.5} = 2239.4 \, \text{ft/sec}$

$\therefore \psi \equiv \rho u = \dfrac{P_M}{R}\left(\dfrac{2239.4}{1.8063 \times 1275.5}\right) = 0.9716 \dfrac{P_M}{R}$

More generally: $\rho = \dfrac{P_M}{R}\left(\dfrac{1}{e}\right)^{\bar{\alpha}} \left(\dfrac{1}{h_M/e - \delta}\right)$ and $u = 146.8 \sqrt{h_M \left(\dfrac{e-1}{e}\right)}$

$\therefore \psi = 146.8 \dfrac{P_M}{R} \sqrt{h_M} \left[\left(\dfrac{1}{e}\right)^{\bar{\alpha}} \left(\dfrac{1}{h_M/e - \delta}\right)\right] \sqrt{\dfrac{e-1}{e}}$ or,

Merging constants into k; $\psi = k \left(\dfrac{1}{e}\right)^{\bar{\alpha}} \dfrac{1}{\dfrac{h_M}{e} - \delta} \sqrt{h_M \left(\dfrac{e-1}{e}\right)}$

or $\quad \dfrac{\psi}{k} = \left(\dfrac{1}{e}\right)^{\bar{\alpha}} \left(\dfrac{e}{h_M - e\delta}\right) \sqrt{h_M \left(\dfrac{e-1}{e}\right)}$ (17-3)

from which it is fairly easy to work out tables for ψ for various values of h_M and e. This is done below for $h_M = 1600$.

e	$\dfrac{h_M}{e}$	$\bar{h}$	$\bar{\alpha}$	δ	$\dfrac{\psi}{k}$	$\dfrac{\psi}{\psi_t}$
1.10	1454.5	1527	3.777	105	.006235	0.9436
1.13	1415.9	1508	3.772	100	.006504	0.9831
1.15	1391.3	1496	3.769	96	.006584	0.9952
1.17	1367.5	1484	3.766	92	.006608	1.000
1.19	1344.5	1472	3.764	88	.006608	0.9988
1.21	1322.3	1461	3.760	85	.006576	0.9940
1.23	1300.8	1450	3.757	82	.006520	0.9855

The values of $\dfrac{\psi}{k}$ are plotted in Fig. 4 as are also similar results for $h_M = 1400$; 1200; and 1000. As may be seen, for $h_M = 1600$, $\dfrac{\psi}{\psi_t} = 1.0$ when $e = 1.17$, i.e. $e_t = 1.17$.

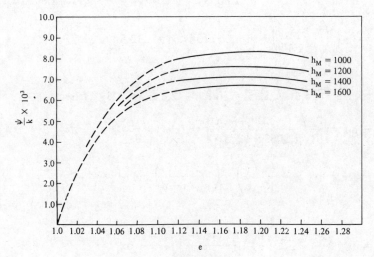

Fig. 17-4

Figure 5 shows a plot of e_t over the range $h_M = 900$ to $h_M = 1600$.

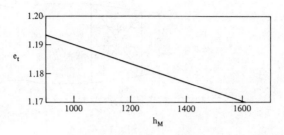

Fig. 17-5

It will be noted how extremely insensitive ψ is to e over quite a wide range of values of e. The variation is only 2% approx. over the range $e = 1.13$ to $e = 1.23$. The implication is that a nozzle designed for non-choking would become a choking nozzle in this range if there is departure from isentropic flow or there is very small nozzle distortion.

It was shown in Section 5 that, with constant specific heat, $\psi_t \dfrac{\sqrt{T_M}}{P_M} = $ const. for a choking nozzle. It is found that, with the h–p method it is approximately true to write

$\psi_t \dfrac{\sqrt{h_M}}{P_M} = $ const. For the range $h_M = 1000$ to $h_M = 1600$.

$\dfrac{\psi_t}{k} \sqrt{h_M}$ varies as follows:

h_M	$\dfrac{\psi_t}{k}\sqrt{h_M}$
1000	0.2611
1200	0.2626
1400	0.2640
1600	0.2643

i.e. a variation of 1.2% over the range.

Choking Nozzles in Series with Energy Extraction between them by the h–p Method

In Section 5 it was shown that, with the FW method, if the choking throats were in fixed section ratio, the energy extraction (for a given efficiency) was proportional to T_M. Also that the temperature ratio (total to total) across the first nozzle and energy extractor (e.g. a turbine) remained constant whatever the value of T_M.

The present purpose is to examine the arrangement using the h–p method.

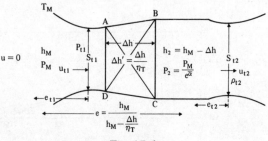

Fig. 17-6

It is represented diagrammatically in Fig. 6 which, as in the case of Fig. 5-6, is representative of a straight turbojet with the compressor turbine having choking nozzles with the final nozzle also choking.

Continuity again requires that $\rho_{t1} u_{t1} S_{t1} = \rho_{t2} u_{t2} S_{t2}$.

The calculation must be dealt with numerically using Figs. 2 and 3.

From equation (3), for the flow through S_{t1} we have

$$Q = S_{t1} \psi_{t1} = S_{t1} k \left(\frac{1}{e_{t1}}\right)^{\bar{\alpha}} \left(\frac{e_{t1}}{h_M - e_{t1}\delta}\right) \sqrt{h_M \left(\frac{e_{t1} - 1}{e_{t1}}\right)} \qquad (17\text{-}4)$$

For the flow through S_{t2}, $u_{t2} = 146.8 \sqrt{(h_M - \Delta h_t)\left(\frac{e_{t2} - 1}{e_{t2}}\right)}$

$$\rho_{t2} = \frac{P_{t2}}{R(h_{t2} - \delta_{t2})} \quad \text{and} \quad P_{t2} = \frac{P_2}{e^{\bar{\alpha}}_{t2}} = \frac{P_0}{e^{\bar{\alpha}} e^{\bar{\alpha}}_{t2}}$$

$$\therefore \quad Q = S_{t2} \psi_{t2} = \frac{146.8}{R} S_{t2} \frac{P_0}{e^{\bar{\alpha}} e^{\bar{\alpha}}_{t2}} \left(\frac{1}{h_{t2} - \delta_{t2}}\right) \sqrt{(h_M - \Delta h)\left(\frac{e_{t2} - 1}{e_{t2}}\right)} \qquad (17\text{-}5)$$

Clearly (4) must equal (5). It is evident, however, that there is no simple formula for $\dfrac{S_{t2}}{S_{t1}}$ so the problem must be handled numerically as in the following example.

Example 17-3

With $h_M = 1600$ $\Delta h = 350$ and $\eta_T = 0.88$ what is the value of $\dfrac{S_{t2}}{S_{t1}}$?

We proceed as follows:

$\Delta h' = \dfrac{350}{0.88} = 397.7$ ∴ the enthalpy ratio e across the first nozzle (which may represent a ring of choking turbine nozzles) and energy extraction ABCD is given by

$$e = \dfrac{1600}{1600 - 397.7} = 1.3308, \; h_2 = 1600 - 350 = 1250,$$

$$\therefore \; \bar{h} = \dfrac{1250 + 1600}{2} = 1425.$$

For this, from Fig. 3, $\bar{\alpha} = 3.75$ ∴ the pressure ratio p is given by $p = e^{\bar{\alpha}}$, i.e.
$1.3308^{3.75} = 2.9203$, ∴ $P_2 = \dfrac{P_M}{2.9203}$.

From Fig. 5, $e_{t1} = 1.175$, ∴ $h_{t1} = \dfrac{1600}{1.175} = 1361.7$

$\therefore \; \bar{h}_{t1} = \dfrac{1600 + 1361.7}{2} = 1480.9$ for which $\bar{\alpha}_{t1} = 3.766$

$\therefore \; p_{t1} = 1.175^{3.766} = 1.8355$, i.e. $P_{t1} = \dfrac{P_M}{1.8355}$

From Fig. 2, for $h_{t1} = 1361.7, \delta = 91$ ∴ $T_{t1} = 1361.7 - 91 = 1270.7$.

Hence $\rho_{t1} = \dfrac{P_M}{R \times 1.8355 \times 1270.7} = \dfrac{P_M}{2332.4 R}$

$\Delta h_{t1} = 1600 - 1361.7 = 238.3$ from which $u_{t1} = 146.8 \sqrt{238.3} = 2266$ ft/sec

$\therefore \; \psi_{t1} = u_{t1} \rho_{t1} = \dfrac{2266.1}{2332.4} \dfrac{P_M}{R} = 0.9716 \dfrac{P_M}{R}$.

For the second nozzle with $h_2 = 1250$ and $P_2 = \dfrac{P_M}{2.9203}$ (from above)

e_{t2} from Fig. 5 is 1.183, so $h_{t2} = \dfrac{1250}{1.183} = 1056.6$

$\therefore \; \bar{h}_{t2} = \dfrac{1250 + 1056.6}{2} = 1153.3$ for which $\bar{\alpha}_{t2} = 3.678$

$\therefore \; p_{t2} = 1.183^{3.678} = 1.8554$ ∴ $P_{t2} = \dfrac{P_2}{1.8554} = \dfrac{P_M}{1.8554 \times 2.9203} = \dfrac{P_M}{5.4183}$.

For $h_{t2} = 1056.6$, $\delta = 48$ (from Fig. 2)

$$\therefore T_{t2} = 1056.6 - 48 = 1008.6, \text{ hence } \rho_{t2} = \frac{P_{t2}}{RT_{t2}} = \frac{P_M}{R} \times \frac{1}{5.4183 \times 1008.6}$$

$$\Delta h_{t2} = 1250 - 1056.6 = 193.4 \quad \therefore u_{t2} = 146.8\sqrt{193.4} = 2041.5 \text{ ft/sec}$$

$$\therefore \psi_{t2} = \frac{P_M}{R} \times \frac{2041.5}{5.4183 \times 1008.6} = 0.3736 \frac{P_M}{R}$$

It follows that $\dfrac{S_{t2}}{S_{t1}} = \dfrac{\psi_{t1}}{\psi_{t2}} = \dfrac{0.9716}{0.3736} = 2.6008$.

Using similar methods and making the assumption that Δh is proportional to h_M, it is found that for $h_M = 1400$ $e = 1.3308$ and $\dfrac{S_{t2}}{S_{t1}} = 2.566$ which is less than $1\frac{1}{2}\%$ smaller than for $h_M = 1600$.

For $h_M = 1200$, again $e = 1.3308$ and $\dfrac{S_{t2}}{S_{t1}} = 2.463$ which is 5.6% smaller than for $h_M = 1600$.

It should be noted that e remains constant if $\dfrac{\Delta h}{h_M} = \text{const}$.

We see that if S_{t2} and S_{t1} are in fixed ratio Δh is not exactly proportional to h_M but is evidently nearly so and could be made exactly so by a slight reduction of S_{t2} as h_M is reduced.

In what follows it will be assumed that Δh is in fact proportional to h_M.

A Comparison between the F.W. Method and the h–p Method for the Straight Jet

For this purpose, three widely different circumstances will be examined. It will be assumed (i) that the design is for supersonic cruise in the stratosphere at 1950 ft/sec with $T_M = 1350°K$ and $\theta_c = 350$; (2) that the efficiency of ram compression η_u is 95% at design and 98% for subsonic conditions; (3) that $\eta_c = 85\%$; (4) $\eta_T = 88\%$; (5) $\eta_j = 98\%$; (6) that for climb at 20,000 ft, gust load considerations limit the speed to 675 ft/sec (approx. 400 knots); (7) that for climb, $T_M = 1400°K$; (8) for take off, i.e. sea level static, $T_M = 1450°K$.

A. The FW Method

The design (i.e. cruise) cycle is the same as Example 16-1 but is repeated below and illustrated in Fig. 7 for ease of reference with additional figures on the diagram.

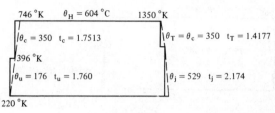

Fig. 17-7

As in Example 16-1 the design specific thrust $\bar{F}_{sD} = 44.54$ sec. The overall temp. ratio of the cycle is $1.76 \times 1.7513 = 3.0822$.

∴ the overall pressure ratio is $3.0822^{3.5} = 51.41$, i.e. the total pressure before the turbine is $51.41\, P_{aD}$ where P_{aD} is the atmospheric pressure at design height. Therefore

$$Q_D = k \frac{51.41}{\sqrt{1350}} P_{aD},\text{ so total thrust } F\,(\equiv F_{SD}\,Q_D)\text{ is } 44.54 \times kP_{aD} \times \frac{51.41}{\sqrt{1350}} = 62.32\,kP_{aD}.$$

Design Efficiency $\eta_{oD} = \dfrac{F_{SD}\,u_D}{\theta_{Ho}\,K_p} = \dfrac{44.54 \times 1950}{604 \times 336} = 0.428$ or 42.8%.

For climb at 675 ft/sec at 20,000 ft, $T_0 = 272°K$ (I.S.A. + 4°) and $P_a = 972.6$ p.s.f.

θ_c, being proportional to T_M; $= \dfrac{1400}{1350} \times 350 = 363°C$, $\theta_u = \left(\dfrac{675}{147.1}\right)^2 = 21.1°C$.

The climb cycle is shown in Fig. 8.

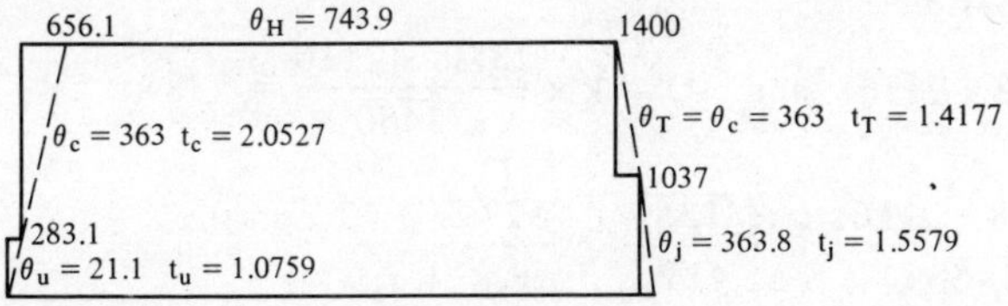

```
       656.1        θ_H = 743.9          1400
       /                                   \
      /θ_c = 363  t_c = 2.0527              \θ_T = θ_c = 363  t_T = 1.4177
     /                                       \1037
    /283.1                                    \
   /θ_u = 21.1  t_u = 1.0759                   \θ_j = 363.8  t_j = 1.5579
```

Fig. 17-8

$u_j = 147.1\sqrt{363.8} = 2806.2$ ft/sec to give $F_S = \dfrac{2806.2 - 675}{32.2} = 66.19$ secs

The overall temp. ratio $t_0 = 1.0759 \times 2.0527 = 2.2085$

∴ the overall pressure ratio $p_0 = 2.2085^{3.5} = 16.01$

$$\therefore Q = \frac{k \times 16.01 \times 972.6}{\sqrt{1400}} = 416.2\,k$$

$$\therefore F = QF_S = 416.2\,k \times 66.19 = 27{,}546\,k$$

$$\frac{F}{F_D} = \frac{Q}{Q_D} \times \frac{F_S}{F_{SD}} = \frac{27{,}546}{62.32\,P_{aD}} = \frac{442}{P_{aD}}$$

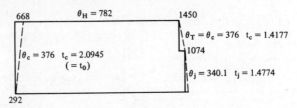

Fig. 17-9

$$\eta_0 = \frac{66.19 \times 679}{743.9 \times 336} = 0.1787 \quad \text{or} \quad 17.87\%$$

$$\therefore \frac{\eta_0}{\eta_{oD}} = \frac{17.87}{42.8} = 0.4175.$$

For sea level static, the cycle is shown in Fig. 9. $T_0 = 292°K$; $P_a = 2116$ p.s.f. θ_c for $T_M = 1450°K$ is now $376°C$.

$$u_j = 147.1\sqrt{340.1} = 2712.8 \quad \therefore F_S = \frac{2712.8}{32.2} = 84.25 \text{ secs}$$

$$\therefore \frac{F_S}{F_{SD}} = \frac{84.25}{44.45} = 1.896$$

The overall pressure ratio corresponding to t_0 (or t_c since there is no ram compression) is $p_0 = 2.0945^{3.5} = 13.28$

$$\therefore P_M = 13.28 \times 2116 \quad \text{and} \quad Q = k \times \frac{13.28 \times 2116}{\sqrt{1450}} = 738\,k$$

$$\therefore \frac{Q}{Q_D} = \frac{13.28 \times 2116}{\sqrt{1450}} \times \frac{\sqrt{1350}}{51.41\,P_{aD}} = \frac{527.3}{P_{aD}}.$$

Hence $\dfrac{F}{F_D} = \dfrac{F_S Q}{F_{SD} Q_D} = \dfrac{1.8916 \times 527.3}{P_{aD}} = \dfrac{997.4}{P_{aD}}.$

The overall efficiency is, of course, zero.

B. The h–p Method

The values of h corresponding to the temperatures of Fig. 7 are derived from Fig. 2 and, for the design cycle, are shown in Fig. 10.

The fuel mass (per lb) m must first be determined for fuel of calorific value of 10,500 C.H.U./lb or 43,800 kJ/lb.

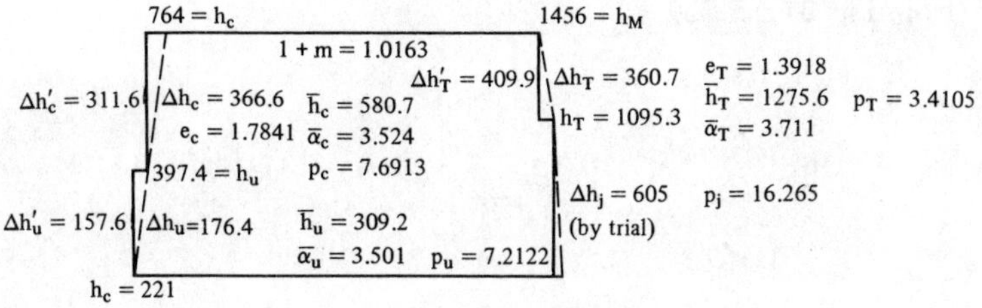

```
                    764 = h_c                           1456 = h_M
                         1 + m = 1.0163
                                        Δh'_T = 409.9   Δh_T = 360.7    e_T = 1.3918
Δh'_c = 311.6  Δh_c = 366.6    h̄_c = 580.7                              h̄_T = 1275.6    p_T = 3.4105
               e_c = 1.7841    ᾱ_c = 3.524              h_T = 1095.3    ᾱ_T = 3.711
                               397.4 = h_u    p_c = 7.6913
                                                        Δh_j = 605      p_j = 16.265
Δh'_u = 157.6  Δh_u = 176.4    h̄_u = 309.2              (by trial)
               h_c = 221       ᾱ_u = 3.501    p_u = 7.2122
```

Fig. 17-10

Using equation (2)

$$m_D = \frac{1456 - 764}{43{,}800 - 1456} = .0163,$$

$$\therefore \Delta h_T = \frac{\Delta h_c}{1+m} = \frac{366.6}{1.0163} = 360.7 \text{ (as shown in Fig. 10).}$$

The values of e, $\bar{h}$, $\bar{\alpha}$ and p for each part of the cycle are shown in Fig. 10.

The overall pressure ratio $p_{oD} = P_{UD} P_{CD} = 7.2122 \times 7.6913 = 55.47$ and since

$$p_T = 3.4105, \quad p_J = \frac{55.47}{3.4105} = 16.265.$$

(The pressure ratios are related by $p_u p_c = p_T p_j$.)

In order to find Δh_j it is necessary to resort to trial to find which value of Δh_j at $\eta_j = 0.98$ gives a pressure ratio of 16.265. Usually about three guesses will suffice though the process takes time because, for each trial value of Δh_j, $\bar{h}$, e and $\bar{\alpha}$ must be found and the value of $e^{\bar{\alpha}}$ plotted to find the required value of p_j.

Having found that $\Delta h_j = 605$, then $u_j = 146.8 \sqrt{605} = 3611$ ft/sec.

So $$F_{SD} = \frac{(1+m)u_j - u_0}{g} = \frac{1.0163 \times 3611 - 1950}{32.2} = 53.41 \text{ secs.}$$

For the h–p cycle it must be borne in mind that the mass flow Q_T in expansion is $1 + m$ times that in compression Q_c.

To find Q_{TD} it is necessary to find ψ_{tD} for the assumed choking turbine nozzle ring.

From Fig. 5, the value of e_t (the critical enthalpy ratio to the nozzle throats) is found to be 1.179 for $h_M = 1456$ to give $h_t = \dfrac{1456}{1.179} = 1234.9$, and $\bar{h}_t = \dfrac{1456 + 1234.9}{2} = 1345.5,$

for which $\bar{\alpha}_t$ (from Fig. 3) is 3.729.

$$\Delta h_t = \begin{cases} h_M = 1456 \\ 221.1 \begin{cases} l_t = 1.179 \\ \bar{h}_t = 1345.5, \bar{\alpha}_t = 3.729, p_t = 1.179^{3.729} = 1.8479 \\ u_t = 146.8\sqrt{221.1} = 2182.8 \text{ ft/sec} \end{cases} \\ h_t = 1234.9, S = 72, T_t = 1234.9 - 72 = 1162.9 \end{cases}$$

Fig. 17-11

The vertical line in Fig. 11 represents isentropic expansion to the throats of the turbine nozzle ring. The figures of Fig. 11 give the information necessary to calculate ψ_{tD}.

$$\rho_{tD} = \frac{P_{tD}}{RT_{tD}} \text{ and, since } P_t = \frac{P_{MD}}{P_{CD}} = \frac{P_{MD}}{1.8479} \text{ and } T_{tD} = 1162.9$$

$$\rho_{tD} = \frac{P_{MD}}{R \times 1.8479 \times 1162.9} \quad \therefore \psi_{CD} = \frac{2182.8}{1.8479 \times 1162.9} \frac{P_{MD}}{R} = 1.0158 \frac{P_{MD}}{R}$$

$$Q_{TD} = \psi_{tD} S_{t1} = 1.0158 S_{t1} \frac{P_{MD}}{R} \text{ or, since } P_{MD} = 55.47 P_{aD}$$

$$Q_{TD} = 56.35 \frac{S_{t1} P_{aD}}{R}$$

$$\text{So } Q_{CD} = \frac{Q_{TD}}{1.0163} = \frac{56.35}{1.0163} \frac{S_t P_{aD}}{R} = 55.442 \frac{S_t P_{aD}}{R}$$

$$\therefore F_D = F_{SD} Q_{CD} = 53.41 \times 55.44 S_t \frac{P_{aD}}{R} = 2961.2 S_t \frac{P_{aD}}{R}$$

$$\eta_0 = \frac{F_{SD} u_D}{[(1+m)1456 - 764] 334.5} = 0.435 \text{ or } 43.5\%.$$

For the climb at $T_M = 1400$ ($h_M = 1515.8$) and $u = 675$ ft/sec, $T_0 = 272$ and $P_a = 972.6$ p.s.f. at 20,000 ft the cycle is shown in Fig. 12.

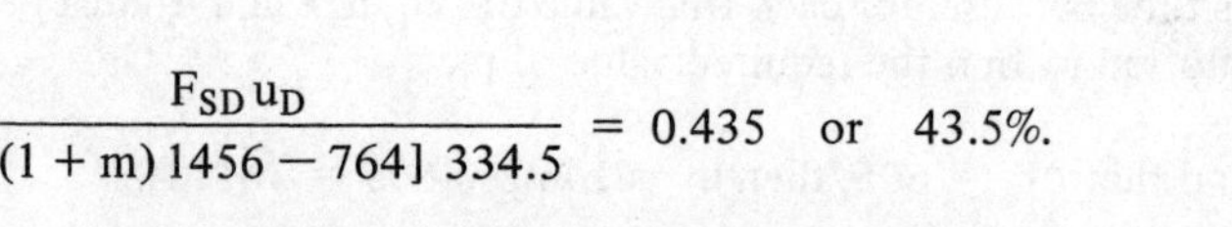

$$\Delta h'_c = 324.4 \begin{cases} 674.8 \\ \Delta h_c = \frac{1515.8}{1456} \times 366.6 = 381.7 \\ e_c = 2.1067 \quad \bar{h}_c = 484 \quad \bar{\alpha}_c = 3.512 \\ p_c = 13.69 \end{cases} \begin{cases} 1 + m = 1.0199 \\ \Delta h'_T = 425.3 \end{cases} \begin{cases} 1515.8 \\ \Delta h_T = 374.3 \quad e_T = 1.3901 \\ \bar{h}_T = 1328.7 \quad \bar{\alpha} = 3.725 \\ 1141.5 \quad p_T = 3.411 \\ \Delta h_j = 413.5 \quad p_j = 5.189 \\ \text{(by trial)} \end{cases}$$

$$\Delta h'_u = 20.7 \begin{cases} 293.1 \\ \Delta h_u = 21.1 \quad \bar{h}_u = 282.6 \quad \bar{\alpha}_u = 3.5 \\ e_u = 1.0761 \quad p_u = 1.2927 \\ 272 \end{cases}$$

Fig. 17-12

The overall pressure ratio $p_0 = 1.2972 \times 13.693 = 17.7$, m from (4) is given by $m = \dfrac{1515.8 - 674.8}{43{,}800 - 1515.8} = .0199$.

Most of the relevant figures are given in Fig. 12.

By trial Δh_j is found to be 413.5 $\therefore u_j = 146.8\sqrt{413.5} = 2985$ ft/sec

$$\therefore F_s = \dfrac{1.0199 \times 2985 - 675}{32.2} = 73.58 \text{ secs.}$$

To find Q_T we might assume that $Q_T \propto \dfrac{P_M}{\sqrt{h_M}}$ but for greater accuracy we proceed as before with the aid of Fig. 13 on which most of the relevant figures are shown for isentropic expansion to the turbine nozzle throats. From Fig. 5, for $h_M = 1515.8$, $e_t = 1.177$ to give $h_t = 1287.9$.

$$\begin{aligned}
h_M &= 1515.8 \qquad P_M = 17.7 \times 972.6 \\
\bar{h}_t &= 1401.8,\ \bar{\alpha}_t = 3.744,\ p_t = 1.8407,\ P_t = \dfrac{17.7 \times 972.6}{1.8407} = 9352 \text{ p.s.f.} \\
&\Delta h_t = 227.9,\ u_t = 2216.1 \\
h_t &= 1287.9 \quad \delta = 79 \quad T_T = 1237.9 - 79 = 1208.9^\circ\text{K.}
\end{aligned}$$

Fig. 17-13

$$\rho_t = \dfrac{P_t}{RT_t} \text{ and } \psi_t = u_t \rho_t = \dfrac{u_t P_t}{RT_t} = \dfrac{2216.1 \times 9352}{1208.9\,R} = \dfrac{17{,}144}{R}$$

$$\therefore Q_T = S_t \psi_t = 17{,}144\,\dfrac{S_t}{R}$$

$$\therefore Q_c = \dfrac{17{,}144}{1.0199}\dfrac{s_t}{R} = 16{,}809\,\dfrac{S_t}{R}$$

Q_{CD} was found to be $55.44\, P_{aD}\, \dfrac{S_t}{R}$

$$\therefore \dfrac{Q_c}{Q_{CD}} = \dfrac{16{,}809}{55.44\, P_{aD}} = \dfrac{303.2}{P_{aD}}$$

F_{SD} was found to be 53.41 sec $\quad \therefore\ \dfrac{F_s}{F_{SD}} = \dfrac{73.58}{53.41}$

$$\therefore \dfrac{F}{F_D} = \dfrac{Q_c F_s}{Q_{CD} F_{SD}} = \dfrac{303.2}{P_{aD}} \times \dfrac{73.58}{53.41} = \dfrac{417.7}{P_{aD}},\ \text{cf.}\ \dfrac{442}{P_{aD}} \text{ for the F.W. method so}$$

the latter overestimates $\dfrac{F}{F_D}$ for the climb by 5.8%.

η_0 for the h–p cycle is given by

$$\eta_0 = \dfrac{73.58 \times 675}{(1.0199 \times 1515.8 - 674.8) \times 334.5} = .17 \quad \text{or} \quad 17\%, \text{cf. } 17.78\%$$

for the F.W. method, the latter therefore overestimates η_c by approx. 4.6%. This, however, is of less importance than it seems in the light of the fact that fuel consumption during the subsonic phase of climb is a very small proportion of total fuel consumption because it is usual to accelerate to supersonic flight at about 30,000 ft.

The take-off (sea-level static) cycle for a T_M of $1450°K$ ($h_M = 1575.8$), $T_0 = 288$ ($j_0 = 289$) and $P_a = 2116$ p.s.f. is shown in Fig. 14,

$$m, \text{ from (4), is given by } m = \dfrac{1575.8 - 685.8}{43,800 - 1575.8} = .0211$$

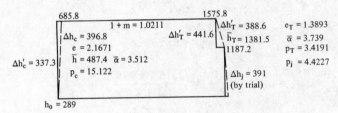

Fig. 17-14

By trial, $\Delta h_j = 391$ ∴ $u_j = 146.8\sqrt{391} = 2902.8$ ft/sec

∴ $F_s = 90.15 \times 1.0211 = 92.05$.

To find Q_T we use Fig. 15 and the fact that from Fig. 4, for $h_M = 1575.8$, $e_t = 1.176$.

$\begin{vmatrix} h_M = 1575.8 \\ \bar{h}_t = 1457.9 \ \bar{\alpha} = 3.759 \ p_t = \dfrac{15.12 \times 2116}{1.8393} \\ \qquad e = 1.176 \text{ (from Fig. 4)} \\ \qquad \Delta h_t = 235.8 \ \therefore \ u_t = 146.8\sqrt{235.8} = 2254.5 \text{ ft/sec.} \\ h_t = 1346 \ \delta = 87 \ \therefore \ T_t = 1340 - 87 = 1253°K \\ \rho_t = \dfrac{17,397}{1253R} \ \therefore \ \psi_c = \dfrac{17,397 \times 22.545}{1253R} = \dfrac{31,302}{R} \\ \therefore \ Q_T = 31,302 \ \dfrac{S_t}{R} \text{ and } Q_c = \dfrac{31,302}{1.0211} \dfrac{S_t}{R} = 30,655 \ \dfrac{S_t}{R}. \end{vmatrix}$

Fig. 17–15

From above, $Q_{CD} = 55.44 \dfrac{S_t}{R} P_{aD}$,

$\therefore \dfrac{Q_C}{Q_{CD}} = \dfrac{30{,}655}{55.44\, P_{aD}}$. Also from above $F_{SD} = 53.41$ secs

$\therefore \dfrac{F}{F_D} = \dfrac{F_s}{F_{SD}} \dfrac{Q_C}{Q_{CD}} = \dfrac{90.15}{53.41} \times \dfrac{30.655}{55.44\, P_{aD}} = \dfrac{933.3}{P_{aD}}$ cf. $\dfrac{997.4}{P_{aD}}$

for the F.W. method. The latter therefore overestimates static thrust by 4.7%. It must be emphasised, however, that these comparisons are based on a rather extreme case. Had the design been for high subsonic cruise, agreement between the two methods would have been much closer. Moreover, for turbo fans where much of the energy interchange involves much lower temperatures, especially in the case of high bypass ratio engines, agreement would be very close indeed.

Agreement between dimensional quantities such as F_s, Q, F, etc., is *not* good, at least in the case of the foregoing comparisons. It is only in the non-dimensional ratios

$\dfrac{F}{F_D}$ and $\dfrac{\eta}{\eta_D}$ that the F.W. method approximates to the h–p method especially in less

extreme cases. These ratios, however, are the most important for performance calculations; the writer therefore claims that he has justified the following procedure for such calculations for $\dfrac{F}{F_D}$ and $\dfrac{\eta}{\eta_D}$:

1) Calculate the design cycle by the h–p method.
2) Calculate the design cycle by the F.W. method using the same value of T_M as for 1).
3) Calculate the values of $\dfrac{F}{F_D}$ and $\dfrac{\eta}{\eta_D}$ for other than the design condition by the F.W. method and assume that they are the same as if calculated by the h–p method, remembering that minor discrepancies are of very little significance having regard to the fact that by far the greater part of the operational time will normally be at the design condition.

SECTION 18
Calculation of Maximum Efficiency

In the two preceding sections, the values of specific compressor energy input used in the examples were arbitrarily chosen and based on what may be achieved in a single compressor, but, as has been shown, for maximum gas generator efficiency, the greater the maximum temperature of the cycle, the greater the optimum pressure ratio.

On the assumption that practical limitations to pressure ratio do not exist, or can be overcome, the effect will now be examined.

Figure 1 shows the cycle diagram used for this purpose, and the table below illustrates the steps in the calculations to find the effect of θ_c. The assumptions used are indicated in Figure 1 for a flight speed of $1950'$/sec. The approximate method is used treating enthalpy as 'pseudo' temperatures, i.e., 1460 is an actual temperature of 1354°K.

Relationships not shown in Figure 1 are: $u_J = 147.0\sqrt{\theta_J}$; $F_S = \dfrac{u_J - 1950}{g}$.

θ_c	t_c	$t_c t_u$	p_o	T_T	t_T	t_J	θ_J	u_J	F_S	θ_H	η_o	η_{GG}	η_p	$p_o F_S$
350	1.749	3.073	50.86	1110	1.374	2.236	601.3	3604	51.38	713	0.419	0.596	0.702	2613
400	1.856	3.261	62.62	1060	1.452	2.246	576.2	3529	49.03	663	0.430	0.604	0.712	3070
450	1.963	3.449	76.19	1010	1.539	2.241	548.1	3442	46.32	613	0.439	0.607	0.723	3529
500	2.071	3.6441	92.38	960	1.637	2.226	518.1	3346	43.36	563	0.447	0.608	0.736	4005
550	2.178	3.8251	109.46	910	1.748	2.188	484.1	3234	39.89	513	0.452	0.601	0.752	4366
600	2.285	4.013	129.50	860	1.876	2.139	448.8	3114	36.15	463	0.454	0.589	0.770	4681

Since maximum cycle temperature is the same for each value of θ_c, the overall pressure ratio p_o is a measure of the mass flow rate *provided that* for a given axial velocity at compressor entry, diameter is increased to accommodate the increase of mass flow based on p_o.

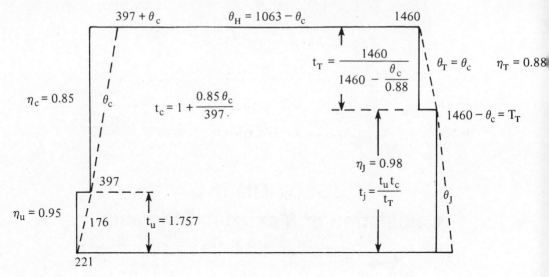

Fig. 18-1

$$\eta_o = \frac{F_S \times 1950}{335.6\,\theta_H} = \eta_{GG} \times \eta_p \quad \text{where } \eta_{GG} = \frac{\theta_J}{\theta_H} \text{ and } \eta_p = \frac{2}{\frac{u_J}{1950}+1} \; ; \; p_o = (t_u t_c)^{3.5} \, ; \, p_c = t_c^{3.5}$$

For a given engine size, given flight speed, and given axial velocity, the mass flow rate would remain constant, so, since F_S falls off substantially as θ_c is increased while η_o increases this latter is at the expense of increased engine size and weight for a given total thrust. As may be seen from the table, the increase in η_o is becoming marginal at $\theta_c = 500$ and above, while F_S is dropping off quite rapidly.

The usual criterion is the minimum value of power plant weight plus fuel weight for the design mission, which suggests that θ_c should be in the range 450–500°C for long range aircraft cruising at Mach 2.0 (1950'/sec in the stratosphere). Moreover, it is in this range that one gets a maximum value of η_{GG} which, as will be seen, is important for turbo-fan type engines.

Some of the quantities obtained from Figure 1 and its accompanying table are plotted in Figure 2, which also includes curves obtained in similar manner for a speed u of 880'/sec and $T_o = 260$ with the same T_{max} as Figure 1. η_u, however, was assumed to be 100% cf. 95% for the supersonic case.

As may be seen from Figure 2 η_o and F_S are very sensitive to u, the former increasing and the latter decreasing with increase of u. The effects of height and speed will be examined in greater detail later.

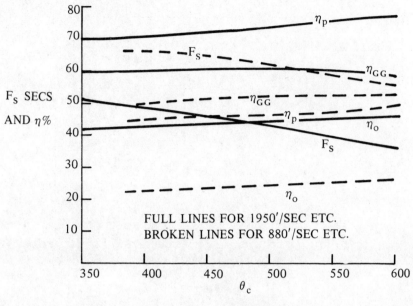

Fig. 18-2

SECTION 19
'Two Spool' Engines

Values of θ_c of the order of 500°C are beyond the capacity of single compressors unless fitted with an elaborate system of variable stators in several stages, hence the evolution of the two spool engine illustrated very diagrammatically in Figure 1. As may be seen, the first ('low pressure' or 'LP') compressor is connected to its driving turbine (the 'LP turbine') by a long shaft assembly 'threaded' through the shaft connecting the second ('HP') compressor to its driving ('HP') turbine.

In 1936 the writer attempted to design a two spool arrangement for an engine with a two stage centrifugal compressor but abandoned the scheme because of the seemingly insoluble problems of assembly and the formidable bearing and other mechanical complications. These difficulties, however, were first overcome by Rolls Royce with the Clyde turbo-prop having an axial flow LP compressor and an HP centrifugal and further developed in later engines with two axial flow compressors in series as in the Olympus 593, four of which power the Concorde.

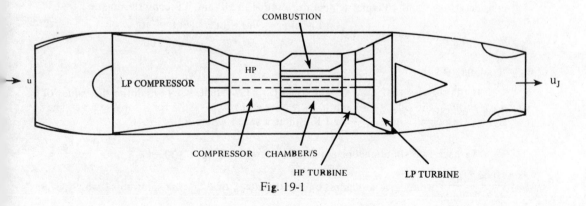

Fig. 19-1

The cycle diagram for the two spool arrangement is now as shown in Figure 2.

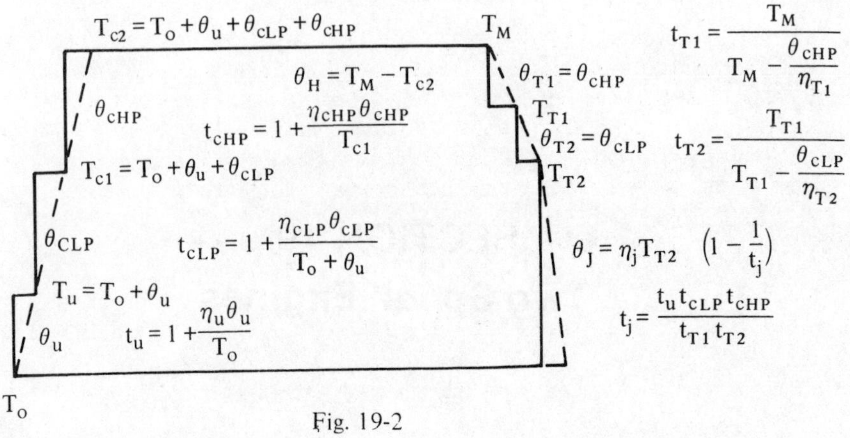

$T_{c2} = T_o + \theta_u + \theta_{cLP} + \theta_{cHP}$

$\theta_H = T_M - T_{c2}$

$\theta_{cHP} \qquad t_{cHP} = 1 + \dfrac{\eta_{cHP}\theta_{cHP}}{T_{c1}}$

$T_{c1} = T_o + \theta_u + \theta_{cLP}$

$\theta_{CLP} \qquad t_{cLP} = 1 + \dfrac{\eta_{cLP}\theta_{cLP}}{T_o + \theta_u}$

$T_u = T_o + \theta_u$

$\theta_u \qquad t_u = 1 + \dfrac{\eta_u \theta_u}{T_o}$

T_o

T_M

$\theta_{T1} = \theta_{cHP}$

T_{T1}

$\theta_{T2} = \theta_{cLP}$

T_{T2}

$\theta_J = \eta_j T_{T2}\left(1 - \dfrac{1}{t_j}\right)$

$t_j = \dfrac{t_u t_{cLP} t_{cHP}}{t_{T1} t_{T2}}$

$t_{T1} = \dfrac{T_M}{T_M - \dfrac{\theta_{cHP}}{\eta_{T1}}}$

$t_{T2} = \dfrac{T_{T1}}{T_{T1} - \dfrac{\theta_{cLP}}{\eta_{T2}}}$

Fig. 19-2

The manner in which the total temperature rise in compression $\theta_{CLP} + \theta_{CHP}$ is distributed between the LP and HP compressors is a matter of choice for the designer and this choice is influenced by several factors which must be weighed against each other. Thus, for a given number of stages, θ_{CHP} can be substantially larger than θ_{CLP} because the entering air is much hotter. This would also mean that the LP turbine blades would be subject to a lower temperature so that, while it may be necessary to cool the HP turbine blades, this could be avoided in the LP turbine. One would also expect that by reducing θ_{CLP} the whole LP assembly — which clearly comprises the bulkiest and heaviest part of the complete assembly — would have reduced weight and bulk at the expense of some increase in weight of the HP assembly. However, such matters as shaft length, bearing and coupling arrangement, etc., must also influence the choice. The writer, however, has had no experience in the detail design of two spool axial flow engines and feels unable to comment adequately on the pros and cons involved. He does feel, however, that thermodynamic considerations point to equal temperature ratios in HP and LP compressors, i.e., that

$$1 + \dfrac{\eta_{CLP}\theta_{CLP}}{T_u} \text{ should equal } 1 + \dfrac{\eta_{CHP}\theta_{CHP}}{T_{c1}} \text{ or } \dfrac{\eta_{CLP}\theta_{CLP}}{T_u} = \dfrac{\eta_{CHP}\theta_{CHP}}{T_u + \theta_{CLP}}.$$

Example 19-1

If the surmise that t_{CHP} should equal t_{CLP} is correct, what would be the values of θ_{CLP} and θ_{CHP} if $\Sigma\theta_c = 500$ and $\eta_{CLP} = 88\%$ and $\eta_{CHP} = 86\%$ for a supersonic engine flying at a height where $T_o = 220°K$ and at a speed where $\theta_u = 176°C$.

We have two simultaneous equations, namely: $\theta_{CHP} = 500 - \theta_{CLP}$ and $\dfrac{0.88\,\theta_{CLP}}{396} = \dfrac{0.86\,\theta_{CHP}}{396 + \theta_{CLP}}$ which give a quadratic for either θ_{CLP} or θ_{CHP} the solution of which gives $\theta_{CLP} = 197$ and $\theta_{CHP} = 303$.

This distribution requires too high a heat drop in the HP turbine for a single stage which a designer might consider undesirable. There are, however, certain advantages (and disadvantages) in using a two stage HP turbine.

A feature of importance in two spool engines must now be discussed; namely that, with a fixed throat area of the propelling nozzle, and with choking in both turbines, there is a 'thermodynamic lock' between the LP and HP assemblies. This arises from the fact that, as we have seen (Section 5) the total temperature before the first of two choking nozzles in series with energy extraction between them is a constant multiple of the energy extraction expressed as a temperature drop. Thus referring to Figure 2, and writing,

$\frac{T_m}{\theta_{T1}} = n$ and $\frac{T_{T1}}{\theta_{T2}} = m$ ∴ since $T_{T1} = T_M - \theta_{T1} = n\theta_{T1} - \theta_{T1} = \theta_{T1}(n-1)$ it follows that

$\theta_{T1}(n-1) = m\theta_{T2}$ or $\frac{n-1}{m} = \frac{\theta_{T2}}{\theta_{T1}}$ or, since $\theta_{T2} = \theta_{CLP}$ and $\theta_{T1} = \theta_{CHP}$,

$\frac{n-1}{m} = \frac{\theta_{CLP}}{\theta_{CHP}} = $ const., therefore, over the speed range where it may be assumed that compressor temperature rise is proportional to the square of the rotational speed, the rpm of the HP assembly is a constant multiple of the rpm of the LP assembly.

Rotor tip speeds may, therefore, be an important factor in deciding the $\theta_{CHP}/\theta_{CLP}$ ratio. Thus, referring to Example 1 above, the shaft power of the smaller HP turbine is greater than that of the LP turbine in ratio 303/197. (Again an argument in favour of a two stage HP turbine).

This thermodynamic lock can be 'broken' in various ways, e.g., by having a variable propelling nozzle throat or by by-passing some of the LP compressor output, thereby increasing θ_{T1}. This latter would be a step in the direction of the turbo fan to be discussed later. A non-choking two stage HP turbine would also serve the purpose.

The cycle of Figure 2 can be simplified to that of Figure 1 by 'presuming' a single compressor of the same total temperature rise and same total temperature ratio and using a value of η_c to conform with this, and by 'combining' the HP and LP turbines in similar fashion (see Section 12). This procedure is useful for finding the variation of performance with height and speed.

SECTION 20
Thrust Boosting

This short section deals with thrust boosting other than by using turbo fans to be discussed later.

It is often desirable, or even necessary, to increase power temporarily at the expense of efficiency. For this there are two main methods, namely (1) coolant injection to reduce T_o 'artificially' and/or (2) after burning. The former is normally used only at take off especially in the tropics while the latter may be used at any flight condition.

Coolant Injection

For cooling the entry air, use is made of the cooling effect of evaporation of various liquids — methanol, a methanol–water mixture, etc., or just plain water.

In the writer's experience the most effective method is liquid ammonia injection. Liquid ammonia has a latent heat of evaporation (about 330 C.H.U./lb), second only to water; but for some reason this method never 'caught on' despite the sensational results in early experiments by the writer and his team in 1942. (In the experiments concerned, a large bottle of liquid ammonia outside the test house supplied a spray of the liquid into the large venturi duct used to measure the air flow into the test cell. It evaporated on injection and caused intense cooling. It was realized that ammonia is combustible so, there being no governor fitted, the operator of the control was warned that it would probably be necessary to cut back the main fuel supply to prevent overspeeding. Unfortunately, the ammonia bottle was not visible from the control room, so a system of hand signals was arranged via a test hand who could pass on a signal from the throttle operator to another test hand operating the ammonia supply valve. The signals were duly given but the operator of the ammonia valve took several seconds to release the very tight valve—or so he claimed. Another version is that he was chasing a rat. The throttle operator, being unaware of the delay, relaxed, believing that the ammonia was being injected and not producing the forecast speed increase, so that when the ammonia *did* feed in he was off guard. The result was spectacular. The tachometer, thrust measuring balance, etc., went 'off the clock' and the mercury in several manometers was blown out; the explosive bang was heard for miles. Surprisingly, the engine (the W.I.) did not disintegrate, but every turbine blade had stretched and fouled the casing. The braking effect of this probably saved the rest of the engine. In a later, more controlled, test, a 40% increase of thrust was obtained but the cooling effect was so drastic that the frost covered centrifugal compressor casing contracted to the point of causing the impellor to seize. It was also found that the ammonia corroded the bronze cages of bearings and other alloys. These problems could have been overcome, but pressure of other work resulted in the abandonment of these experiments.) A disadvantage of ammonia is its strong affinity for water and its high heat of solution therein, which, in hot humid air, can 'neutralize' the latent heat of evaporation of quite a large amount of ammonia hence the amount necessary to achieve a given cooling effect depends very much on the relative humidity. (Air at 303°K and 2110 p.s.f. with 100% relative humidity contains about 3.5% of water vapor by weight which would not only be capable of dissolving nearly its own weight of ammonia but would have a substantial warming effect due to its own condensation, the effect of which would be difficult to calculate because the latent heat of condensation into minute droplets is less than that of complete condensation to liquid by the amount of energy contained in the surface tension of the droplets. There would, of course, be a cooling effect on re-evaporation, but this would occur rather late in the compression process and so be comparatively ineffective).

After Burning

The exhaust from a jet engine contains ample oxygen for further combustion before final expansion through the jet nozzle, hence thrust can be increased considerably by after burning in the jet or 'tail' pipe at the expense of the efficiency of final expansion due to the pressure loss caused by the after burning equipment — usually two or more annular 'flame holders' of $\subset$ or $<$ section.

As we have seen, u_j is given by $u_j = 147.1 \sqrt{\eta_J T_T \left(1 - \frac{1}{t_J}\right)}$ where t_J is the total to static temperature ratio of final expansion and T_T is the total temperature before expansion. By after burning, T_T can be increased by burning up to the limit of the available oxygen, but, in practice, the amount of after burning is far less than this because the penalty in reduction of jet pipe efficiency would be considerable (indeed intolerable) for long range cruise without the after burner in operation. Moreover, with after burning, a variable jet nozzle is necessary to avoid decreasing mass flow by choking and for complete or near complete expansion before final exhaust. Hence, the weight and mechanical complexity of the mechanism necessary tends to put a limit on the amount of after burning.

If the axial velocity at turbine exhaust is very high as it usually is, then to avoid high pressure loss at the point of location of the after burning flame holders, etc., it is very desirable to reduce velocity upstream of the after burner by a diffusing divergence. This, of course, also entails pressure loss so the optimum is the combination which gives the least total jet pipe pressure loss with the after burner off.

After burning is normally required, for a short line only, e.g., for take off; in combat; for transition through the 'sound barrier,' etc. Its effectiveness is obviously very dependent on the pressure ratio available for final expansion, the percentage thrust gain is therefore least at take off and greatest at maximum engine rpm and highest air speed. Also the greater the final expansion ratio the smaller the effect on overall efficiency, i.e., the adverse effect on fuel consumption.

Example 20-1

The sea level static specific thrust without after burning is 80 secs and the temperature (static) of final exhaust is 690°K; what is the amount of after burning necessary to increase the static thrust by 20%? assuming complete mixing and also assuming that the expansion in the final nozzle is isentropic.

To give F_s = 80 secs, u_j must be 32.2 x 80 = 2576'/sec, the temperature equivalent of which θ_j is $\left(\frac{2576}{147.1}\right)^2$ = 307.1°C ∴ the total temperature T before and through the final nozzle = 307.1 + 690 = 997.1°K.

To increase the static thrust by 20% requires that u_j be increased by 20% and ∴ θ_j by $(1.2)^2$ = 1.44 ∴ with after burning θ_j = 1.44 x 307.1 = 442.2°C.

The temperature ratio t_j across the nozzle is unchanged, i.e., $t_j = \frac{997.1}{690}$ = 1.445

∴ the total temperature T to give θ_j = 442.2 is found from $T\left(\frac{0.445}{1.445}\right)$ = 442.2 from which T = 1436. Thus the temperature increase necessary by after burning is 1436 − 997.1 = 438.9°C.

Example 20-2

Taking the figures of Example 1, what increases in nozzle throat and exist areas would be necessary for complete expansion with the same mass flow rate Q?

(a) The throat section S_c: Without after burning the critical velocity u_c = $147.1\sqrt{\frac{997.1}{6}}$ = 1895'/sec and the static temperature $T_c = \frac{997.1}{1.2}$ = 830.9; with after burning $u_c = 147.1\sqrt{\frac{1436}{6}}$ = 2274'/sec and $T_c = \frac{1436}{1.2}$ = 1196.7.

The throat pressures will be the same with and without after burning ∴ the critical density ρ_c will be inversely proportional to T_c.

Hence, $Q = k\frac{S_{c1} u_{c1}}{T_{c1}} = k\frac{S_{c2} u_{c2}}{T_{c2}}$ where suffixes 1 and 2 denote without and with after burning.

∴ $\frac{S_{c2}}{S_{c1}} = \frac{T_{c2}}{T_{c1}} \cdot \frac{u_{c1}}{u_{c2}} = \frac{1196.7 \times 1895}{830.9 \times 2274}$ = 1.2, i.e., a 20% increase of throat section is necessary.

(b) The exit section S_E: Since the exit static pressure is the same, exit density ρ_E is inversely proportional to exit static temperature T_E.

∴ $\frac{S_{E2}}{S_{E1}} = \frac{T_{E2}}{T_{E1}} \cdot \frac{u_{j1}}{u_{j2}} = \left(\frac{1436 - \theta_{J2}}{690}\right)\frac{2576}{1.2 \times 2576}$ (see Example 1)

$$\therefore \frac{S_{E2}}{S_{E1}} = \frac{1436 - 442.2}{690 \times 1.2} = 1.2.$$

(This result could have been assumed from the findings of Section 5 where it was shown that S_E/S_c for a con-di nozzle is dependent only on the temperature ratio for complete isentropic expansion.)

Example 20-3

Assuming the same degree of after burning as in Example 1, what would be the effect on specific thrust and overall efficiency at a speed of 2000'/sec if, without after burning $F_S = 41$ secs, $\eta_o = 0.43$, and the total temperature of the jet is $1000°K$?

To give $F_S = 41$, u_j is given by $41 = \frac{u_j - 2000}{32.2}$ from which $u_j = 3520'/\text{sec}$ and

$$\theta_J = \left(\frac{3320}{147.1}\right)^2 = 510.1 \therefore t_j = \frac{1000}{1000 - 510.1} = 2.041.$$

If $\eta_o = 0.43$ then $\frac{F_S u}{336 \theta_H} = 0.43$, i.e., $\frac{41 \times 2000}{336 \times \theta_H} = 0.43$ from which θ_H (the heat addition in the main combustion) is $567.6°C$. From Example 1, the after burner increases the temperature before final expansion by $438.9°C$, i.e., to $1438.9°K$

$\therefore \theta_{J2}$ with after burning is given by $\theta_{J2} = 1438.9 \left(1 - \frac{1}{t_j}\right) = 1438.9 \left(\frac{1.041}{2.041}\right) = 733.9°C$

$\therefore u_{J2} = 147.1\sqrt{733.9} = 3982'/\text{sec} \therefore F_{S2} = \frac{3982 - 2000}{32.2} = 61.56$ secs, i.e., an increase of 50.2%.

With after burning the total heat added $\Sigma \theta_H = 567.6 + 438.9 = 1006.5°C$

$$\therefore \eta_o = \frac{F_{S2} u}{336 \times 1006.5} = \frac{61.56 \times 2000}{336 \times 1006.5} = 0.364.$$

Example 20-4

Taking the figures of Example 3, what increase in nozzle section areas would be necessary for complete expansion with the same mass flow rate?

We have (as in Example 2) $\frac{S_{c2}}{S_{c1}} = \frac{T_{c2}}{T_{c1}} \frac{u_{c1}}{u_{c2}}$

$T_{c1} = \frac{1000}{1.2} = 833.3°K$; $u_{c1} = 147.1\sqrt{\frac{1000}{6}} = 1898'/\text{sec}$

$T_{c2} = \frac{1438.9}{1.2} = 1199°K$; $u_{c2} = 147.1\sqrt{\frac{1438.9}{6}} = 2276'/\text{sec}$

$$\therefore \frac{S_{c2}}{S_{c1}} = \frac{S_{E2}}{S_{E1}} = \frac{1199}{833.3} \times \frac{1898}{2276} = 1.2$$

(The fact that this answer is the same as for Example 2 is because the total temperatures in the two examples differ only by $2.9°$.)

If, in Example 3, the after burning had increased the total temperature by 1000°C, F_s would be 83.7 secs and η_o would be 0.318, i.e., the thrust would be more than doubled. However, though there would be ample oxygen for such a large amount of after burning, the problem of nearly complete mixing before final expansion within a reasonable length would be quite acute. How serious could this be? Much less than one might suppose, as will now be demonstrated.

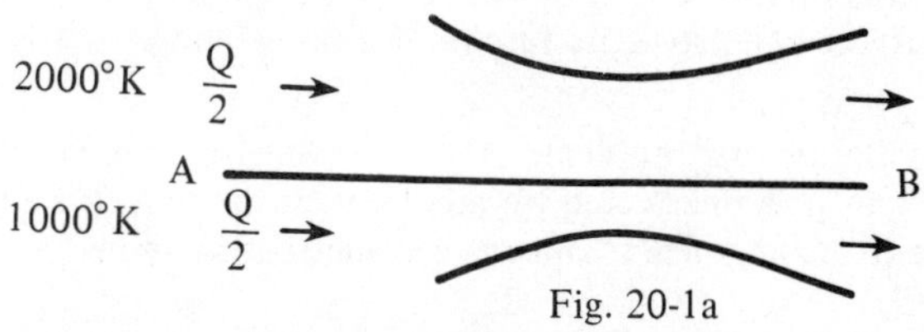

$$2000°K \quad \frac{Q}{2} \rightarrow \qquad \rightarrow \quad u_j = 147.1\sqrt{\frac{2-1}{2}(2000)} = 4648.5'/\text{Sec}$$

$$1000°K \quad \frac{Q}{2} \rightarrow \qquad \rightarrow \quad u_j = 147.1\sqrt{\frac{2-1}{2}(1000)} = 3287'/\text{Sec}$$

Fig. 20-1a

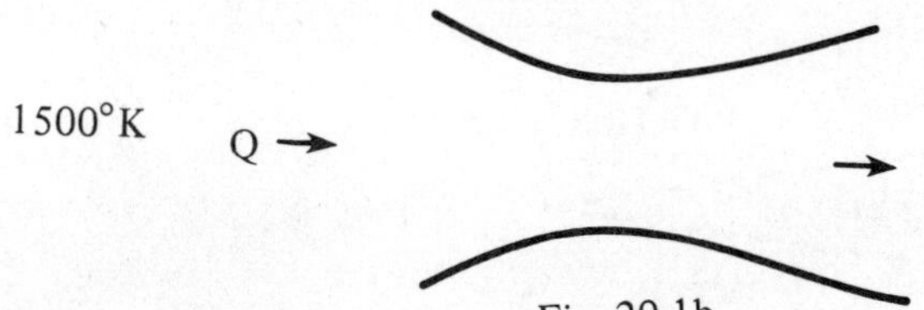

$$1500°K \quad Q \rightarrow \qquad \rightarrow \quad u_j = 147.1\sqrt{\left(\frac{2-1}{2}\right)(1500)} = 4025'/\text{Sec}$$

Fig. 20-1b

Figure 1a represents a con–di propelling nozzle having a total to static temperature ratio of $t_j = 2.0$ when air speed u is 2000'/sec. AB is a flexible diaphragm separating the total flow Q into two equal flows of $\frac{Q}{2}$. Above AB the total temperature is 2000°K and below AB it is 1000°K. The density of the lower flow is twice that of the upper, hence the eccentric position of AB.

Figure 1b represents a similar nozzle with a flow Q at a total temperature of 1500°K with the same temperature ratio.

In Figure 1a, the diaphragm (being flexible) ensures that the pressure along it are the same on each side.

The specific thrust of the upper flow in Figure 1a is $\frac{4648.5 - 2000}{32.2} = 82.25$ secs and for the lower flow is $\frac{3287 - 2000}{32.2} = 19.98$ secs. The total thrust, therefore, equals $\frac{Q}{2} \times 82.25 + \frac{Q}{2} \times 19.98 = 61.11$ Q lbs.

For Figure 1b the corresponding thrust is $Q \times \frac{4025 - 2000}{32.2} = 62.91$ Q lbs, i.e., about 3% greater — an astonishingly small difference for such a drastic case.

It can easily be shown that the throat and exit sections of the 'divided' nozzle would need to be about 1-1/2% larger than the undivided one for the same total flow.

The foregoing is not intended to imply that mixing is unimportant. A loss of thrust of 3% would be most undesirable. However, such an extreme case would be most improbable in practice.

If, in Figure 1a, the upper and lower total temperatures had been 1600 and 1400°K the thrust would be reduced by only 0.14%.

Where inadequate mixing is most undesirable is in the main combustion chamber otherwise the non-uniformity of velocity at exit from the turbine nozzles would seriously upset the flow pattern.

A vapor trail is visible evidence of the mixing problem. Also, experiments with stationary jet engines indicate that the axis jet flow is unaffected by mixing with the external air for some distance downstream of the exit plane. What happens is somewhat as shown in Figure 2; AB being about 6-10 diameters.

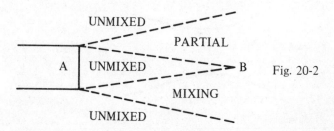

Fig. 20-2

To obtain mixing within a reasonable distance, it is necessary to use turbulence generators with their associated pressure loss, or to greatly increase the 'interface' by subdividing the flows it is desired to mix by means of fluted devices, again with pressure loss due to skin friction. The narrower the widths of the flute channels, the shorter the mixing distance needed — at the expense of increased skin friction.

The problem is obviously affected by the difference of velocity of the mixing streams. For example, two streams in contact at the same velocity and without turbulence; even with a large temperature difference, would take a very long time to mix since mixing would virtually be solely a matter of molecular diffusion.

The greater the velocity difference, the greater the mixing rate due to turbulence at the interface. But, since the mixing takes place without change of momentum, there is an inevitable conversion of kinetic energy into internal energy, part of which, however, is 'recoverable' in expansion.

It is not intended herein to go into the mixing problem at length so a simple example is used to indicate what happens.

Figure 3 illustrates a core flow of 1 lb/sec mixing with an annular flow of 1.2 lb/sec, the two flows having the velocities and temperatures shown. It is assumed that the static pressure is uniform across the plane AB and that downstream of AB, the static pressure remains constant. This would; in fact, require a slight change in section (an increase):

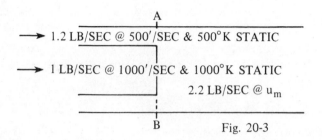

Fig. 20-3

Since there will be no change in total momentum before and after mixing

$$2.2 \times u_m = 1000 + 1.2 \times 500 \quad \therefore u_m = 727.3'/\text{sec}$$

The kinetic energy of the core stream before mixing is $\dfrac{1 \times 1000^2}{2g}$ and that of the annular stream is $\dfrac{1.2 \times 500^2}{2g}$. After mixing the total ke is $\dfrac{2.2}{2g} \times 727.3^2$ ∴ the loss Δke is given by $\Delta\text{ke} = \dfrac{1}{2g}[1000^2 + 1.2 \times 500^2 - 2.2 \times 727.3^2] = 2116.1$ ft/lbs. ∴ the loss per unit of total flow = $\dfrac{2116.1}{2.2} = 961.9$ ft/lbs, equivalent to $\dfrac{961.9}{336} = 2.86°$. This figure, therefore, represents the conversion of ke into internal energy per unit mass. It may be seen that there will be no great error in neglecting it in cycle calculations, as will be done later.

The pressure loss in a fluted mixing device may well be largely offset by reduced skin friction in the rear end of the jet pipe and final nozzle.

SECTION 21
Turbo Fans

Introductory

During the 1930s the writer was much concerned with the problem of increasing the propulsive efficiency of jet engines at the flight speeds which then seemed to be limited to about 500 m.p.h. owing to the high drag rise as the speed of sound was neared. It was then the general belief that the 'sound barrier' could never be passed. The writer did not share this belief and, indeed, his 1936 notebooks contained performance calculations for the straight jet up to 1500 m.p.h. which showed very clearly that efficiency could be greatly increased at supersonic speeds. (At this time the first experimental engine was under construction but had not yet been completed – its first run was not until April 12th, 1937). However, there then seemed little hope that a supersonic airplane could have a lift/drag ratio better than about 4.0 c.f. 16 or more for high/subsonic speeds. It thus seemed improbable that increase of engine efficiency could compensate for the serious drop in L/D. It would, in any event have been most unwise to have pressed the concept of jet powered supersonic aircraft in the light of the formidable wall of scepticism about the practicability of jet engines for any purpose. So the writer concentrated on the problem of improving power plant efficiency at moderate and high subsonic speeds.

It was obvious that means to 'gear down' the jet were needed, i.e. to aim for a high mass low velocity jet instead of the low mass high velocity jet of the simple jet engine.

The writer first considered the use of an ejector for the purpose but rejected this scheme when calculations showed that, though some improvement in thrust could be obtained at take off and low speed, an ejector actually reduced thrust at higher speeds. It soon became clear that mechanical means would be necessary to extract part of the energy of the jet and impart it to additional 'by-pass' air. At modest speeds the turbo-prop would have been the answer and, indeed, the Royal Aircraft Establishment was working on an axial flow gas turbine to drive a propellor in parallel with Power Jets' development of the centrifugal type jet engine. (Later, when the first experimental engine began to show real promise, the R.A.E., in co-operation with Metropolitan-Vickers, converted their turbo-prop scheme into an axial flow jet engine—the F-2, later known as the Beryl). The writer, however, remained convinced that the propellor was not the answer for speeds of 500 m.p.h. and above, so turned his attention to the concept of using part of the gas generator output energy to drive a high mass flow low pressure compressor or 'fan'.

The primary problem was the mechanical arrangement, i.e. how to connect the fan to its driving turbine without using the two spool arrangement which, at that time, seemed to present formidable bearing, coupling and assembly problems. His first solution is shown in Figure 1 which is the principle drawing of U.K. Patent 471,368 filed in March 1936. This patent was, in fact, the master patent for all turbo fans. When it expired in 1952 it was granted the maximum allowable extension of 10 years. Even so, it expired before turbo-fans

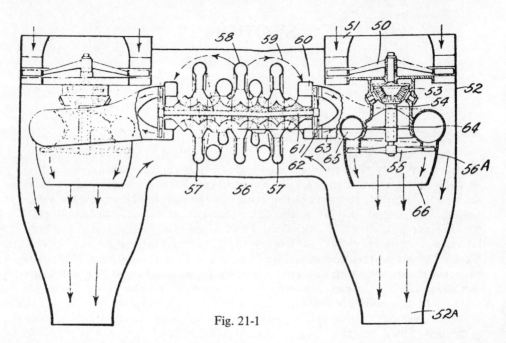

Fig. 21-1

began to supersede straight jet engines. (The Rolls-Royce Conway was going into limited production at the time but it had the very low bypass ratio of about 0.6). All modern turbo-fans fall within the first claim of this patent.

As may be seen in Figure 1, the gas generator was athwartships in a plenum chamber (56) and supplied energised gas to two turbines (55) driving fans (50) in lateral nacelles. The two fans 'supercharged' the plenum chamber as well as supplying the bypass air which rejoined the exhaust from the fan turbines before final expansion.

The gas generator is shown with two first stage centrifugal compressors (57) in parallel and a single second stage centrifugal compressor (58) all driven by two turbines (62) in parallel via combustion chambers (not shown).

No attempt was made at detail design because of pressure of work on the experimental jet engine. Moreover there was no money available for such an ambitious project.

Figure 2 shows another of the several 'tricks' proposed by the writer to circumvent threaded shafts. It is the principal drawing of U.K. Patent 593,403.

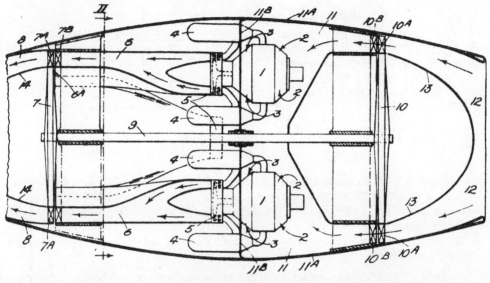

Fig. 21-2

In this arrangement three or four simple jet engines (1) were symmetrically disposed about the shaft (9) connecting the fan (10) with its driving turbine (7) driven by the energised gas from the jet engine gas generators.

This scheme also was never followed up for lack of time and money, but one might imagine that a modern form of it using axial flow type gas generators might prove interesting.

Eventually, when the simple jet had reached an advanced stage of development and was in limited production, Power Jets did start to develop a turbo-fan—the L.R.1. Unfortunately the company had been nationalised in 1944 at which time the L.R.1 was in the initial design stage. In the spring of 1946 when the L.R.1 was half built, the Company (except for a small 'rump') was merged with the R.A.E. gas turbine unit to form the National Gas Turbine Establishment. Under pressure from the aircraft industry in the U.K., which by then had been almost entirely converted to aircraft gas turbines, the Terms of Reference of the N.G.T.E precluded the design and development of engines. The functions of the N.G.T.E. were restricted to research and assisting the industry. In consequence the contracts for the L.R.1 and other projects in progress were cancelled. (This deprivation of the right to continue the work they had initiated resulted in the break-up of the greater part of the writer's very able team.)

The L.R.1, illustrated in Figure 3 very diagrammatically, also avoided the two spool arrangement. There was no mechanically independent gas generator. The power for both the main compressor (an axial flow compressor with a final centrifugal compressor stage) and the three stage fan was provided by a two stage turbine, the fan being driven through an epicyclic reduction gear.

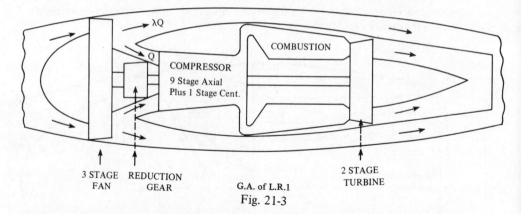

3 STAGE FAN   REDUCTION GEAR   G.A. of L.R.1   2 STAGE TURBINE

Fig. 21-3

The L.R.1 was primarily intended to power a 4 engined bomber having a 4000 miles still air range with a 10,000 lb. bomb load but was also visualised as suitable for a trans-Atlantic civil aircraft. The core unit was also designed to be applicable to turbo-props and different bypass ratios by changing the gear ratio. It was, however, basically a 3.1 bypass ratio turbo fan. Its premature demise undoubtedly put back the turbo-fan era by about twenty years. (In Figures 3 and 4 λ is the bypass ratio).

Meanwhile, three years earlier, a short lived attempt was made to develop an aft fan along the lines illustrated roughly in Figure 4. At the time it was known as a 'thrust augmentor' because it was a device which could be 'tacked on' to existing jet engines without any modification to the main engine forward of the compressor turbine exhaust.

In Figure 4, (1) is the nozzle ring of the aft fan turbine (2), (3) and (4) are the rotor and stator blades of the fan. As may be seen, the fan rotor blades were formed as extension of the turbine blades but the shroud ring (5) separated the hot and cold flows.

Only one stage is shown in Figure 4, though normally two or more would be necessary—especially for the turbine component.

The first design—the No. 1 Thrust Augmentor—had no stator blading at all. It comprised a pair of contra rotating turbine-fan wheels. Seemingly the acme of simplicity, but it had to

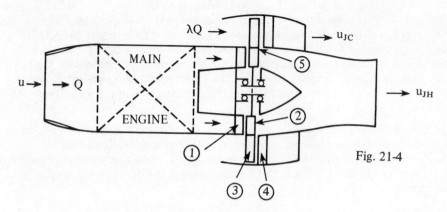

Fig. 21-4

be abandoned when detail design revealed that the necessary blade profiles and relative velocities were quite impracticable.

The second design—the No. 2 Thrust Augmentor was a more conventional two stage arrangement. It was never built by Power Jets but the drawings were shipped to the General Electric Company and became the basis of the aft fans later developed by G.E.

The aft fan, however, has several disadvantages as compared with the now conventional front fan, namely:—

1) The turbine blade speed is necessarily lower than that of the fan which is opposite from what is desirable.

2) The hot portion of the blading is the more highly stressed—again opposite from what is desirable.

3) The blades are integrally cast and so the fan blading has to be of the same material as the turbine, hence they are far heavier than they could be from the point of view of stress and temperature.

4) There is a leakage problem through the clearances. The higher pressure hot stream leaks into the fan flow thereby reducing fan efficiency.

5) There is a boundary layer 'build-up' along the core casing which also reduces fan efficiency. This can be reduced to a large extent by extending the fan inlet ducting the full length of the engine—at the expense of a substantial increase in nacelle drag.

6) The fan contributes nothing to the overall pressure ratio hence the main engine is larger than it would be if 'supercharged' by a front fan.

It would clearly be desirable to reverse the positions of the turbine and fan blades, i.e., to have the former external to the latter. This removes the first and second of the disadvantages enumerated above, but requires some very awkward ducting since the ducts carrying the hot gas supply from the main engine to the 'tip turbine' must pass through the inlet flow to the fan.

The arrangement is shown in Figure 5 which is the principal drawing of U.K. Patent 588,085.

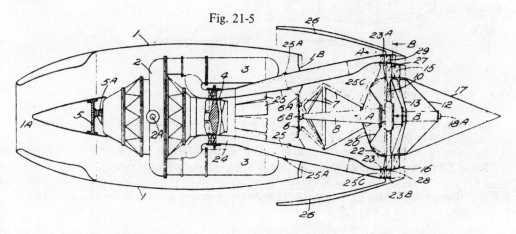

Fig. 21-5

The No. 3 Thrust Augmentor designed and built by the writer and his team in 1943 was in accordance with the arrangement shown in Figure 5. On test, a 40% gain of static thrust was achieved but this was far lower than expectations because of the interference of the hot ducting with the fan inlet flow. No doubt this interference could have been greatly reduced by redesign of the gas generator ducting, but again pressure of other work ruled out further development. However, the fan with a tip turbine has shown some promise as a lift fan.

Later, as Power Jets' capacity expanded, the 'No. 4 Thrust Augmentor' was designed for addition to the Power Jets W2/700 engine. It was basically similar to the No. 2 but the scheme included after burning aft of the aft fan assembly. The purpose of the whole arrangement was to power the small Miles M52 supersonic airplane.

Again, when both power plant and aircraft were nearing completion they fell victims of the consequences of Ministry policy. The contracts were cancelled.

High Bypass Ratio Turbo-Fans

Figure 6 is a diagrammatic representation of the type of turbo-fan with high bypass ratio λ which has now largely replaced straight jet engines in subsonic aircraft.

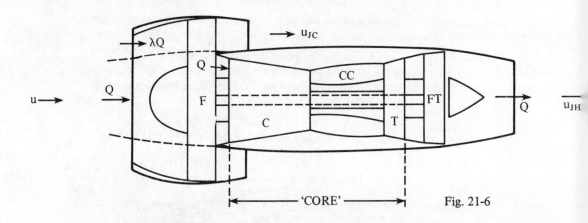

Fig. 21-6

A two spool arrangement is shown in which the shaft connecting the fan turbine FT to the fan F is threaded through the shaft of the compressor—combustion chamber—turbine unit C–CC–T which comprises the 'core engine'. The core engine itself may be a two spool assembly as in the Rolls Royce R.B. 211 which is thus a three spool engine. Another alternative arrangement is that used in the Pratt and Whitney J.T.9.D in which the core compressor has an LP section and HP section, the former being integral with the fan.

Bypass ratios have increased greatly in latter years and are now in the range 5:1 to 6:1 or more.

As indicated by the broken lines in the intake of Figure 6, part of the fan flow passes to the core compressor and thus contributes to the overall compression. Indeed, for purposes of calculation it is desirable to assume that the inner part of the fan (within the broken lines) is part of the core cycle.

If λ is the bypass ratio and the core flow is Q lb/sec. than the bypass flow is λQ and total flow is $(1 + \lambda)$Q, hence if θ_F represents the specific work done by the fan and θ_{FT} the specific work of the fan turbine, then $\theta_{FT} = (\lambda + 1) \theta_F$.

The specific thrust F_s is the thrust per unit core flow and is given by

$$F_s = \frac{1}{g} [u_{JH} - u + \lambda (u_{JC} - u)] \tag{21-1}$$

u_{JC} being the 'cold jet' velocity and u_{JH} the 'hot jet' velocity.

As may be seen in Figure 6 there is no mixing of the bypass and core flows before their respective final expansions.

The bypass cycle comprises compression and expansion only. The cycle (or cycles) may be represented as in Figure 7. The displacement to the left of CE (bypass expansion) and

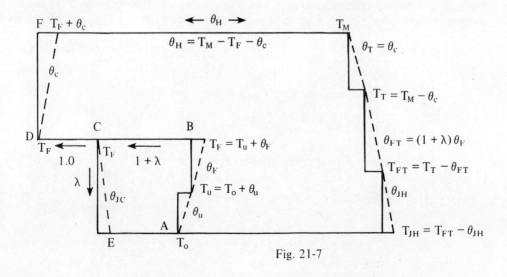

Fig. 21-7

D.F. (core compressor) as compared with A.B. (ram plus fan compression) is to indicate the different mass flows involved. T_F, the total temp. at fan discharge, is the same at B, C and D.

Alternatively the core and bypass cycles may be represented independently as in Figure 8 as long as it is remembered that *for each unit of flow in the core cycle there are λ units of flow in the bypass cycle.*

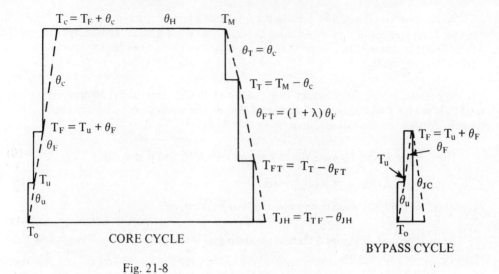

Fig. 21-8

As compared with the straight jet, the expansion efficiency is somewhat reduced, and therefore the thermal efficiency, because a much greater part of the expansion is subject to turbine losses, but this is much more than compensated for by the gain in propulsive efficiency.

Apart from the overall temp. ratio t_o there are seven temp. ratios to be taken into account, namely:—

1) Ram temp. ratio t_u given by $t_u = 1 + \dfrac{\eta_u \theta_u}{T_o}$ \hfill (21-2)

2) Fan temp. ratio t_F given by $t_F = 1 + \dfrac{\eta_F \theta_F}{T_u}$ \hfill (21-3)

3) Compressor temp. ratio t_c given by $t_c = 1 + \dfrac{\eta_c \theta_c}{T_F}$ \hfill (21-4)

4) Core Comp. Turbine temp ratio t_T given by $t_T = \dfrac{T_M}{T_M - \dfrac{\theta_c}{\eta_T}}$ \hfill (21-5)

5) Fan Turbine temp. ratio t_{TF} given by $t_{TF} = \dfrac{T_T}{T_T - \dfrac{\theta_F(1+\lambda)}{\eta_{FT}}}$ \hfill (21-6)

6) Core Jet temp. ratio t_{JH} given by $t_{JH} = \dfrac{t_u t_F t_c}{t_T t_{TF}} \left(= \dfrac{(1) \times (2) \times (3)}{(4) \times (5)} \right)$ \hfill (21-7)

7) Bypass Jet temp. ratio t_{JC} given by $t_{JC} = t_u t_F \; (= (1) \times (2))$ \hfill (21-8)

Other relationships between temp. ratios can be obtained by combining some of the above, thus t_T and t_c may be related by substituting for θ_c from (4) in (5) and so on.

The cold jet velocity u_{JC} is found from $u_{JC} = 147.1 \sqrt{\eta_{JC} T_F \left(1 - \dfrac{1}{t_{JC}}\right)}$

or $\quad u_{JC} = 147.1 \sqrt{\eta_{JC} T_F \left(1 - \dfrac{1}{t_u t_F}\right)} \qquad (21\text{-}9)$

and the hot jet velocity u_{JH} is found from $u_{JH} = 147.1 \sqrt{\eta_{JH} T_{TF} \left(1 - \dfrac{1}{t_{JH}}\right)}$

or $\quad u_{JH} = 147.1 \sqrt{\eta_{JH} T_{TF} \left(1 - \dfrac{t_T t_{TF}}{t_u t_F t_c}\right)} \qquad (21\text{-}10)$

∴ Substitution for u_{JC} and u_{JH} in (1) gives

$$F_s = \frac{1}{g}\left[147.1\left(\lambda \sqrt{\eta_{JC} T_F \left(1 - \frac{1}{t_u t_F}\right)} + \sqrt{\eta_{JH} T_{TF}\left(1 - \frac{t_T t_{FT}}{t_u t_F t_c}\right)}\right) - u(\lambda + 1)\right] \qquad (21\text{-}11)$$

Though η_u, the ram efficiency, is indicated as less than unity in Figures 7 and 8 so much of the ram compression occurs ahead of the intake orifice that η_u may be taken as 1.0 with negligible error (since large bypass ratio engines are suitable only for subsonic speeds) in which case $t_u = 1 + \dfrac{\theta_u}{T_o}$.

The design value of the bypass ratio λ is a matter of choice for the designer. Among the factors he must take into account are:—

1) The much greater specific thrust at take-off and low speeds with high bypass ratio at the expense of a rapid fall off at high speed (of specific thrust).

2) The effect on power plant weight. The weight of the fan-fan turbine assembly is offset to some extent by the reduction in size of the core assembly but this offset is greater for low bypass ratio engines which in any case have lighter fan-fan turbine assemblies.

3) The effect on fan cowling drag.

4) The effect of the drag of the annular cold jet (which may well be supersonic) on the core casing.

Strictly speaking, the two items of drag 3) and 4) above should be deducted from engine thrust to give the true net thrust. The drag over the core casing can be much reduced by extending the length of the fan exhaust cowling but at the expense of increased drag both internally and externally of the latter. In general it pays to have a long fan exhaust duct in the case of moderate and low bypass ratios. Moreover, the higher pressure in the exhaust duct of a low bypass ratio engine makes 'duct burning' more worthwhile for thrust boosting.

Example 21-1.

Figure 9 shows the cycle for the design condition for $u = 900'/\text{sec.}$ and $T_o = 221°K$ and $\lambda = 5.0$. Assumptions not shown in the diagram are:— 1) $\eta_u = 1.0$, 2) $\eta_F = .88$, 3) $\eta_c = .85$, 4) $\eta_T = .87$, 5) $\eta_{FT} = .90$, 6) $\eta_{JC} = \eta_{JH} = .98$.

```
Tc = 768.5            θH = 687.5
                                          TM = 1456
  θc = 450    tc = 2.201                  θT = θc = 450    tT = 1.551
                                          TT = 1006
         318.5        TF = 318.5          θTF = (1 + λ) θF = 360    tTF = 1.
318.5      λ + 1 = 6.0   θF = 60    tF = 1.204
           tJC = 1.409                    TTF = 646
 λ = 5.0   θJC          Tu = 258.5
                        θu = 37.5°  tu = 1.170   θJH    tJH = 1.204
           TJC          To = 221                 TJH
```

Fig. 21-9

(t_{JH} in Figure 9 from $t_{JH} = \dfrac{t_u t_F t_c}{t_T t_{TF}}$ and t_{JC} from $t_{JC} = t_u t_F$)

$$\theta_{JH} = 646 \times .98 \left(\dfrac{.204}{1.204}\right) = 107.3 \therefore u_{JH} = 147.1 \sqrt{107.3} = 1522.7'/\text{sec}.$$

$$\theta_{JC} = 318.5 \times .98 \left(\dfrac{.409}{1.409}\right) = 90.54 \therefore u_{JC} = 147.1 \sqrt{90.54} = 1399'/\text{sec}.$$

$$F_s = \dfrac{1}{g} \ (\lambda (1399 - 900) + 1522.7 - 900) = 96.79 \text{ secs}.$$

$$\therefore \eta_o = \dfrac{96.79 \times 900}{336 \times 687.5} = .377 \quad \text{(This compares with about .25 for a straight jet at}$$
the same speed).

Using the same methods and varying λ one finds the following results:—

λ	u_{JH}	u_{JH}	F_s
3	2276	1399	89.18
4	1940.2	1399	94.27
6	910.6	1399	93.27

The variation of F_s with λ for u = 900 ft/sec. is shown in Figure 10.

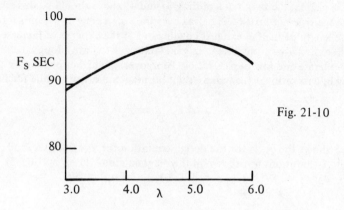

Fig. 21-10

Since the same value of θ_F was used for each value of λ it follows that u_{JC} is not affected by variation of λ though the size of the cold jet nozzle would, of course, depend very much on the choice of λ.

θ_H was also the same for each case, so peak F_s corresponds to peak overall efficiency.

As may be seen, $\lambda = 5$ is very close to the optimum and F_s is insensitive to λ in the range $4.75 - 5.25$. As flight speed is reduced, the theoretical optimum value of λ increases very substantially, e.g. for a freighter designed to cruise at 400 m.p.h., it may well be of the order of 15 or more, subject to practical difficulties of weight, installation, etc.

Several years ago, the writer found a proof that the optimum value of η_o occurred when $u_{JC} = \sqrt{\eta_F \eta_{TF}}\; u_{JH}$ but the proof was very lengthy and it is not proposed to repeat it here. (It is possible that there is a much shorter proof). Thus, in Example 1, λ should have been such that $u_{JH} = \dfrac{1399}{\sqrt{.88 \times .9}} = 1572'/\text{sec.}$ whereas for $\lambda = 5$ u_{JH} was found to be $1523'/\text{sec.}$ approx. in Example 1, so the choice of 5 for λ was a 'very near miss'.

It may also be noted from the example that the temp. ratio across the cold jet nozzle is above the critical ratio 1.2, while the temp. ratio across the core jet nozzle is just above 1.2, so both nozzles are choking but only the cold jet justifies a con-di nozzle.

If one accepts that $u_{JH} = \dfrac{u_{JC}}{\sqrt{\eta_F \eta_{TF}}}$ then $\theta_{JH} = \dfrac{\theta_{JC}}{\eta_F \eta_{TF}}$. Using this

$$t_{JH} = \frac{T_M - \theta_c - (\lambda + 1)\theta_F}{T_M - \theta_c - (\lambda + 1)\theta_F - \dfrac{\theta_{JC}}{\eta_{JC}\eta_F\eta_T}} \qquad (21\text{-}12)$$

Using this equation in combination with equations (2), etc., and given the values of θ_u, θ_F, θ_c, T_M and the various stage efficiencies, it is possible to obtain a somewhat complicated quadratic equation for λ^*, the optimum value of λ.

Example 21-2.

What is the value of λ^* if $\theta_F = 50$ and $\theta_c = 460$ given that all other assumptions are as in Example 1?

$$t_u = 1 + \frac{37.5}{221} = 1.168 \quad \text{and} \quad T_u = 258.5$$

$$t_F = 1 + \frac{.88 \times 50}{258.5} = 1.17 \quad \text{and} \quad T_F = 308.5$$

$$t_c = 1 + \frac{.85 \times 460}{308.5} = 2.2674 \quad \therefore \quad t_{JC}t_c = t_u t_F t_c = 3.098$$

$$t_T = \frac{1456}{1456 - \dfrac{460}{.87}} = 1.570 \quad \text{and} \quad T_T = T_M - \theta_c = 996$$

$$t_{TF} = \frac{996}{996 - \frac{(\lambda+1)50}{.9}}$$

$$\theta_{JC} = .98 T_F \left(1 - \frac{1}{t_u t_F}\right) = .98 \times 308.5 \left(1 - \frac{1}{1.168 \times 1.17}\right) = 81.09$$

∴ from (12), $\dfrac{3.098}{1.57\left[\dfrac{996}{996-\dfrac{(\lambda+1)50}{0.9}}\right]} = \dfrac{996-(\lambda+1)50}{996-(\lambda+1)50-\dfrac{81.09}{0.98 \times 0.88 \times 0.9}}$

which boils down to $1.862 - 0.11\lambda = \dfrac{946 - 50\lambda}{841.5 - 50\lambda}$ from which λ^* is found to be 6.07.

This does not, in fact, quite agree with λ^* as found from the cycle diagram which gives the following results:—

λ	u_{JC}	F_{sJC}	u_{JH}	F_{sJH}	ΣF_s	$\dfrac{u_{JH}}{u_{JC}}$
6.0	1326.6	79.49	1524.6	19.40	98.89	1.149
6.1	1326.6	80.81	1484	18.13	98.94	1.119
6.2	1326.6	82.13	1442	16.84	98.97	1.086
6.3	1326.6	83.46	1399	15.50	98.96	1.054

(F_{sJC} is the bypass specific thrust and F_{sJH} is the core specific thrust; each per unit core flow).

Whereas $\lambda = 6.07$ gives $\Sigma F_s = 98.93$. Clearly an insignificant difference. As the table shows, ΣF_s is very insensitive to λ and the value of $\dfrac{u_{JH}}{u_{JC}}$ in the neighborhood of the optimum.

As may be seen, u_{JC} is unaffected by λ for the given values of u, θ_F, η_u, η_F and η_{JC} ∴ the drag over the core casing would be the same, but the fan cowling drag would, of course, increase as λ increases so that, for maximum net thrust, i.e., thrust less drag, the 'real' optimum value of λ would be somewhat less than that calculated from the cycle alone.

Example 21-3.

This example is an energy balance check on the figures of Example 1.

Referring to Figure 9, $\theta_{JC} = 90.54$ ∴ $T_{JC} = 318.5 - 90.54 = 227.96$, ∴ the heat rejected by the cold jet = $\lambda(227.96 - 211) = 5 \times 6.96 = 34.8$. $\theta_{JH} = 107.3$ ∴ $T_{JH} = 646 - 107.3 = 538.7$ ∴ the heat rejected = $538.7 - 221 = 317.7$ ∴ Total heat rejected = $34.8 + 317.7 = 352.5$. The energy increase of the bypass flow = $\lambda(\theta_{JC} - 37.5)$ = $5(90.54 - 37.5) = 265.2$. The energy increase of the core flow = $\theta_{JH} - 37.5 = 107.3 - 37.5$ = 69.8 ∴ the total energy increase = $69.8 + 265.2 = 335$. Adding this to the heat rejected we have $335 + 352.5 = 687.5$ which is equal to the heat added, which is how it should be.

Off Design Conditions

The design parameters will normally be those which are the optimum for the cruise condition, but it is necessary to examine the conditions necessary for satisfactory performance at take-off and other flight conditions, e.g., take-off, climb, 'stand off', etc.

Taking the design figures of Example 1 for θ_F, θ_c, T_M, η_F, η_c, η_T, η_{TF}, η_{JC} and η_{JH}, let us find out what the sea level static cycle/s would be. The cycle diagram is shown in Figure 11 assuming $T_o = 288°K$.

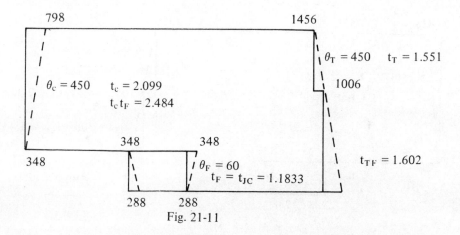

Fig. 21-11

It is found that t_{TF} cannot exceed 1.602 even if the final nozzle is opened up to the point where u_{JH} equals or is less than the axial velocity from the fan turbine.

θ_{TF} from $\dfrac{1006}{1006 - \dfrac{\theta_{TF}}{.09}} = 1.602$ is found to be 340.2°C $\therefore \lambda + 1 = \dfrac{340.2}{60} = 5.67$

$\therefore \lambda = 4.67$. Assuming that the hot exhaust energy cannot exceed 4% of θ_{TF}, $\theta_{JH} = .04 \times 340.2 = 13.6°$ which gives $u_{JH} = 147.1\sqrt{13.6} = 542.2'$/sec. which would give a thrust of 16.8 secs.

$u_{JC} = 147.1\sqrt{.98 \times 348 \times \dfrac{.1833}{1.1833}} = 1068'$/sec. to give a thrust of $33.18 \times \lambda$

i.e., $33.18 \times 4.67 = 155$, hence the total specific thrust is $155 + 16.8 = 171.8$ secs.

The cold jet velocity is nearly twice that of the hot jet which is far from the optimum ratio $\sqrt{.9 \times .88} = .89$. Moreover the reduction of λ is undesirable since it means reduced axial velocity in the fan.

For take-off, however, it is usual to allow an increase in T_M, and therefore an increase of θ_c. Let us examine the effect of increasing T_M to 1550°K for the case corresponding to the design cycle of Example 1.

As has been shown in Section 10, the condition for compressor axial velocities to match blade speeds is that the maximum cycle temp. T_M should be a constant multiple of the total temp. at core compressor entry. For the design cycle of Figure 9 (Example 1)

$\dfrac{T_M}{T_F} = \dfrac{1456}{318.5} = 4.57$ ∴ T_F should be $\dfrac{T_M}{4.57}$ so that for $T_M = 1550°K$, $T_F = 339$

∴ θ_F (for $T_o = 288$) is reduced to $339 - 288 = 51°C$ while θ_c is increased to $\dfrac{1550}{1456} \times 450$
$= 479°C$. The cycle for take-off now becomes as shown in Figure 12.

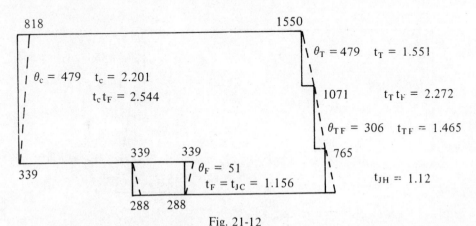

Fig. 21-12

$t_{TF} \times t_{JH} = \dfrac{2.544}{1.551} = 1.640$ which is still below the design value of t_{TF} (= 1.66) but if t_{JH} were to be reduced to 1.0 by opening up the final nozzle (as in the case of Figure 11) θ_{TF} would be 376.3 to give $\lambda + 1 = \dfrac{376.3}{51} = 7.378$ and $\lambda = 6.378$. Such a large value of λ in proportion to θ_F would be most undesirable, so let us restore λ to 5.0 to make $\theta_{TF} = 6 \times 51 = 306$ as shown in Figure 12. $t_{JH} = \dfrac{2.544}{2.272} = 1.12$ ∴

$u_{JH} = 147.1 \sqrt{.98 \times 765 \times \dfrac{.12}{1.12}} = 1315.5'$/sec. to give 40.85 secs.

For the cold jet, $u_{JC} = 147.1 \sqrt{.98 \times 339 \times \dfrac{156}{1.156}} = 984'$/sec. to give

$5 \times \dfrac{984}{322} = 152.84$ secs. ∴ Specific Thrust (per lb. of core flow) = 152.84 + 40.85 = 193.7 secs. This is a substantial increase (nearly 13%) of specific thrust as compared with the cycle of Figure 11. Moreover, on the basis that $Q \propto \dfrac{P_M}{\sqrt{T_M}}$, Q would be increased by 5.4% (remembering that Q is the core flow). The combined effect of the increases of specific thrust and of Q would give an overall increase of total thrust of 18.8% (for the cycle of Figure 12 as compared with that of Figure 11.)

In the discussion of axial flow compressors in Section 12 it was pointed out that an increase of axial velocity, i.e., of Q (or $1 + \lambda Q$ in the case of a bypass engine fan) reduces the work capacity but the effect was not examined. It is, in fact, quite drastic as may be seen from Figure 13 which shows the vector diagram for the exit conditions at mean radius for a design θ_F of 30°C, i.e. $\dfrac{u_b u_\omega}{gK_p} = 30$. As may be seen, an increase of leaving axial velocity. From 550'/sec. to 600'/sec. reduces u_ω to 186'/sec. and θ_F to $\dfrac{1200 \times 186}{gK_p} = 20.63°C$.

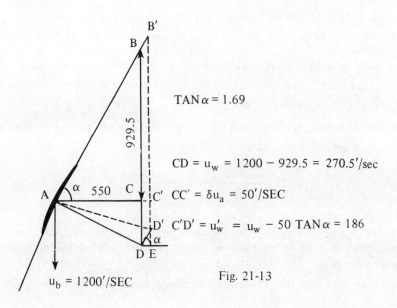

Fig. 21-13

More generally, $\delta u_\omega = - \delta u_a \tan\alpha$.

For the design condition, as has been shown in Section 12, the specific work (and therefore θ_F) can be uniform from root to tip only if the induced whirl is that of a free vortex with constant axial velocity.

Any departure from the design condition gives rise to a very complex situation. Tan α increases with r and hence the reduction of θ_F becomes greater. Indeed when $u_\omega = u_b - u_a \tan\alpha = 0$ then $\theta_F = 0$ and, since u_b also increases with r, much depends on how u_a varies with r if Q, and therefore u_a, is varied from design. The inter-relationship of u_a and u_ω in off design conditions is very complex because, as shown in Section 3, radial equilibrium requires that $2\dfrac{dr}{r} + \dfrac{d\theta_\omega}{\theta_\omega} + \dfrac{d\theta_a}{\theta_\omega} = 0$ (if there is no radial velocity component).

This equation may be converted to $2\frac{dr}{r} + \frac{du_\omega^2}{u_\omega^2} + \frac{du_a^2}{u_\omega^2} = 0$

or $\frac{dr}{r} + \frac{u_\omega du_\omega}{u_\omega^2} + \frac{u_a du_a}{u_\omega^2} = 0$ (see Equation 3-7)

or $\frac{dr}{r} + \frac{du_\omega}{u_\omega} + \frac{u_a du_a}{u_\omega^2} = 0$

By using $u_\omega = u_b - u_a \tan\alpha$ either u_ω and du_ω or u_a and du_a can be substituted to convert equation (3-7) into a differential equation connecting either r and u_ω or r and u_a. If we want to get rid of u_ω and du_ω then $du_\omega = du_b - \tan\alpha du_a - u_a d\tan\alpha$, or, since $u_b = \omega r$ and $\tan\alpha = \frac{\omega r - u_\omega}{u_a}$, with a certain amount of juggling, equation (3-7) can be converted into a complicated differential equation connecting r and u_a. It helps to note from Figure 13 that $du_a \tan\alpha = -du_\omega$.

No attempt is made here to obtain the differential equations which connect r and u_ω or r and u_a, but one can arrive at the following equation:—

$$\delta u_a = \frac{u_\omega}{u_a} \frac{(u_a \delta \tan\alpha - u_\omega \frac{\delta r}{r} - \delta u_b)}{1 - \frac{u_\omega}{u_a}\tan\alpha} \quad (21\text{-}13)$$

which, for the design blade speed, can be used to find the variation of u_a, u_ω, etc., if sufficiently small intervals of δr are chosen. Tan α, of course, is determined by design condition at each value of r.

Figure 14 illustrates the results of an iterative process applied to a case where the root

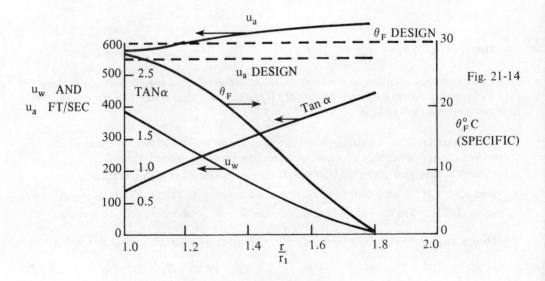

Fig. 21-14

blade speed $\left(\frac{r}{r_1} = 1.0\right)$ was 800'/sec. and the design condition was for a 30°C fan with an exhaust axial velocity of 550'/sec. at all radii; i.e. a free vortex variation of whirl velocity.

This case is somewhat extreme but possibly underestimates the effect of an increase of Q since no allowance was made for the density changes which would result from the variation of u_a and u_ω across the radius. A reduction of θ_F implies a reduction of pressure ratio and both this and the increase of u_a would reduce density, Figure 14 should therefore be regarded as very approximate, nevertheless, it illustrates the drastic effect on θ_F. As may be seen, an increase of u_a from 550'/sec. to 631'/sec. at $\frac{r}{r_1} = 1.4$ causes θ_F to fall off from 28.4° at $\frac{r}{r_1} = 1.0$ to zero at $\frac{r}{r_1} = 1.8$ approx. Beyond $\frac{r}{r_1} = 1.8$, u_ω becomes negative. i.e., the rotor would be producing torque thereby reducing the net driving torque.

Figure 14 makes no allowance for blade losses and assumes that the angle of flow at exit conforms to the blade exit angle whereas, in practice, secondary flows within the blade channel could cause some flow deviation.

An increase of u_a from 550 to 631'/sec. implies an increase of Q less than, but of the order of 14.7% which is far beyond the increase of 5.4% found for the cycle of Figure 12 as compared with that of Figure 11.

Figures 13 and 14 explain why the pressure ratio falls off so drastically with only small increase of mass flow from design, even with a single stage. The effect is 'compounded' with multi stage compressors.

In Figure 13, AD represents the angle of flow relative to the stator blade entry at design while AD' shows the change in this which would result from increasing u_a from 550 to 600'/sec. As may be seen the change is considerable and could very well be more than enough to cause a flow breakaway within the stator blade channels and greatly reduce the pressure recovery within them.

The drastic effect of flow changes can be mitigated by using guide vanes ahead of the rotor to give a whirl in the direction of rotation at the expense of a reduction in work capacity for a given blade speed.

What about the effect of a change of speed? Fortunately, if u_a is reduced in proportion to blade speed then, neglecting the small change in density which would result, exit blade angles remain unchanged to give free vortex flow with constant axial velocity. So, if Q is closely proportional to r.p.m. the flow is compatible with design and θ_F is proportional to the square of r.p.m.

At this point it is worth pointing out that centrifugal compressors with back swept blades do not suffer from the complexities of axial flow compressors because the flow at rotor exit would be substantially uniform across the blade tips. Unfortunately their size and stress considerations prevent their use in aircraft, except, possibly, as the final stage of a compound compressor.

Reverting to the design case of Example 1 and Figure 9, and remembering that for 'correct' core compressor flow T_M should be a constant multiple of the total temp. at compressor entry $(= T_F)$ then $\frac{T_M}{T_F}$ should be constant, i.e., since $T_F = T_o + \theta_u + \theta_F$, $\frac{T_M}{T_o + \theta_u + \theta_F}$ should be constant.

Thus, in Example 1, $\frac{1,456}{318.5} = 4.57$ ∴ the core compressor would be suited to all flight conditions with $T_m = 1,456$ if $\frac{1,456}{T_o + \theta_u + \theta_F} = 4.57$. e.g., for a speed of 600'/sec. ($\theta_u = 16.7$) at a height where $T_o = 248°K$ (about 20,000') then $T_o + \theta_u + \theta_F$ should be 318.5 ∴ $\theta_F = 318.5 - 16.7 - 248 = 53.8°$. This implies that the fan speed should be $\sqrt{\frac{53.8}{60}} \times$ design, i.e., 95% of design r.p.m.

Referring to the cycle of Figure 12, it will be noted that the temp. ratio across the final nozzle t_{JH} is well below the critical ratio 1.2 so there is no thermodynamic lock since the final nozzle is not choking. The same is true for moderate speeds and heights. Thus, in the immediately preceding case of u = 600'/sec. $T_o = 248°K$ and $\theta_F = 53.8$ and with $\lambda = 5.0$, t_{JH} works out at 1.149.

We have now reached the point where it is possible to calculate the full performance of an aircraft designed to cruise at maximum lift drag ratio in the stratosphere and powered by high bypass ratio turbo-fans. This lengthy process will not be attempted here, but a number of points which must be borne in mind are as follows:—

1) For take-off and initial climb, T_M may be allowed to be higher than for continuous climb and cruise.

2) On the climb the indicated air speed may be limited by gust load considerations (400 knots I.A.S. in the case of Concorde up to about 30,000').

3) Fuel consumed on the climb decreases the weight (as it also does for cruise).

4) The height gained on climb is a gain of potential energy which will be partially recovered on descent. ("partially" because of weight reduction by the fuel consumed).

5) The form drag is proportional to I.A.S. and the induced drag (or 'lift drag') is inversely proportional to I.A.S. for subsonic aircraft.

6) The optimum rate of climb (subject to the above mentioned gust load limitations) is that which yields the best m.p.g. taking into account that the potential energy gained on the climb is partially recoverable on descent.

7) The fuel required for stand off and/or diversion, i.e., reserve fuel, depends to a large extent on air traffic control requirements which are rarely of the kind best suited to the aircraft.

8) Fuel must be allowed for taxiing and awaiting clearance for take-off (which can be quite lengthy if, due to weather and other things, there is a large number in line at and near the runway threshold. On more than one occasion the writer has been a passenger when the airplane had to return to take on more fuel after an excessively long delay. One would expect

that this sort of thing may one day be a thing of the past with improvements in airport layouts and ground control).

9) The slow climb during cruise as fuel is consumed, needed for optimum range, may not be allowed by air traffic control regulations. (Fortunately this restriction—so far— does not apply to S.S.T.s flying well above the main traffic lanes).

10) Headwinds and/or the longer flight path of 'weather pattern' flying must be allowed for. (Again S.S.T.s are less 'vulnerable' to these sources of extra fuel consumption).

For the engine to be consistent with design at flight speeds and heights other than those on which the design is based and operating at design T_M, the following rules apply:—

1) θ_c = constant = that for design *if the compressor turbine is choking*.

2) t_T = constant = that for design (which follows from 1).

3) $\dfrac{T_M}{T_F}$ = constant *if the cold nozzle is choking*.

4) $(\lambda + 1)\dfrac{Q}{N_F}$ = constant (where N_F = fan r.p.m.)

5) Subject to 4) λ should be such that $u_{JC} = \sqrt{\eta_F \eta_c}$ approx.

As has been shown there are conditions where rules 1–5 cannot simultaneously apply, e.g., take-off and initial climb where it is desirable to increase T_M. If T_M is different from design then the following additional rules apply:—

6) $\dfrac{T_M}{\theta_c}$ = constant = that for design *if the compressor turbine is choking*.

7) t_T = constant = that for design (which follows from 6).

Though it is not proposed to attempt a full set of performance calculations, it seems desirable to illustrate the application of the foregoing rules. For this purpose the design cycle of Example 1 will be taken for an engine to produce 10,000 lbs. thrust for an aircraft cruising at a lift/drag ratio of 18 (i.e., the weight per engine = 180,000 lbs.) at a height where the atmospheric pressure is 392 p.s.f. (about 40,000′).

To avoid reference back, the following figures from the example are repeated as follows:—
$u = 900'/\text{sec.}$ ($\theta_u = 37.5°C$); $T_o = 221°K$;

For the ram compression $\eta_u = 1.0$, $t_u = 1.170$ and $T_u = 258.5$

For the fan:— $\theta_F = 60°$, $t_f = 1.204$, $T_F = 318.5$, $\eta_F = .88$ and $\lambda + 1 = 6.0$

For the core compressor:— $\theta_c = 450$, $t_c = 2.201$, $T_c = 768.5$ and $\eta_c = .85$

For the core compressor turbine:— $T_M = 1456$ $\theta_T = 450°C$, $t_T = 1.551$ and $\eta_T = .87$

For the fan turbine:— $T_T = 1006$, $\theta_{TF} = 360°C$ $t_{TF} = 1.66$ and $\eta_{TF} = .9$

For the hot jet:— $T_{TF} = 646°K$, $\theta_{JH} = 107.3$, $t_{JH} = 1.204$, $u_{JH} = 1522.7'/\text{sec.}$ and $\eta_{JH} = .98$

For the cold jet:— $T_F = 318.5$, $\theta_{JC} = 90.5$, $t_{JC} = 1.409$, $u_{JC} = 1399'/\text{sec}$. and $\eta_{JC} = .98$

These figures gave $F_s = 96.79$ secs. (per unit core flow).

For the purpose of the following examples some further assumptions are necessary, namely 1) at entry to the fan the axial velocity $u_a = 550'/\text{sec}$.

2) after descent to 5000' (after the end of cruise) the aircraft weight is 70% of that at beginning of cruise, i.e., $.7 \times 180,000 = 126,000$ lb/engine

3) at 5000', $T_o = 278°K$ and atmospheric pressure $= 1,761$ p.s.f.

We also need to know:— 4) Q for the design condition, 5) the annulus area S at fan entry, 6) the overall pressure ratio for the design condition.

For 4) $Q = \dfrac{10,000}{96.79} = 103.1$ ∴ the fan flow $= 6 \times 103.1 = 618.7$ lb/sec.

For 5) the atmospheric density $\rho_o = \dfrac{392}{96 \times 221} = .01848$; the temp. equivalent of $550'/\text{sec}. = 14°C$ ∴ density ρ_a at fan entry $= \left(\dfrac{T_u - 14}{T_u}\right)^{2.5} \times .01848 \times \left(\dfrac{T_u}{T_o}\right)^{2.5}$

∴ $\rho_a = \left(\dfrac{258.5 - 14}{221}\right)^{2.5} \times .01848 = .0238$ ∴ $S = \dfrac{(\lambda+1)Q}{u_a \rho_a} = \dfrac{618.7}{.0238 \times 550} = 47.3$ ft²

For 6) the overall pressure-ratio $p = (t_u t_F t_c)^{3.5} = (1.17 \times 1.204 \times 2.201)^{3.5} = 52.48$

For the calculation of drag we assume that maximum lift/drag-ratio occurs when form drag equals induced (i.e. lift) drag and that form drag is proportional to ρu^2 and induced drag is inversely proportional to ρu^2 (neglecting high Mach No. increase of form drag)

i.e. $D = k_1 \rho u^2 + \dfrac{k_2}{\rho u_2}$. For the design condition assumed, $k_1 \rho u_2 = \dfrac{k_2}{\rho u^2} = 5000$ lb

and $\rho u^2 = .01848 \times 900^2 = 14,969$, ∴ $k_1 = \dfrac{5,000}{14,969} = .334$ and $k_2 = 5,000 \times 14,969 = 74.845 \times 10^6$ ∴ $D = .334 \rho u^2 + \dfrac{74.845 \times 10^6}{\rho u^2}$ (flight at speeds less than max L/D is 'unstable' in the sense that any drop in speed results in increased drag which causes a further drop in speed. This difficulty, however, can be avoided when necessary by the use of spoilers, wing flaps, etc.).

Example 21.4.

What should be the engine cycle conditions for steady cruise at $500'/\text{sec}$. air speed at a height of 5000', using the assumptions, etc. from above?

The answer to this could be dealt with by tabulation, but the writer feels that the reader will find it easier to follow by cycle diagrams.

At 5000', $\rho = \dfrac{1761}{96 \times 278} = .066$ and at 500'/sec. $\rho u^2 = 16{,}500$

$\therefore D = .334 \times 16{,}500 + \dfrac{78.845 \times 10^6}{16{,}500} = 5{,}511 + 4{,}778 = 10{,}289$ lbs. (thus, as the induced drag component is the smaller, 500'/sec. is higher than minimum drag speed).

Case 1. $T_M = 1456$. This being the same as design, T_F should also be the same, i.e., $T_F = 318.5$. $\theta_u = \left(\dfrac{500}{147}\right)^2 = 11.6°C$ $\therefore T_u = 278 + 11.6 = 289.6°K$ so $\theta_F = T_F - T_u = 318.5 - 289.6 = 28.9°C$.

The cycle becomes as shown in Fig. 15.

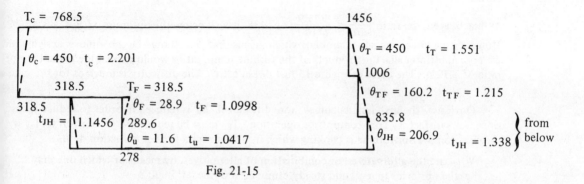

Fig. 21-15

The overall temp. ratio = $1.0417 \times 1.0998 \times 2.201 = 2.5214$ $\therefore$ pressure ratio = 25.45

$\therefore$ since T_M is unchanged, $Q = \dfrac{p_o}{p_{oD}} \times \dfrac{25.45}{52.48} \times Q_D$. At 5,000', $p_o = 1{,}761$ p.s.f.

$\therefore Q = \dfrac{1761}{392} \times \dfrac{25.45}{52.48} \times 103.1 = 224.6$ lb/sec.

θ_F having dropped from 60° to 28.9°, the fan speed should reduced in the ratio $\sqrt{\dfrac{28.9}{60}} = .694$ $\therefore$ the axial velocity u_a should be $.694 \times 550 = 382'$/sec. for which

$\theta_{ua} = \left(\dfrac{382}{147}\right)^2 = 6.74°C$ $\therefore \rho_a = \left(\dfrac{289.6 - 6.74}{278}\right)^{2.5} \times \rho_o = 1.0443\, \rho_o$

At 5,000', $\rho_o = \dfrac{1{,}761}{96 \times 278} = .066$ $\therefore \rho_a = .069$ $\therefore (\lambda + 1)Q = \rho_a u_a S = .069 \times 382 \times 47.3 = 1{,}245$ and, since $Q = 224.6$, $\lambda + 1 = \dfrac{1{,}245}{224.6} = 5.54$. Hence $\theta_{TF} = 5.54 \times 28.9 = 160.2$

and $t_{TF} = \dfrac{1,006}{1,006 - \dfrac{160.2}{9}} = 1.215$ ∴ $t_{JH} = \dfrac{t_u t_F t_c}{t_T t_{FT}} = \dfrac{1.0417 \times 1.0998 \times 2.201}{1.551 \times 1.215} = 1.338$

from which $\theta_{JH} = .98 \times 835.8 \left(\dfrac{.338}{1.338}\right) = 206.9°C$, ∴ $u_{JH} = 147\sqrt{206.9} = 2,114.5'/\text{sec}$

$$\theta_{JC} = 318.5 \times .98 \left(\dfrac{.1456}{1.1456}\right) = 39.7°C, \therefore$$

$u_{JC} = 147\sqrt{39.7} = 925.9'/\text{sec}$.

$F_s = \dfrac{1}{g}[\lambda \times (925.9 - 500) + 2114.5 - 500]$ which, since $\lambda = 4.54$, gives $F_s = 110.2$ sec

∴ Thrust $= QF_s = 224.6 \times 110.2 = 24,748$ lbs.

This is well over twice the drag (= 10,289) so the aircraft would climb or accelerate or both.

As may be seen, the ratio $\dfrac{u_{JC}}{u_{JH}} = \dfrac{925.9}{2114.5} = .438$ is well below the optimum $\sqrt{.9 \times .88} = .89$. Moreover the hot jet nozzle is choking which means that θ_{TF} should be the same as design which isn't possible (inter alia, the products of the turbine temp. ratios would exceed the overall temp. ratio of 2.521). The cycle of Figure 15 just doesn't 'fit'. The difficulty is much as for the take-off cycle.

Obviously the matching becomes more difficult if T_M is reduced in order to reduce thrust, since, for satisfactory core compressor operation, T_F must be reduced in proportion to T_M if the core compressor turbine is choking which in turn means a further reduction of θ_F.

What are the alternatives, or combination of alternatives, by means of which one may 'escape' the obstacles to level and steady cruise at 500'/sec. at 5,000'?

They are one or more of the following:—

1) Spoilers and/or flaps may be used to increase the drag.

2) Core compressor efficiency can be reduced by operating well down the pressure ratio—mass flow parameter characteristic thus reducing overall pressure ratio and therefore Q.

3) λ may be increased by opening up the final nozzle—at the expense of fan efficiency

4) The link between T_M and T_F may be broken by using a core compressor turbine which is non-choking, i.e., has two or more stages in which there is no choking in either nozzle or rotor blades.

5) Ditto for the fan turbine.

6) The use of variable intake guide vanes (i.e. stator blades) ahead of the fan first stage rotor to reduce the work capacity of the fan without seriously affecting its efficiency.

7) The use of a two spool core compressor with the first stage coupled to the fan (as in the J.T. 9D).

8) The use of a two spool core compressor mechanically independent of the fan, i.e., a three spool engine (as in the R-R R.B.211).

9) The bypassing of a portion of the core compressor delivery so that the mass flow rate in expansion is less than Q. i.e. using a high pressure bypass in addition to the low pressure bypass. ('Bleeding' the core compressor to provide turbine blade cooling is a step in this direction, but becomes less necessary for reduced T_M).

Of these alternatives, 1), 2), and 3) are very undesirable on the grounds of excessive fuel consumption. 9) is also undesirable for the same reason plus the considerable mechanical complication it would entail.

Of the remaining five, various combinations need to be considered. In the writer's opinion, 6) i.e., the use of variable intake guide vanes, is the best single method whether or not one or more of 4), 5), 7) and 8) are also used.

Intake guide vanes (again in the writer's opinion) are desirable as at least a partial protection against bird strikes—one of the most common hazards, especially at take-off and landing. Their use is often opposed on the grounds of fan noise due to the 'siren effect' as rotor blades pass through their wakes, and the need for anti-icing means.

Strictly speaking, for the energy increase to be uniform from root to tip, the intake guide vanes should be shaped for free vortex flow in the direction of rotor rotation so as to give a constant value of $u_b \times \Delta u_\omega$ but this refinement is scarcely necessary. Variable intake guide vanes are so potent in reducing work capacity that the deflection needed is very small. For example, a 30° fan with a mean u_b of 1200'/sec. must produce a whirl of

$$\frac{30 \times 336 \times 32.2}{1200} = 270.5'/\text{sec.}$$ at mean radius without inlet guide vanes. If it is desired to reduce θ_F to 20° then $\frac{\Delta u_\omega \times 1,200}{32.2 \times 336} = 20$ from which $\Delta u_\omega = 189.3'/\text{sec}$. Thus, the inlet guide vanes must generate a pre-whirl of $270.5 - 180.3 = 90.2'/\text{sec}$. at mean radius. With an axial velocity of, say, 550'/sec. the exit angle (of the guide vanes) is

$\tan^{-1} \frac{90.2}{550} = 9.3°$. At this small angle a constant angle produces a whirl very little different from a free vortex. It is found that over a radius range of 1 to 2 (root to tip) the variation in θ_F is about from 20.1° at the root to 19.96 at the tip.

The writer has not examined the matter but there *might* be a case for having guide vane exit angles which would generate a whirl greater than that of a free vortex at the root and less at the tip so that the fall off of θ_F from root to tip with increase of Q, i.e., of axial velocity would be far less drastic than the sort of thing illustrated in Figure 14. On the other hand, any variation in θ_F must take the form of a variation in axial velocity with axial discharge from the first stage stators. Thus the increased axial velocity, at the root at entry to a second stage, would reduce θ_F at the root of the second stage so that the distribution of θ_F in the second stage would tend, to some extent at least, to compensate for the distribution of θ_F for the first stage.

Case 2. Having got nowhere in Case 1 in meeting the requirements for steady cruise at 500'/sec. at 5000', beyond demonstrating that the result is unsatisfactory if the design cycle of Example 1 has the built-in 'locks' between T_M and T_F, between T_F and T_u, etc., it is now necessary to look at the problem with those locks avoided by one or more of the methods discussed in Case 1.

Referring to the design cycle of Example 1 and the cycle of Figure 9 it may be seen that choking in the compressor turbine is unavoidable in a single stage turbine with an energy drop equivalent to 450°C. A single stage turbine *could* be designed for this having a large residual whirl after the rotor which contributed to the power of the fan turbine, but only at the expense of choking in the rotor blading, whether or not there was choking in the nozzle blading. It is therefore clearly desirable to use a multi-stage core compressor turbine, in which there is no choking either in nozzle or rotor blading. The same applies even more to the fan turbine with its design temp. ratio of 1.66 which is, in fact, greater than that of the core compressor turbine (t_T = 1.551). Having regard to the fact that the choking velocity is proportional to the square root of the total temperature relative to the blading, even a two stage turbine would probably be marginal from the choking point of view for a specific energy drop equivalent to 450°C. Without examining the matter in detail, the writer suspects that the blade speed necessary to avoid choking would be excessive and the blading and rotor disc would be overstressed.

With three stage core compressor and fan turbines, choking can be avoided in all stators and rotors and the design becomes much more 'flexible', especially as there are no serious problems in providing variable jet nozzles for both the core and bypass jets.

After two quick 'tryouts' it is found that the cycle of Figure 16 meets the requirement specified, namely a thrust of 10,290 lbs. at 500'/sec. at 5000'.

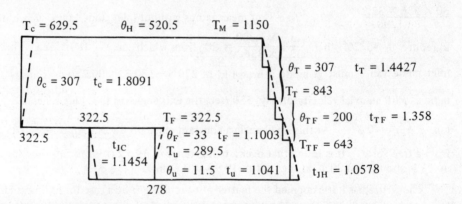

Fig. 21-16

For the cycle of Figure 16 the same component efficiencies as those in Example 1 were used, namely:— η_u 1.00, η_F = .88, η_c = .85, η_T (for the two stages) = .87, η_{TF} (for the two stages) = .9, η_{JH} = η_{JC} = .98.

$$\lambda + 1 = \frac{\theta_{TF}}{\theta_F} \text{ so } \lambda \text{ is seen to be 5.061}$$

From $\theta_{JH}' = .98 \times 643 \left(\frac{.0578}{1.0578}\right)$, θ_{JH} = 34.4 ∴ u_{JH} = 147.1 $\sqrt{34.4}$ = 863'/sec.

From $\theta_{JC} = .98 \times 322.5 \left(\frac{.1454}{1.1454}\right)$, θ_{JC} = 40.1 ∴ u_{JC} = 147.1 $\sqrt{40.1}$ = 931.7'/sec.

$$\therefore F_s = \frac{1}{g} [\lambda(931.7 - 500) + 863 - 500] \text{ and, for } \lambda = 5.061, F_s = 79 \text{ secs.}$$

The pressure ratio of the design cycle was 52.48; the pressure ratio for the cycle of Figure 16 is given by $p = (t_u t_F t_c)^{3.5} = (1.041 \times 1.1003 \times 1.8091)^{3.5} = 12.81$.

The atmospheric pressure for the design cycle (P_{oD}) was 392 p.s.f. and P_o for 5,000' is 1,761 p.s.f. $\therefore \frac{pP_o}{p_D P_{oD}} = \frac{P_{max}}{P_{maxD}} = \frac{12.81}{52.48} \times \frac{1761}{392} = 1.0966$, i.e. the cycle of Figure 16 at 5,000' has virtually the same maximum pressure as the design cycle at 40,000'.

If one may assume that one of the temp. ratios in expansion is sufficiently near the choking value of 1.2 so that $Q \propto \frac{P_{max}}{\sqrt{T_M}}$ then $\frac{Q}{Q_D} = \frac{P_{max}}{P_{maxD}} \sqrt{\frac{T_{MD}}{T_M}} = 1.0966 \sqrt{\frac{1,456}{1,150}} = 1.234$

Q_D for the design cycle was found to be 103.1 lb/sec/lb of core flow $\therefore$ for this case $Q = 1.234 \times 103.1 = 127.2$ lb/sec. $\therefore$ Thrust $= QF_s = 127.2 \times 79 = 10,050$ lbs.

This is a little short of the required 10,290 but is near enough considering the element of uncertainty in the assumptions, especially in the case of Q which, in practice, would be somewhat greater with the temperature ratios indicated in Figure 16 for two 2 stage turbines in series.

The cycle of Figure 16 is not necessarily the optimum for the conditions. For example, the hot jet velocity u_{JH} is lower than the cold jet velocity u_{JC} whereas it should be higher.

The overall efficiency η_o is given by $\frac{F_s u}{K_p \theta_H}$ $\therefore \eta_o = \frac{79 \times 500}{336 \times 520.5} = .226$, which is very reasonable for the low speed and height. It could probably be improved by juggling with θ_F, θ_c, λ and T_M but not by any significant amount. In any event the flight condition corresponds to flight in a holding pattern pending landing clearance. For diversion to an alternate airfield it would pay (subject to air traffic control) to open up and climb for approximately half the distance.

A further look at Figure 16 indicates that fan speed would be 74% of design and core assembly speed about 83% of design (N being approx. proportional to $\sqrt{\theta}$) while the reduction of volumetric flow is greater. Variable stators for both fan and core compressor would assist in obtaining better matching between blade speeds and mass flow rate. A full investigation into this would require a knowledge of fan and core compressor characteristics for which testing would be necessary. In all probability it would be found necessary to run at points on the characteristics well below those for optimum efficiency of fan and core compressor. To that extent the cycle of Figure 16 is altogether too optimistic.

Low Bypass Turbo Fans with Exhaust Mixing

This type of engine, illustrated diagrammatically in Figure 17, is likely to become of great importance for S.S.T.s and other supersonic aircraft flying at Mach Nos. up to about 2.5 (above which 'straight' jets and even ram jets begin to be competitive except for take-off, acceleration and climb).

λQ
MIXING
Q
(λ + 1)Q
(λ + 1)Q
u
uJ
FLUTES
'CORE'

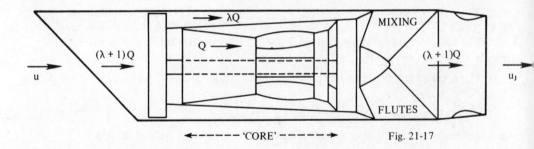

Fig. 21-17

The core engine is much the same as for the high bypass engine, but the bypass flow λQ rejoins and mixes with the core flow immediately aft of the fan turbine instead of discharging through an independent 'cold' nozzle.

It follows that the static pressures of the bypass and core flows must be equal, or very nearly equal, at exhaust from the fan turbine.

In deciding on the value of λ, the designer must weigh the 'pros and cons' of several factors including:–

1) effect on wave drag
2) effect on nacelle skin friction
3) effect on power plant weight
4) the effectiveness of mixing.

In theory, neglecting nacelle drag, etc., the efficiency of any turbo-fan engine would benefit by mixing the core and bypass flows before final expansion, but, in practice, adequate mixing becomes increasingly difficult with increase of λ.

The cycle diagram for this type of engine is shown in Figure 18.

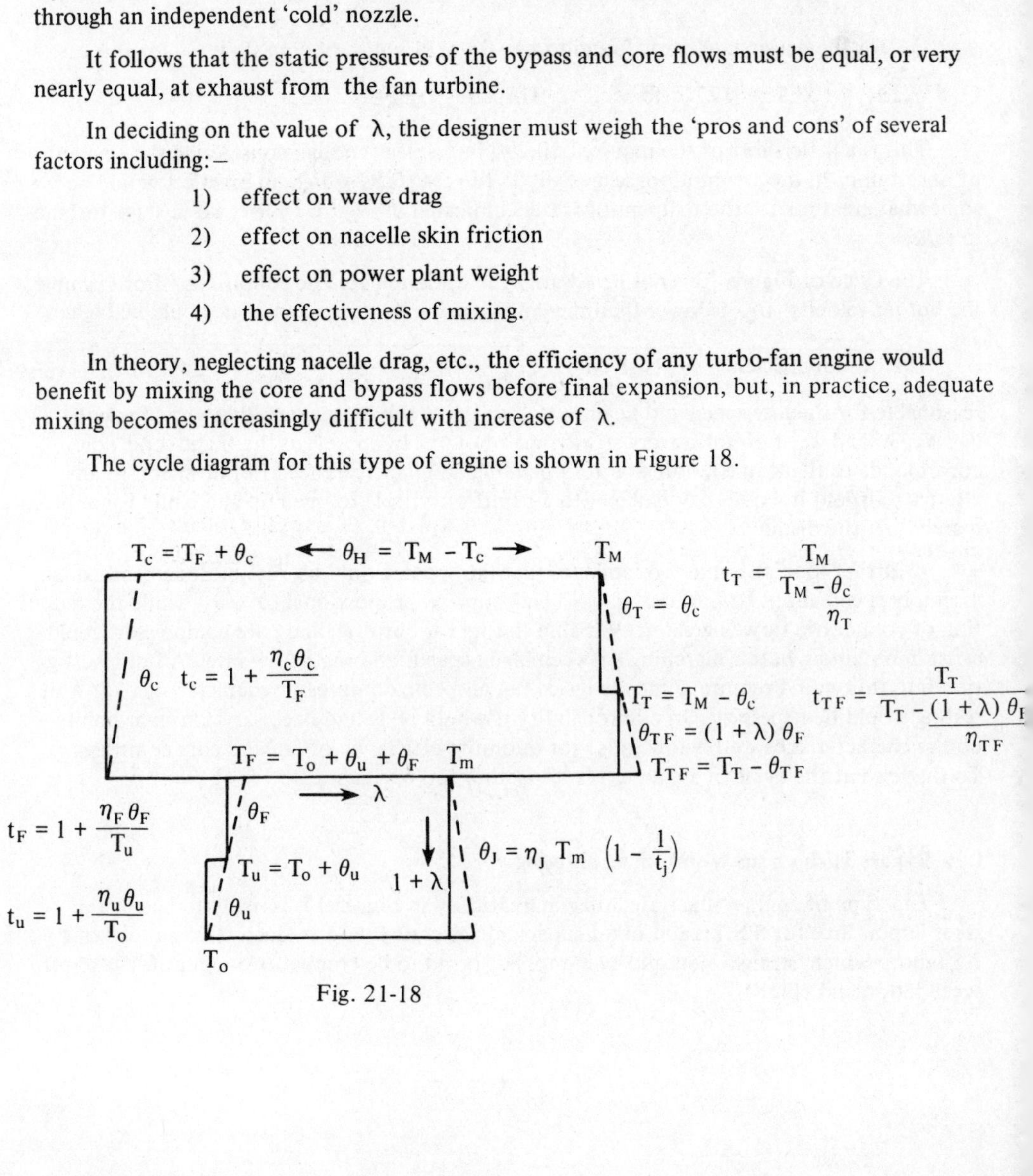

$T_c = T_F + \theta_c$
$\theta_H = T_M - T_c$
T_M
$t_T = \dfrac{T_M}{T_M - \dfrac{\theta_c}{\eta_T}}$
$\theta_T = \theta_c$
θ_c
$t_c = 1 + \dfrac{\eta_c \theta_c}{T_F}$
$T_T = T_M - \theta_c$
$t_{TF} = \dfrac{T_T}{T_T - \dfrac{(1+\lambda)\,\theta_F}{\eta_{TF}}}$
$\theta_{TF} = (1 + \lambda)\,\theta_F$
$T_F = T_o + \theta_u + \theta_F$
T_m
$T_{TF} = T_T - \theta_{TF}$
λ
θ_F
$t_F = 1 + \dfrac{\eta_F \theta_F}{T_u}$
$\theta_J = \eta_J\, T_m \left(1 - \dfrac{1}{t_j}\right)$
$T_u = T_o + \theta_u$
1 + λ
$t_u = 1 + \dfrac{\eta_u \theta_u}{T_o}$
θ_u
T_o

Fig. 21-18

The constant pressure line through T_F to T_{TF} is a total pressure line. More accurately it should be a static pressure line, but the final temp. ratio t_j is so high at supersonic speeds that any error due to this inaccuracy is found to be negligible in relation to other assumptions.

In this compound cycle, as before, λ is the bypass ratio and T_m (not to be confused with T_M) is the total temp. after mixing of the bypass and core flows.

In addition to the relationships shown in the diagram, several others may readily be seen, e.g.:—

$$T_m = \frac{\lambda T_F + T_{TF}}{\lambda + 1} \tag{21-14}$$

$$t_J = t_u t_F \tag{21-15}$$

$$t_c = t_T t_{TF} \quad \text{or} \quad t_{TF} = \frac{t_c}{t_T} \tag{21-16}$$

From Equation (16) and Figure 18 it may be seen that

$$\frac{T_T}{T_T - \frac{(1+\lambda)\theta_F}{\eta_F}} = \left(\frac{T_M - \frac{\theta_c}{\eta_T}}{T_M}\right)\left(1 + \frac{\eta_c \theta_c}{T_u + \theta_F}\right)$$

or

$$\frac{1}{1 - \frac{(1+\lambda)\theta_F}{\eta_F T_T}} = \left(1 - \frac{\theta_c}{\eta_T T_M}\right)\left(1 + \frac{\eta_c \theta_c}{T_u + \theta_F}\right) \tag{21-17}$$

From (17), if T_u (i.e. $T_o + \theta_u$), θ_c, T_M and the various efficiencies are known, θ_F may be found if λ is predetermined and vice versa. If λ is given, θ_F can be found as the solution of a somewhat untidy quadratic, but if θ_F is given, λ is readily calculated as indicated in the table below for u = 2,200'/sec. (i.e. θ_u = 223.7) T_o = 220°K; T_M = 1650°K; θ_c = 450°C; η_u = .95; η_F = .88; η_c = .85; η_T = .87 and η_{FT} = .89.

From this data $\frac{1}{t_T}$ = 0.687; T_T = 1200°K and T_u = 443.7°K.

θ_F	T_F	t_c	t_{TF}	θ_{FT}	$\lambda + 1$	λ
60	503.7	1.759	1.209	184.4	3.07	2.07
70	513.7	1.745	1.199	176.9	2.53	1.53
80	523.7	1.730	1.189	169.6	2.12	1.12
90	533.7	1.717	1.179	162.4	1.80	0.80

As may be seen, λ drops very rapidly with increase of θ_F. (It is desirable to obtain a relationship between θ_F and λ in any case).

An example of a cycle for θ_F = 83°C using the values of u, θ_c, T_M, etc. given above is shown in Figure 19.

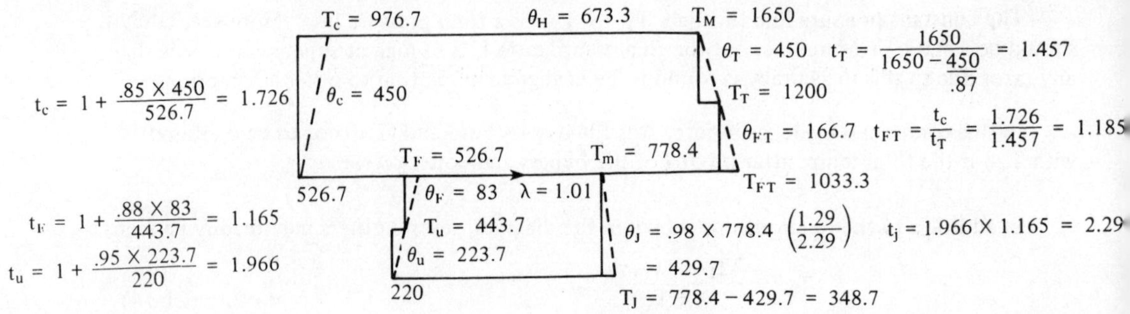

$$t_c = 1 + \frac{.85 \times 450}{526.7} = 1.726$$

$$t_F = 1 + \frac{.88 \times 83}{443.7} = 1.165$$

$$t_u = 1 + \frac{.95 \times 223.7}{220} = 1.966$$

$T_c = 976.7 \qquad \theta_H = 673.3 \qquad T_M = 1650$

$\theta_c = 450$

$T_F = 526.7 \qquad T_m = 778.4$

$\theta_F = 83 \quad \lambda = 1.01$

$T_u = 443.7$

$\theta_u = 223.7$

526.7

220

$\theta_T = 450 \quad t_T = \dfrac{1650}{1650 - 450} = 1.457$

$T_T = 1200 \qquad \qquad .87$

$\theta_{FT} = 166.7 \quad t_{FT} = \dfrac{t_c}{t_T} = \dfrac{1.726}{1.457} = 1.185$

$T_{FT} = 1033.3$

$\theta_J = .98 \times 778.4 \left(\dfrac{1.29}{2.29}\right) \quad t_j = 1.966 \times 1.165 = 2.29$

$= 429.7$

$T_J = 778.4 - 429.7 = 348.7$

Fig. 21-19

It will be noted that Figure 19 includes the additional assumption that $\eta_J = .98$.

$\lambda = 1.01$ is obtained from $\theta_{FT} = (\lambda + 1)\theta_F$ i.e. $\lambda + 1 = \dfrac{166.7}{83} = 2.01$

T_m is obtained from $\dfrac{1.01 \times 526.7 + 1{,}033.3}{2.01} = 778.4$

t_J is obtained from $t_j = t_u t_F = 1.966 \times 1.165 = 2.290$

Heat balance check:— The jet energy equivalent = $\theta_J \times 2.01 = 863.7$

The ram energy equivalent = $\theta_u \times 2.01 = 449.6$

∴ the equivalent of the energy increase is $863.7 - 449.6 = 414.1$

The equivalent of the heat rejected is $2.01(348.7 - 220) = 258.7$

$414.1 + 258.7 = 672.8$ c.f. $\theta_H = 673.3$. The agreement would have been exact if a greater number of significant figures had been used.

The jet velocity $u_J = 147.1\sqrt{429.7} = 3{,}049'/\text{sec}$. ∴ the specific thrust (per lb/sec of core flow) $F_s = \dfrac{2.01}{g}(3{,}049 - 220) = 53$ secs.

Overall efficiency $\eta_o = \dfrac{F_s u}{K_p \theta_H} = \dfrac{53 \times 2{,}200}{336 \times 673.3} = .515$

This type of engine is thus seen to be very efficient at flight Mach Nos. of about 2.2.

The thermal efficiency $\eta_{th} = \dfrac{(\lambda + 1)(\theta_J - \theta_u)}{\theta_H} = \dfrac{2.01(429.7 - 223.7)}{673.3} = .612$

The propulsive (or Froude) efficiency $\eta_F = \dfrac{2}{\frac{u_j}{u}+1} = \dfrac{2}{\frac{3049}{2200}+1} = .838$.

$\eta_{th} \times \eta_P = \eta_o = .838 \times .612 = .513$ which substantially agrees with the figure found from $\eta_o = \dfrac{F_s u}{K_p \theta_H}$.

The improvement in overall efficiency obtained by mixing before final expansion is in the range 4 – 5 %. This may not seem impressive but it could mean a possible increase in payload (for a given range) of the order of 30%.

Clearly such high efficiencies go a long way to offsetting the low lift/drag ratios of supersonic aircraft.

In Figure 19, t_j is seen to have the high figure of 2.29 so one might expect that it would pay to increase T_m by additional heating in the bypass ('duct burning') or by after burning but, though F_s could be increased very substantially it can easily be shown that overall efficiency would be reduced. For example, if, in the cycle of Figure 19, T_m is increased to 1000°K, the equivalent of the extra heat addition is $2.01 (1000 - 778.4) = 445.4$ to give a total heat energy addition of $445.4 + 673.3 = 1118.7$. θ_j would be $.98 \times 1000 \left(\dfrac{1.29}{2.29}\right) = 552°C$.

$u_j = 147.1 \sqrt{552} = 3456'/\text{sec}$. and $F_s = \dfrac{3456 - 2200}{32.2} \times 2.01 = 78.42$ secs.

$\therefore \eta_o = \dfrac{78.42 \times 2200}{336 \times 1118.7} = .459$. So F_s is increased from 53 to 78.42 secs. at the expense of a drop in η_o from .515 to .459. Nevertheless, for comparatively short range missions, it may be worthwhile to sacrifice overall efficiency to gain such a large increase in thrust, especially if the increase is needed for temporary boost only. The above reduction in efficiency does not take into account any pressure loss which would result from the additional combustion apparatus, and which would cost some loss in efficiency even when not in use. This loss, however, would be very small.

SECTION 22
Effect of Height and Speed on Performance

For a preliminary look at this question the supersonic engine cycle of Figure 18 of Section 21 will be used for a height where $T_o = 221°K$ since it seems likely that straight jets are a thing of the past for other than high supersonic speeds.

In calculations on this basis, the writer found himself up against an aspect of the subject which was new to him and which must now be explained. It was found that for design speeds of the order of 2000'/sec. where the ram pressure ratio is a high proportion of the total pressure ratio, the axial velocity at compressor entry would become a limiting factor.

As has been stressed earlier, the axial flow u_a at the compressor face should not differ appreciably from a constant proportion of blade speed, i.e. $\dfrac{\theta_a}{\theta_c}$ should be approximately constant. To satisfy this condition it was found that reduced flight speeds meant reduced engine r.p.m. below flight speeds of the order of 1600'/sec., otherwise u_a became much too high to conform with the condition that the mass flow rate Q is proportional to the maximum pressure of the cycle and inversely proportional to the square root of the maximum cycle temp.

Let it be assumed that θ_a, the temp. equivalent of axial velocity u_a at compressor entry, is given by $\theta_a = x\theta_c$. It is, of course, desirable that x should be constant but unfortunately this is not possible for varying flight conditions if θ_c is constant especially for supersonic engines operating at subsonic speeds, as may be seen from the following reasoning:

Considering first the flow at compressor entry; $Q \propto \rho_a u_a$ or, since $u_a = 147.1\sqrt{\theta_a}$ and $\theta_a = x\theta_c$, $\quad Q \propto \rho_a \sqrt{x\theta_c}$ \hfill (22-1)

If ram efficiency is less than 100% then ρ_a, the density at compressor entry, is found as shown in Figure 1.

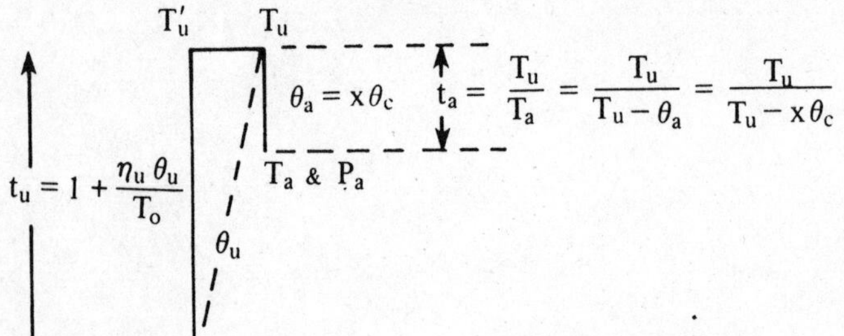

Fig. 22-1

The total to static pressure-ratio $p_u = t_u^{3.5}$. The total to static pressure ratio to give $u_a \ (= 147.1\sqrt{x\theta_c})$ is $t_a^{3.5}$ ∴ the static pressure at compressor entry $P_a = P_o \left(\dfrac{t_u}{t_a}\right)^{3.5}$

$$\therefore \rho_a = \frac{P_a}{RT_a} = \frac{P_o}{RT_a}\left(\frac{t_u}{t_a}\right)^{3.5} \quad \text{or, since } T_a = T_u - x\theta_c, \ \rho_a \propto \frac{1}{T_u - x\theta_c}\left(\frac{t_u}{t_a}\right)^{3.5} \quad (22\text{-}2)$$

Substituting for ρ_a in (1) one may write

$$Q = k_1 \frac{1}{T_u - x\theta_c}\left(\frac{t_u}{t_a}\right)^{3.5} \sqrt{x\theta_c} \quad (22\text{-}3)$$

Now considering Q as governed by the expansion:—as has been shown, $Q \propto \dfrac{P_{max}}{\sqrt{T_{max}}}$ and $T_{max} \propto \theta_c$, ∴ since $P_{max} \propto (t_u t_c)^{3.5}$,

$$Q = k_2 \frac{(t_u t_c)^{3.5}}{\sqrt{\theta_c}} \quad (22\text{-}4)$$

It follows from (3) and (4) that $k_1 \left(\dfrac{1}{T_u - x\theta_c}\right)\left(\dfrac{t_u}{t_a}\right)^{3.5} \sqrt{x\theta_c} = k_2 \dfrac{(t_u t_c)^{3.5}}{\sqrt{\theta_c}}$

or

$$\frac{\theta_c \sqrt{x}}{(T_u - x\theta_c)(t_a t_c)^{3.5}} = \text{const.} \quad (22\text{-}5)$$

using $t_a = \dfrac{T_u}{T_u - x\theta_c}$ (5) becomes $\dfrac{\theta_c \sqrt{x}(T_u - x\theta_c)^{2.5}}{(T_u t_c)^{3.5}} = \text{const.}$ (22-5a)

Example 22-1.

The value of the constant in (5a) can be found by using the design values for the quantities in the equation. Assuming that θ_a is equivalent to Mach 0.6 then $\theta_a = x\theta_c = 0.6\dfrac{T_u}{6} = 0.1\,T$ If other assumptions for the design condition are 1) $T_o = 221$, 2) $u = 2100'/\text{sec.}$; 3) $\eta_u = .95$; 4) $\theta_c = 500$ and 5) $\eta_c = .84$ what is the value of the constant in equation (5a)?

$$T_u = T_o + \theta_u = 221 + \left(\frac{2100}{147}\right)^2 = 425.1°K \quad (\theta_u = 204.1)$$

$$500 \, x = 0.1 \; T_u = 42.51 \quad \therefore \quad x = .085$$

$$t_u = 1 + \frac{\eta_u \theta_u}{T_o} = 1 + .95 \times \frac{204.1}{221} = 1.877$$

$$t_c = 1 + .84 \, \frac{\eta_c \theta_c}{T_u} = 1 + \frac{.84 \times 500}{425.1} = 1.988$$

$$\therefore \quad \frac{\theta_c \sqrt{x} \, (T_u - x\theta_c)^{2.5}}{(T_u t_c)^{3.5}} = \frac{500\sqrt{.085} \, (425.1 - .085 \times 500)^{2.5}}{(425.1 \times 1.988)^{3.5}} = .0238$$

It is clearly undesirable that $x = \frac{\theta_a}{\theta_c}$ should vary much from the design value with axial flow compressors. If it is too small there is a risk that the angle of attack will be reduced to the stalling point and surging of the compressor will occur. It it is too large then the characteristics of an axial flow compressor are such that there is a rapid fall off in pressure ratio and efficiency, i.e. θ_c and η_c are reduced. A reduction in θ_c means a proportionate reduction in T_{max}; which all adds up to reduction in thrust and overall efficiency.

In fact it is impossible to calculate overall performance unless the effect of x on θ_c, η_c, etc. are known. But if x is assumed constant then it is reasonable to calculate the way θ_c etc. vary with height and speed.

Example 22-2.

If the flight speed is 900'/sec. ($\theta_u = 37.5°$) and $T_o = 250°K$ (about 20,000') and $\frac{\theta_a}{\theta_c} = x = .085$ (as in Example 1) what is the corresponding value of θ_c from equation (5a) assuming that η_u and η_c are the same as in Example 1?

Since $t_c = 1 + \frac{\eta_c \theta_c}{T_u}$, θ_c must be found from $\frac{\theta_c \sqrt{.085} \, (250 + 37.5 - .085 \, \theta_c)^{2.5}}{(T_u t_c)^{3.5}} = .0238$

$$T_u = 287.5 \quad \therefore \quad \frac{.292 \, \theta_c \, (287.5 - .085 \, \theta_c)^{2.5}}{\left[287.5 \left(1 + \frac{.84 \, \theta_c}{287.5} \right) \right]^{3.5}} = .0238$$

By trial θ_c is found to be 338°C.

On completing Example 2, the writer made a discovery which was novel to him but which is probably a blinding glimpse of the obvious to a competent mathematician. He was struck by the fact that the value of $\frac{\theta_c}{T_u}$ was the same in both Examples 1 and 2. This proved to be no fluke as will now be shown.

Writing a (a constant) instead of x, equation (5a) may be rewritten

$$\frac{T_u \frac{\theta_c}{T_u} \sqrt{a} \, \left[T_u \left(1 - a \frac{\theta_c}{T_u} \right) \right]^{2.5}}{\left[T_u \left(1 + \frac{\eta_c \theta_c}{T_u} \right) \right]^{3.5}} = \text{Const.}$$

which reduces to
$$\sqrt{a}\,\frac{\frac{\theta_c}{T_u}\left(1 - a\frac{\theta_c}{T_u}\right)^{2.5}}{\left(1 + \eta_c\frac{\theta_c}{T_u}\right)^{3.5}} = \text{Const.} \qquad (22\text{-}5b)$$

from which it is clear that if η_c is constant $\frac{\theta_c}{T_u}$ must be constant and the same as the design value.

Therefore for constant axial velocity Mach number at compressor intake θ_c is a constant multiple of T_u.

Since it has been shown that T_M is a constant multiple of θ_c it follows that T_M *is also a constant multiple of* T_u.

Another curious result is obtained as follows:— Figure 2 illustrates a method of finding how the axial velocity at compressor exit varies.

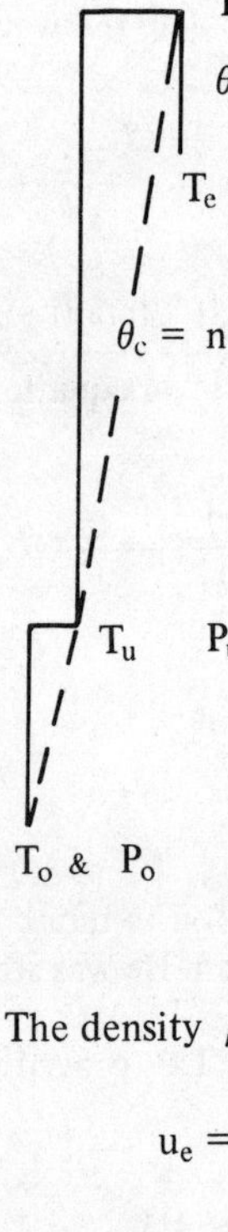

$T_c = T_u(1+n) \qquad P_c = P_o(t_u t_c)^{3.5}$

$\theta_e = xT_c \qquad t_e = \frac{T_c}{T_e} = \frac{1}{1-x}$

$T_e = T_c(1-x) = T_u(1+n)(1-x) \qquad P_e = P_o(t_u t_c)^{3.5}(1-x)^{3.5}$

$\theta_c = nT_u$

Fig. 22-2

$T_u \qquad P_u = P_o t_u^{3.5}$

$T_o\ \&\ P_o$

The density ρ_e at T_e is given by $\rho_e = \frac{P_e}{RT_e}$ and the axial velocity u_e is

$$u_e = 147\sqrt{xT_u(1+n)} \quad \therefore\quad Q = k_1\rho_e u_e = k_2\frac{P_e}{T_e}\sqrt{xT_u}$$

Substituting for P_e and T_e from the diagram and merging constants

$$Q = k_3\frac{P_o(t_u t_c)^{3.5}(1-x)^{3.5}\sqrt{xT_u}}{T_u(1-x)}$$

or $\quad Q = \dfrac{k_3 \, P_o (t_u t_c)^{3.5} \, (1-x)^{2.5} \, \sqrt{x}}{\sqrt{T_u}}$ (22-6)

For expansion, $Q = k_4 \dfrac{P_o (t_u t_c)^{3.5}}{\sqrt{m T_u}}$ (22-7)

(since it has been shown that T_M is multiple of T_u for constant Mach No. for axial velocity at compressor entry and assuming constant η_c)

Equating (6) and (7) gives:—

$$\dfrac{k_3 \, P_o (t_u t_c)^{3.5} \, (1-x)^{2.5} \, \sqrt{x}}{\sqrt{T_u}} = \dfrac{k_4 \, P_o (t_u t_c)^{3.5}}{\sqrt{m T_u}}$$

which reduces to $(1-x)^{2.5} \sqrt{x}$ = const. from which x *must be constant*, i.e. over the range of operation for which the assumptions are valid, *the axial velocity at compressor discharge is a constant multiple of $\sqrt{T_c}$ and therefore of $\sqrt{T_u}$*. Thus axial velocities at both entry and exit remain 'in step' with rotational speed. It is a reasonable presumption that axial velocities throughout the compressor are proportional to rotational speed over the range where mass flow rate Q may be assumed $\alpha \, \dfrac{P_{\max}}{\sqrt{T_{\max}}}$, i.e. over the range where the turbine nozzles are choking or near choking.

Considering Examples 1 and 2, a change of T_u from 425.1 to 287.5°K means a change of θ_C from 500°C to 338°C. The corresponding variation in rotational speed would be in the ratio $\sqrt{\dfrac{338}{500}}$ = .822 — not drastic for the large change in flight conditions.

Evidently the variation in T_u would be far smaller (and hence the variation in θ_c, T_M, etc.) if the design condition had been for high subsonic speed. Thus, for example, if the design condition had been for 900'/sec. at a height where $T_o = 221°K$ then T_u would have been 258.5°K which implies that at sea level static ($T_u = T_o = 288°K$) the speed of rotation should be about 5½% *greater* than design.

Having determined the proportionality of θ_c and T_M to T_u it is now possible to revise the non-dimensional diagram for the straight jet engine as shown in Figure 3, which is valid only for the range where η_c, η_T and η_u may be assumed constants (η_j is normally so near 1.0 that it may be presumed constant). This, however, is a reasonably safe assumption over the normal operating range other than take off.

It will be noted that $1 + \phi_u$, i.e. $\dfrac{T_u}{T_o}$, occurs in nearly every quantity and one might be tempted to base a non-dimensional diagram on $\dfrac{T_u}{T_o}$ but, since $T_u = T_o + \theta_u$ it is a very variable quantity. So is T_o but it is determined by I.S.A., i.e. by the height up to the tropopause above which it is constant.

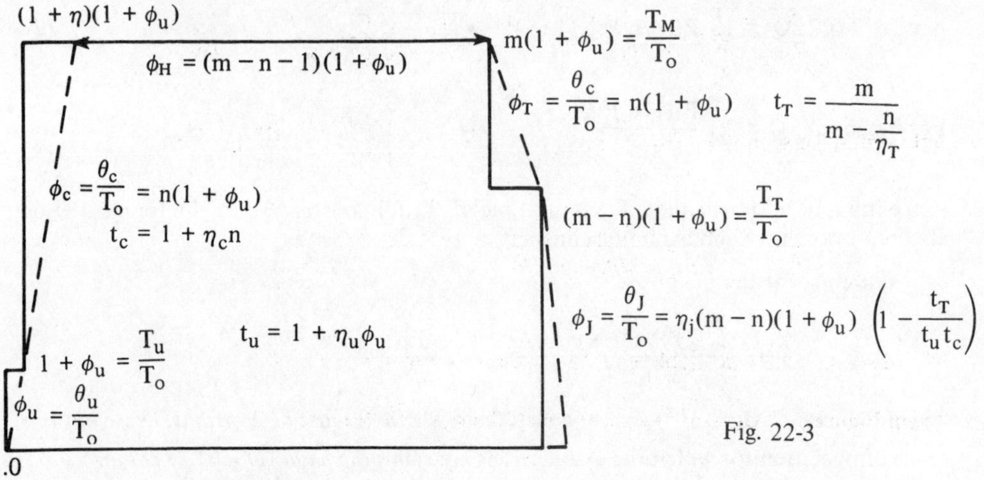

Fig. 22-3

It will be noted that the temperature ratios t_c and t_T are constants, but t_u and t_j depend upon flight speed.

Figure 3 gives all the information necessary to find the effect of speed and height on performance for a given design cycle, though it is perhaps less confusing to work in temps. and temp. differences.

We have information as follows:—

$$F_s \propto \sqrt{\phi_J} - \sqrt{\phi_u} \quad \text{and} \quad F_{sD} \propto \sqrt{\phi_{JD}} - \sqrt{\phi_{uD}}$$

where the suffix D implies the design condition

$$\therefore \quad \frac{F_s}{F_{sD}} = \frac{\sqrt{\phi_J} - \sqrt{\phi_u}}{\sqrt{\phi_{JD}} - \sqrt{\phi_{uD}}} \quad \text{or} \quad \frac{\sqrt{\theta_J} - \sqrt{\theta_u}}{\sqrt{\theta_{JD}} - \sqrt{\theta_{uD}}} \tag{22-8}$$

$$Q \propto \frac{(t_u t_c)^{3.5}}{\sqrt{m(1+\phi_u)}} P_o \quad \text{and} \quad Q_D \propto \frac{(t_{uD} t_c)^{3.5}}{\sqrt{m(1+\phi_{uD})}} P_{oD} \quad \therefore \quad \frac{Q}{Q_D} = \sqrt{\frac{1+\phi_{uD}}{1+\phi_u}} \left(\frac{t_u}{t_{uD}}\right)^{3.5} \frac{P_o}{P_{oD}} \tag{22-9}$$

$$\text{or} \quad \frac{Q}{Q_D} = \sqrt{\frac{T_{oD} + \theta_{uD}}{T_o + \theta_u}} \left(\frac{t_u}{t_{uD}}\right)^{3.5} \frac{P_o}{P_{oD}} \tag{22-10}$$

$$\frac{F}{F_D} = \frac{QF_s}{Q_D F_{sD}} \quad \therefore \text{ from (8) and (10)} \quad \frac{F}{F_D} = \frac{\sqrt{\theta_J} - \sqrt{\theta_u}}{\sqrt{\theta_{JD}} - \sqrt{\theta_{uD}}} \sqrt{\frac{T_{oD} + \theta_{uD}}{T_o + \theta_u}} \left(\frac{t_u}{t_{uD}}\right)^{3.5} \frac{P_o}{P_{oD}} \tag{22-1}$$

$$\text{or} \quad \frac{F}{F_D} = \frac{\sqrt{\theta_J} - \sqrt{\theta_u}}{\sqrt{\theta_{JD}} - \sqrt{\theta_{uD}}} \sqrt{\frac{T_{uD}}{T_u}} \left(\frac{t_u}{t_{uD}}\right)^{3.5} \frac{P_o}{P_{oD}} \tag{22-1}$$

$$\eta_o \propto \frac{F_s \sqrt{\theta_u}}{\theta_H} \quad \text{and} \quad \eta_{oD} \propto \frac{F_{sD} \sqrt{\theta_{uD}}}{\theta_{HD}} \quad \therefore \text{ using (8)}$$

$$\therefore \frac{\eta_o}{\eta_{oD}} = \frac{F_s}{F_{sD}} \sqrt{\frac{\theta_u}{\theta_{uD}}} \frac{\theta_{HD}}{\theta_H} = \frac{\sqrt{\theta_J} - \sqrt{\theta_u}}{\sqrt{\theta_{JD}} - \sqrt{\theta_{uD}}} \sqrt{\frac{\theta_u}{\theta_{uD}}} \frac{\theta_{HD}}{\theta_H} \qquad (22\text{-}13)$$

but, as may be seen from Figure 3, $\dfrac{\theta_{HD}}{\theta_H} = \dfrac{\phi_{uD}}{\phi_H} = \dfrac{1+\phi_{uD}}{1+\phi_u} = \dfrac{T_{oD}+\theta_{uD}}{T_o+\theta_u}$

$$\therefore \frac{\eta_o}{\eta_{oD}} = \frac{\sqrt{\theta_J} - \sqrt{\theta_u}}{\sqrt{\theta_{JD}} - \sqrt{\theta_{uD}}} \sqrt{\frac{\theta_u}{\theta_{uD}}} \frac{T_{oD}+\theta_{uD}}{T_o+\theta_u} \qquad (22\text{-}13)$$

It is obviously desirable to find an expression for θ_J as follows:—

From Figure 3, $\phi_J = \dfrac{\theta_J}{T_o} = \eta_j (m-n)(1+\phi_u) \dfrac{(t_u t_c - t_T)}{t_u t_c}$ \qquad (22\text{-}14)

$$\phi_{JD} = \frac{\theta_{JD}}{T_{oD}} = \eta_j (m-n)(1+\phi_{uD}) \left(\frac{t_{uD} t_c - t_T}{t_{uD} t_c}\right) \qquad (22\text{-}14a)$$

remembering that t_T and t_c have been found to be constants.

Dividing (14) by (14a) gives

$$\frac{\theta_J}{\theta_{JD}} = \frac{T_o}{T_{oD}} \left(\frac{1+\phi_u}{1+\phi_{uD}}\right)\left(\frac{t_{uD}}{t_u}\right)\left(\frac{t_u t_c - t_T}{t_{uD} t_c - t_T}\right)$$

or $\quad \theta_J = \theta_{JD} \dfrac{T_u}{T_{uD}} \dfrac{t_{uD}}{t_u} \left(\dfrac{t_u t_c - t_T}{t_{uD} t_c - t_T}\right)$

or $\quad \theta_J = \theta_{JD} \dfrac{T_u}{T_{uD}} \dfrac{t_{uD}}{t_u} \left(\dfrac{\dfrac{t_u t_c}{t_T} - 1}{\dfrac{t_{uD} t_c}{t_T} - 1}\right)$ \qquad (22\text{-}15)

Example 22-3.

Using the design cycle shown in Figure 4 and assuming that $F_D = 10,000$ lbs. at $u_D = 2,100'/\text{sec}$. How does thrust F and overall efficiency η_o vary with speed near sea level where $T_o = 290$, over the speed range $0 - 900'/\text{sec.}$? Assume $P_o = 2100$ p.s.f.

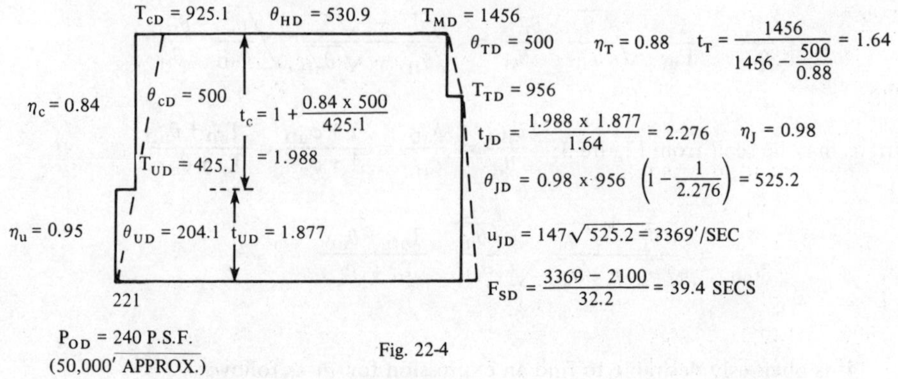

$T_{cD} = 925.1$ $\theta_{HD} = 530.9$ $T_{MD} = 1456$

$\theta_{TD} = 500$ $\eta_T = 0.88$ $t_T = \dfrac{1456}{1456 - \dfrac{500}{0.88}} = 1.64$

$\eta_c = 0.84$ $\theta_{cD} = 500$ $t_c = 1 + \dfrac{0.84 \times 500}{425.1} = 1.988$

$T_{TD} = 956$

$T_{UD} = 425.1$

$t_{JD} = \dfrac{1.988 \times 1.877}{1.64} = 2.276$ $\eta_J = 0.98$

$\theta_{JD} = 0.98 \times 956 \left(1 - \dfrac{1}{2.276}\right) = 525.2$

$\eta_u = 0.95$ $\theta_{UD} = 204.1$ $t_{uD} = 1.877$

$u_{JD} = 147\sqrt{525.2} = 3369'/\text{SEC}$

$F_{SD} = \dfrac{3369 - 2100}{32.2} = 39.4$ SECS

221

$P_{OD} = 240$ P.S.F.
(50,000' APPROX.)

Fig. 22-4

$$Q_D = \dfrac{10,000}{39.4} = 253.8 \text{ lb/sec.}$$

$$\eta_{oD} = \dfrac{39.4 \times 2100}{530.9 \times 336} = .464$$

The following tabulation results:—

u'/sec	θ_u°C	T_u°K	t_u	θ_J°C	u_j'/sec	F_s secs	Q lb/sec	QF_s lb	θ_H°C	η_o
0	0	290	1.0	111.8	1554	48.27	296.8	14,326	362.2	0
200	1.8	291.8	1.006	115.7	1581	42.89	302.1	12,958	363.2	.07
400	7.4	297.4	1.024	127.4	1659	39.10	318.4	12,451	371.5	.125
600	16.16	306.2	1.055	147.1	1783	36.74	348.4	12,788	383.1	.171
800	29.6	319.6	1.097	174.7	1943	35.49	390.9	13,873	399.2	.212
900	37.5	327.5	1.123	191.4	2034	35.21	419.1	14,758	409	.231

To obtain θ_J equation (15) was used but F_s was found from $\dfrac{147\sqrt{\theta_J} - u}{g}$.

Equation (10) in the form $Q = Q_D \sqrt{\dfrac{T_{uD}}{T_u}} \left(\dfrac{t_u}{t_{uD}}\right)^{3.5} \dfrac{P_o}{P_{oD}}$ was used for Q. For θ_H we have

$T_M = \dfrac{T_u}{T_{uD}} \times 1456$ and $\theta_c = \dfrac{T_u}{T_D} \times 500$ and

$$\theta_H = T_M - T_u - \theta_c = \dfrac{T_u}{425.1}(1456 - 500) - T_u = T_u \left(\dfrac{956}{425.1} - 1\right).$$

For η_o we have $\eta_o = \dfrac{F_s u}{336\, \theta_H}$

Thus, for the purpose of the example, equations (8), (10) and (13), etc. are not really necessary.

The values of F ($= QF_s$) and η_o are plotted in Figure 5.

Example 22-4.

Using the same design cycle as Example 3, find the variation of F and η_o for the speed range 900 – 2100'/sec. at a height where $T_o = 221°K$ and $P_o = 390$ p.s.f. (about 40,000').

The tabulation is as follows:—

u	θ_u	T_u	t_u	θ_J	u_j	F_s	Q	QF_s	θ_H	η_o
900	37.5	258.5	1.161	164.8	1887	30.66	98.37	3016	322.7	.254
1200	66.6	287.6	1.286	227.2	2216	31.55	133.38	4208	359.2	.314
1500	104.1	325.1	1.447	308.1	2580	33.54	189.80	6366	406.0	.369
1800	149.9	370.9	1.644	402.5	2949	35.69	277.70	9909	463.3	.413
2100	204.1	425.1	1.877	525.2	3369	39.40	412.40	16,248	530.9	.464

The values of F and η_o are also plotted in Figure 5.

The foregoing discussion and Examples 3 and 4 bring out some very important points. Perhaps the chief of these is that the designer is faced with something of a dilemma if the design cycle of a high supersonic engine is based on its design cruising height and speed—

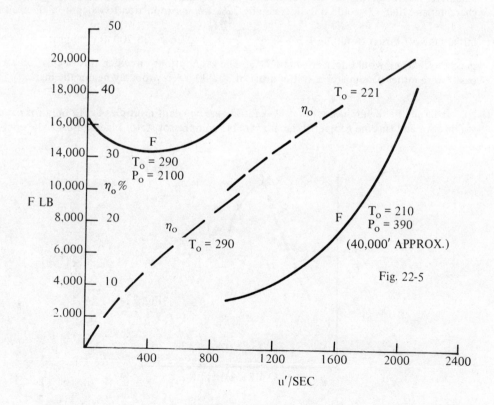

Fig. 22-5

as it should be for long cruising range—in that at take-off and low heights and speeds the thrust is far too low if the conditions that T_{max} and θ_c are multiples of T_u are to be observed.

The reason for this is that if T_u, the total temp. at compressor entry is low, so are the appropriate values of θ_c and T_{max}. Thus, in Example 3, at $u = 0$ $T_u = T_o = 290°C$ and θ_c would be $\frac{290}{T_{uD}} \times \theta_{cD}$ i.e. $\frac{290}{425.1} \times 500 = 341.1°C$, and T_{max} would be $\frac{290}{425.1} \times 1456 = 993.3°$

and the rotational speed N would be $\sqrt{\frac{290}{425.1}}$ $N_D = .83$ N_D approximately.

For θ_c and T_M to be the same for the design condition and for take-off then the design condition would need to be based on the same T_u (for fixed compressor and turbine geometry), thus, if for sea level $T_o = 290°K$, and $u = 0$, the appropriate design condition would require that $T_u = T_o + \theta_u = 290°K$. Thus for a height at which $T_o = 221°K$ the matching value of $\theta_u = 290 - 221 = 69°C$ which corresponds to a speed of $147\sqrt{69} = 1221'/sec$.

Clearly a very much greater thrust for take off, etc. could be obtained by increasing N to N_D or more but at the expense of departing from optimum flow conditions in the compressor. A rough calculation shows that an increase of T_{max} to $1500°K$ should increase F_s from 48.27 secs to 73.71 secs. and N from .83 N_D to N_D approx. but, with a vertical pressure ratio v mass flow parameter characteristic for the compressor, the working point would drop so far down the characteristic that Q would only increase by 22%, i.e. in proportion to N approx., hence the initial take-off thrust would be $1.22 \times \frac{73.71}{48.27} \times 14,326 = 26,700$ lbs. approx. The compressor efficiency would drop to about 71%. The assumptions on which this calculation was based were rather pessimistic; a static thrust of 30,000 lbs. is probably nearer the mark.

It should be re-emphasised that θ_c and T_M are constant multiples of T_u only for *fixed* compressor and turbine geometry; i.e. no variable compressor stator blades, etc. If the com-

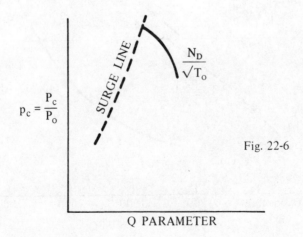

Fig. 22-6

pressor characteristic could be made to be more in accordance with Figure 6 then a much better performance could be obtained when T_u is much lower than the design condition because the working point would not drop so far down the characteristic.

Variable compressor stators help to obtain the kind of characteristics shown in Figure 6. So would rotor blades designed for whirl velocity equal to blade speed at mean radius at rotor exit (i.e. the relative leaving velocity at mean radius would be axial)—if one could 'get away' with the large deflection this would entail both in rotors and stators. This might prove possible if the blade pitch could be sufficiently small. This, however, even if aerodynamically possible, would almost certainly be structurally impossible except for very short blading.

SECTION 23
A Super-Thrust Engine

By way of conclusion of this work it may be of interest to take a look at what might be possible if the requirement was for a very large thrust regardless of fuel consumption. Such an engine would, of course, be for very specialised purposes, e.g. for boosting a space shuttle vehicle to speeds where rockets could take over with reasonable efficiency and thus save the large amount of rocket fuel needed for launch and acceleration to speeds of the order of 3,000 m.p.h. After rocket takeover the air breathing engines would become 'dead weight' for the remaining ascent into orbit, but would at least be available for use on return and so alleviate the great accuracy needed at present to bring the engineless vehicle to a safe landing.

Super-thrust engines could also be of use for single mission very high performance missiles.

As we have seen, in all engines discussed herein, the fuel necessary is far below that corresponding to the stoichiometric air fuel ratio of about 15:1. It follows that the means for obtaining great thrust would be to have additional combustion wherever this is possible, e.g. between the core compressor turbine and the fan turbine in the case of a low bypass ratio engine, combined with after burning before final expansion.

A low bypass ratio engine is indicated as being the most suitable for the purpose in order to obtain a high final expansion ratio and thus increase the effectiveness of after burning.

Figure 1 shows the take-off cycle for a super thrust engine based on the following assumptions: $T_o = 288°K$, $\theta_T = 200°C$ (i.e. the 'fan' is a multi-stage L.P. compressor); that T_{max} can be as high as $1850°K$ for short periods with considerable blade cooling; that the fan turbine can also operate with a turbine entry temp. of $1850°K$; that after burning is permissible up to $2500°K$; that the L.P. turbine is spaced sufficiently far from the core compressor turbine for supplementary combustion chambers to be installed between them; that $\theta_c = 500°C$; and that component efficiencies are as follows:— $\eta_F = .86$, $\eta_c = .84$; $\eta_T = .86$; $\eta_{FT} = .86$; $\eta_J = .96$.

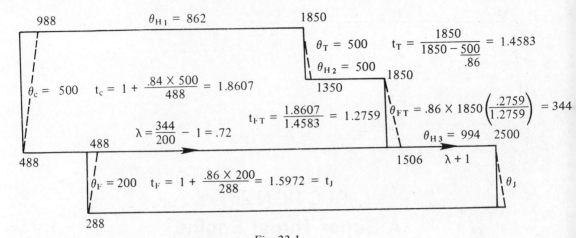

Fig. 23-1

$$\theta_J = .96 \times 2500 \left(\frac{.5972}{1.5972}\right) = 897.4 \quad \therefore \quad u_J = 147.1\sqrt{897.4} = 4406.5'/\text{sec}.$$

$F_s = \frac{1+\lambda}{g} \times 4406.5 = 235.4$ secs., i.e. of the order of 3 times that of a straight jet and comparable with the specific thrust of a rocket. The total heat energy added $\propto$ $862 + 500 + 994 + .72\,(2500 - 488) = 3805$. The equivalent of the exhaust energy = $1.72\,\theta_J = 1.72 \times 897.4 = 1543.5$ $\therefore \eta_{th} = \frac{1543.5}{3805} = .406$ which is surprisingly high considering that such a large proportion of the heat is added at quite a modest temp. ratio.

The specific fuel consumption in lbs/sec. for a calorific value of 10,500 C.H.U/lb. would be $\frac{3805 \times .24}{10,500} = .087$ lb/sec. for 1.72 lb/sec. of air $\therefore$ the air/fuel ratio $= \frac{1.72}{.087} = 19.8$ approx., i.e. appreciably weaker (or 'leaner') than stoichiometric.

In the above example no attempt was made to determine the optimum value of λ.

Unfortunately there is a serious 'matching problem' as between low speeds at low heights and very high speeds in the stratosphere—hence the several attempts to solve the problem of variable bypass ratio.

At very high speeds there is little or no advantage in using any bypass at all. Indeed, even with the straight jet, the higher the speed the lower the compressor temp. rise should be (otherwise, *in the limit*, the ram temp. rise plus the compressor temp. rise would approach the max cycle temp. to the point of eliminating the heat addition in the main combustion).

The cycle of Figure 2 illustrates the point for a speed of 3000 m.p.h. (u = 4400'/sec. and $\theta_u = 896°C$) with a straight jet plus after burning with component efficiencies as indicated in the diagram.

```
1316            θ_H1 = 534           1850
  θ_c = 200   η_c = .87   t_c = 1.1559   θ_T = 200   η_T = .86   t_T = 1.1438
                                          θ_H2 = 850
  1116                          1650                                2500

  θ_u = 896  η_u = .92  t_u = 3.7469    η_J = .96  t_J = 3.7865    θ_J = 1766

220                                                               734
```

Fig. 23-2

For $\theta_J = 1766$ $u_J = 6182'/\text{sec.}$ and $F_s = \dfrac{6182 - 4400}{32.2} = 55.34$ secs.

$\theta_{H1} + \theta_{H2} = 534 + 850 = 1384$ ∴ $\eta_o = \dfrac{55.34 \times 4400}{336 \times 1384} = .524$

In this example, complete expansion in the jet nozzle would require that $\dfrac{S_E}{S_o} = 2.375$ which would obviously be very undesirable. For $S_E = S_o$ the under expansion would be such that the pressure at the nozzle exit would be a little over 3 atmospheres. The same difficulty occurs with the ramjet case below.

A ramjet at the same speed and same ram and jet efficiencies and $T_{max} = 2500$ would have $F_s = 54.98$ and $\eta_o = .520$. So under such conditions the turbojet is merely a 'supercharged ramjet'. It is evident that the utility of a turbojet ends at about 3000 m.p.h. and possibly below, having regard to power plant weight.

It would clearly be desirable to convert from the cycle of Figure 1 to that of Figure 2 and then switch to ramjets if air breathing engines are to be used for speeds greater than 3000 m.p.h.

For take off, initial acceleration and climb, a further substantial increase of thrust could be obtained by coolant injection. As mentioned earlier in this work, liquid ammonia is the most potent agent for this purpose. Coolant injection, by reason of its effect on initial cycle temp. T_o, could also greatly alleviate the problem of matching as between sea level static and high speed in the stratosphere.

An engine with a core mass flow of 500 lb/sec. operating with the cycle of Figure 1 would have a take-off thrust of 117,700 lbs. With ammonia injection this might be increased to 160,000 lbs. or more by virtue of the double effect of increasing F_s and increasing mass flow rate.

Twisted Obsession

PDotson

P.Dotson

P.O.B 252

Preston, MD 21655

Copyright @ 2010 P.Dotson

Printed in the United States of America

This is a work of fiction. Names, characters, places, and incidents either are the product of the author's imagination or are used fictitiously, and any resemblance to the actual persons, living or dead, events, or locales is entirely concidental

This book is dedicated to my beautiful mother who inspired me to write and told me to continue to strive for my dreams. My sister Puroney who pushed me to finish this book. The TLC crew Pam, Dani, and Kara how we loved to swap books. My big brother Sterling LaPrince Dotson Sr. I love you always keep ya head up

LOCATION: LIBRARY.
COLLECTION. LOAN.
CLASS MARK: DOT.
BARCODE No: 537009
DATE: 03/12/12.

Twisted Obsession

Prologue

The cold winter air caused Allexis to shiver as she stood outside on her front porch at the break of dawn. She vigorously rubbed her hands together before rubbing her arms in hopes of raising her body temperature. The brisk weather stung her nostrils causing her eyes to water just a little. Birds chirped loudly in the background while a rooster crowed in the distance. Everything seemed to be in their rightful places except for her.

The taste of salt on her lips caused her to realize that she was crying. She cursed herself, wanting to be strong for the long journey ahead of her. But she couldn't control the broken bridge her emotions were riding on. Tears of anguish, hurt humiliation and betrayal streamed down the side of her face and down onto her sweat shirt. She sniffled a little before wiping her face with the back of her hand. Hurtful memories played in her mind as she thought back to the reason's she why she was standing outside at the break of dawn all by her lonesome.

"You know yo' mama won't believe you," he said to her as she lay curled up in the fetal position on her bed. "So you might as well step up, be a woman and enjoy what *Daddy* 'bout to give you," he smirked.

Allexis slowly sat up while tears continued to steadily flow from her eyes. The emotions she was experiencing at that moment were clearly evident on her face. But she knew that didn't matter to him. The day she had been dreading her entire life had finally come. "Please, don't do this," she begged as she looked at him with pleading eyes, just hoping that just for once he would look at her as his daughter and not an sexual object. The way he use to when she was young and he would give her piggy back rides and play hide and go seek, well that's how she use to remember him.

But that was false hope, because her pleas fell on deaf ears, as he grabbed her by her long wavy black hair and kissed her violently. He roughly fondled her breast and caressed her untouched treasure. Allexis wept and prayed in silence while her father continued to violate her body. She could taste the liquor that lingered on his tongue and the stench of his unwashed body made her gag.

Deep within himself he knew he was wrong for touching his daughter in such an unthinkable way. But he couldn't contain the strong urges that forced him to do this evil

Twisted Obsession

thing. He needed to have that sense of control, the way his father had control over him.

The distinct sound of the door slamming downstairs caused them both to jump. "Alexis ... Daddy ..." she heard her mother yell. "Would one of you mind giving me a hand?"

"Shit," she heard him mutter.

Neither one of them was expecting her to be home this early. She was a school teacher and usually Monday's were her busiest days. Allexis breathed a huge sigh of relief for now. She could hold onto her innocence for one more day, she hoped.

"You better not say anything'" he threatened as he threatened on previous occasions after violating her body. He gave her a warning look, before leaving her room.

Once he was gone a wave of emotions took over her body as she began to cry uncontrollably. Her chest heaved rapidly up and down as she began panting for air. She quickly got up and ran into her bathroom and threw up, until her stomach was completely empty. Afterwards she managed to pull herself to her feet. Turning on the cold water she splashed some on her face before grabbing her tooth brush to brush her teeth. She brushed them until her gums were sore.

But no matter how hard she brushed she still couldn't get rid of the terrible taste that he had left in her mouth. Just the thought of his tongue in her mouth and his hands touching her in places a father shouldn't, made her feel extremely dirty. She wanted to take a shower but decided against it. She couldn't help but to look at her pitiful reflection in the mirror. Her hair was a mess, her eyes were puffy and red which stood out against her now pale and ghastly looking skin.

Hugging her body tight she timidly walked back out into her spacious bedroom. Most girls would kill to have what Allexis had. She had any and every named brand clothing, with shoes to match. She even had her own car before the age of sixteen. But people only saw half of what Allexis was living. They never knew the hell that she went through behind closed doors, or experienced the pain and humiliation she endured at the hands of her father.

The picture of her and her older sister Tina that sat on her dresser caught her attention. She picked up the frame and hugged it tightly to her chest. "Why didn't you take me with you?" she whispered as fresh tears lined her face. "You promised you would come back

Twisted Obsession

for me."

But Tina had made that promise over four years ago when Allexis was twelve and Tina was sixteen. At the time Allexis was oblivious to the skeletons in her family's closet. But now there was no hiding. Everything that her father had kept hidden was now out in the open. It all started making sense to Allexis now. The sudden change in her sister's behavior, how Tina went from being A straight student to completely dropping out of school, hanging out with the wrong crowd and eventually she began to experiment with drugs It had gotten to the point where Allexis didn't even recognize her sister anymore.

As Allexis had flashbacks of her sister's troubled past, she decided right then and there she wouldn't allow her father the opportunity to shatter her innocence. She refused to live another day in a house built solely on lies. Taking deep breaths she built up enough courage to walk downstairs. As she made her way downstairs she could hear both her mother and father laughing in the kitchen. Just the sound of his voice made her sick to her stomach.

"Hi, sweetheart," her mother greeted her as she walked in the kitchen. "Can you help me put some of these groceries away?"

"Mama, we need to talk," Allexis said ignoring the fact that her father was staring at her. She could feel the beads of sweat starting to form on her forehead and upper lip.

"What is it sweet heart?" her mother asked with a look of concern on her face.

At the age of forty-six her mother was still stunningly beautiful. She had long thick beautiful black hair that fell just below her waist, but she always wore it in a bun at the nape of her neck. She was built like a model, tall, slim built with legs for days. Beautiful green eyes, her signature beauty mark that sat just above her right eyebrow and that one dimple in her left cheek that always appeared when she smiled. She could have easily been on any top notch magazine cover Allexis thought.

Allexis took a deep breath and wiped her now clammy hands on her jeans before responding. "Ummmm, can we talk in private," she asked nervously biting her bottom lip.

"You have something to say, Allexis?" her father asked in an underlying tone. "You can say it to the both of us. There are no secrets in this house."

Twisted Obsession

Allexis gasped as soon as the words escaped his lips. *The audacity* she thought to herself. But she refused to crumble under his menacing gaze. Allexis took a step backwards, "You know exactly what I have to say," she muttered defiantly through clenched teeth. "You've been sneaking into my room for the past five months. Kissing me and touching me in a way a father should *never* touch his daughter."

"Now that's a got dame lie," her father roared causing Allexis to tremble just a little.

"Mama it's true. You do believe me don't you?" Allexis asked taking ahold of her mother's hand. "Mama please, why would I lie about his?" Allexis could only watch in awe as her mother withdrew her hand from her grasp and stood by her husband.

"I won't stand for it Allexis," her mother said shaking her head. "You know your father loves you. Why would you say such terrible things?"

Still hoping for the best and refusing to give up Allexis reached out to her mother again. "Mama, what are you saying? Why do you think Tina left? He's sick, he tried to rape-" But Allexis words were cut short with a hard slap to the face. In shock Allexis looked at her mother the pain evident in her eyes, but she refused to cry. But the pain of the blow didn't hurt nearly as much as her mother's rejection. She was always told mothers were supposed to protect their children. She guessed her mother was the exception.

"How dare you say that about your father Allexis?" her mother yelled grabbing her chest. "He is your father and he loves you," she whispered as if she were trying to convince herself.

This was all déjavu to her same plot different character. She remembered that day as if it were yesterday, when a broken Tina came to her for help. But she pushed her daughter away, because she didn't want to believe that a father would do this his own daughter, to his own flesh and blood. Besides she knew Tina had gotten in with the wrong crowd and was on drugs. She tried to tell herself that it was the drugs and Tina's friends that were talking in her ear. As she stood before Allexis she kept her head down unable to make eye contact with her daughter, in fear of what she might see. Afraid to face the same painful look Tina gave her right before she left.

"She just thinks she is better than me 'cause she got white blood runnin' through her

Twisted Obsession

veins. But you ain't no betta than me girl," he taunted. "You got some of my blood runnin' through dem viens too," he said pointing his bony finger in Allexis face.

"Well you remember that the next time you get the urge to sneak into my room," Allexis muttered.

"You little bitch," he growled through clenched teeth as he reached for her.

But Allexis was too fast as she darted out of harm's way just in the knick of time, running upstairs to her bedroom and slamming the door hard in his face. Thinking quick on her toes she mustered up enough strength to move her dresser in front of the door. Her father ranted, yelled and pounded on the door for what seemed like hours but in actuality was just a couple of minutes.

"Let her be," she heard her mother say in what sounded like a defeated voice.

"I know you're not going to listen to that *bullshit*," he asked with a hint of fear in his voice. "There lies, got dame lies."

"I'm not saying that. It's obviously something's bothering her and she needs an outlet," she said trying to convince herself but not really believing her own *bull shit*. "I'll make some calls in the morning, okay," she said just wanting some peace and quiet.

The night's events had sent her in a whirl wind. All she wanted to do was run a nice warm bath, pour herself a nice glass of wine and pop in her Tina Marie cd and hope in the morning that when she woke up things would be back to normal.

"But she has to learn how to respect me. I am her father. I'm the man of this house got dammit," he continued to rant.

What a joke, Allexis thought to herself. He hadn't worked in years and from the looks of things he didn't plan on ever working again. Not with her mother making a pretty decent living and making every aspect of his life so convenient.

"I know," she heard her mother say rocking her out of her thoughts. "Come on let's go to bed. Things will be better in the morning."

But Allexis wasn't about to stick around and find out. Clutching her duffel bag tightly Allexis looked at her childhood home one last time before spinning around on her heels. With only half her worldly possessions and twenty-five hundred dollars to her name, she was ready to face the world. She figured the world couldn't be any worse than the hell she's already experienced.

* * *

Twisted Obsession

Chapter 1

"Hey sis what you doin'?" Tina asked flopping down on the couch beside Allexis.

"Nothing, just studying," Allexis replied never looking up from her book.

"Nothing, studying," Tina mocked. "You need to put that shit down and get the fuck up outta this house for a minute," she said over her shoulder as she got up and walked into the kitchen. She opened up the refrigerator and pulled out a corona before walking back out into the living room.

Allexis rolled her eyes. *"Whatever!"*

She wasn't the least bit worried about going out. In Allexis mind she had plenty of time for that. Right now she was more focused on finishing school and getting her degree. She was more than eager to help other troubled teens out there in the world. But her sister didn't see it that way, often referring to her as a book worm.

"It ain't no whatever! Get that pussy wet," Tina said as she smacked her ass and dropped it like it was hot.

"Girl, whatever, you know I don't get down like that," Allexis laughed.

"I know you don't," Tina said taking a sip from her drink before placing it on the stand. "But we 'bout to change all dat, please believe sweet heart, because we goin' out tonight," she stated emphatically grabbing Allexis's books without warning and throwing it across the room.

"Girl," Allexis squealed. "Do you know how much these books cost?" she asked as she retrieved her book and carefully inspected it to make sure no damage was done. Once she was satisfied she sat back down on the couch and picked up where she left off. "You need to lay off of those," Allexis said referring to the Corona sitting on the stand.

"Come on Alleixis, you need to loosen up a little. Get some meat up in you," Tina joked. Allexis gave her a sideways glance. "Nah, I'm just playin'. I'm glad you're still a virgin. Lord knows I wasn't at your age. But you do you and if you want to wait until that special one comes along, go head. Just don't let the kitty get to dry."

"Girl, shut-up."

"You know I love you. But in the mean time you could have a little fun while you waitin'. *Damn!*" Tina could only shake her head because her sister always opted to stay

Twisted Obsession

cooped up in the house. Sure she was happy Allexis was focused on her studies, but she felt her sister should loosen up and have some fun once in a while.

"But I don't have anything to wear," Allexis whined.

Tina waved Allexis off with her hand. "Quit playin' you can wear somethin' of mine."

Alliexis smirked and folded her arms across her chest. Tina looked down at her own full figured frame before checking out Allexis's petite frame.

"On second thought, Allexis you know damn well you can find something to wear as full as your closet is. And if not, I'm sure Doricka can find you something to wear," Tina smiled.

Both girls burst into laughter.

"Okay first of all who the hell would name their child Doricka? And second you know Doricka wears that off the wall mess," Allexis smiled as she thought about how fashionably challenged Doricka was.

"Okay, don't be talkin' 'bout my friend," Tina laughed as she playfully mushed Allexis in the head. "But your right, that name is fucked up and her wardrobe is even more fucked up." They shared another laugh. "But know seriously I'll tell her to bring you something decent to wear. So be ready by eight. Okay."

"Okay we will see," Allexis said still not convinced.

She watched all smiles as Tina bounced back to her room. The reunion between the two was a rather awkward one at first. Both were at a loss for words, and simply stared at each other for what seemed like hours before either one spoke. Allexis was the first to close the gap that held them apart by embracing her older sister in a hug.

The two were able to pick up where they left off and Allexis had learned a lot about her sister that day and was proud of how far she had come considering her circumstances. When Tina left home for New York at the tender age of sixteen she was pregnant, hurt, scared and alone. Rationalizing with her conscious she knew there was no way she could bring a child into this world. So with the help of a friend she was able to get an abortion.

With minimum education and no GED, Tina retorted to the next best thing, *stripping*. To cope with the pain of exposing herself to men once again, she eventually turned to drugs. What started out as something to help her get through a night of dancing soon

Twisted Obsession

turned into a habit of getting her through her childhood memories of the physical, emotional, mental and sexual abuse that she suffered at the hands of their father. If it weren't for Dom, Tina's current boyfriend, Tina says she knows she would have been dead. That's how badly the drugs had taken over her life. But Dom was her hero. He came in and relocated her to Virginia putting her in rehab where she was able to get the help she needed. She's been clean for 4 years, got her GED and currently works part time as a receptionist at a dentist office.

But the move to Virginia didn't sit well with Allexis and Tina's parents. They wanted Allexis to come back home. Even from jail her father was trying to control her life. But with Dom's help, he was able to get them one of the best lawyers in Virginia. Their father was bought up on charges of sexual abuse, but not only against Allexis and Tina, but also on some of the local girls in their neighborhood. The girls felt some kind of way about putting their father in jail, but Dom and their lawyer assured them it was for the best so that he wouldn't be free to hurt another little girl. So their father was sentenced to 15 years in prison and Tina was awarded sole custody of Allexis.

All of those events transpired three years ago. Allexis still has troubles trusting people, especially men. But her past didn't stop her from excelling in school, graduating in the top five of her class. She is currently in her second year of college, hoping to obtain a degree in psychology. But despite everything Allexis still longed for a relationship with her mother, after their father's sentencing she's spoken to her mother only twice, but Tina doesn't want to have anything to do with their mother. Allexis just hoped and prayed that one day they all could reunite.

* * *

Knock! Knock!

Allexis knew that it was Doricka. She did her best to suppress the laughter in her throat as she walked towards the front door. She already knew what to expect and when she opened the door she received nothing less than the ordinary from Doricka.

Now she knows she looks a mess, Allexis thought to herself, as she stepped aside and allowed Doricka to *model* her outfit.

"Sup?" Doricka asked as she sauntered through the door, thinking she was *cute*. "This

Twisted Obsession

is my cousin Passion."

"Hi," Passion said with a wave of her hand as she walked in behind Doricka.

"Oh, I'm sorry. Hello," Allexis replied. She had been so wrapped up in Doricka's wardrobe that she hadn't even noticed Passion sitting back in the cut.

But Allexis still could only shake her head at Doricka's ensemble. She had on a pair of zebra striped leggings, a black and red halter-top, and tan suede boots. To put icing on the cake she had the nerve to have on a blonde wig with lime green bangs. Then she had the nerve to strut in like she was beautiful.

"*Oh, hell naw!*" Tina yelled as soon as her eyes landed on Doricka's outfit. "Girl, take that shit off," Tina said walking up to Doricka and snatching off her wig, and waving it the air to keep it out of Doricka's reach.

"Girl, you betta give me back mah shit," Doricka said as she chased Tina around the living room.

But Tina was too fast jetting out Doricka's reach and running into the guest room slamming the door in Doricka's face. Both Allexis and Passion could only sit back and laugh at the two.

"Tina you betta quit playin'" Doricka yelled as she pounded on the door. "Ain't nothin' wrong wit what I got on is it Allexis?" Doricka asked seriously looking down at her tan suede boots before looking back up at Allexis.

"Ummm," were the only words that Allexis could find in her vocabulary at that moment.

Doricka just simply rolled her eyes and waved her hand. She didn't really care what they thought because she thought she looked good. "Whatever," she said rolling her eyes. "Passion goes ahead and sit on down. You don't have to be shy. Allexis doesn't bite. At least I don't think she does anyway," Doricka joked.

"I'm sorry. Where are my manners? Would you like something to drink?" Allexis offered her eyes still on Doricka's pitiful ensemble. She didn't know where that girl came up with her ideas or style of fashion.

"No thanks," Passion replied taking a seat on the lounge chair.

"Doricka, bring yo' ass in here and put on somethin' else," Tina yelled. "You are not

Twisted Obsession

going anywhere with me looking like that," Tina stated matter of factly with her hands on her hips.

At first Doricka started to object but when they went out Tina usually paid for everything, courtesy of Dom of course. "Y'all don't know how to be different," Doricka sulked as she trotted back into the guest room with Tina, carrying her tote bag over her shoulder.

"Man your cousin is a mess," Allexis said plopping down on the couch opposite Passion. She thought Passion was a cute dark-skinned girl with hazel eyes. Her hair was styled in a cute bob that stopped just above her shoulders.

"Humph, don't I know it. You can't tell that chick that she can't dress." They both laughed. "So are you from around here?"

"No' I'm from Maryland. What about you?"

The two ladies sat and conversed nicely with each other. Passion went on to tell Allexis that she was originally from Virginia but grew up in Miami. Allexis also learned that she was twenty-one years old, and enrolled in beauty school. She was currently on summer vacation and decided to come back home and visit family. Passion continued to talk freely while Allexis listened attentively, grateful that Passion loved to talk so that she wouldn't have to, only sharing few details of her own life. She wasn't quite ready to open up about her past just yet, but she liked Passion and could tell that they were going to be good friends.

"You ain't even dressed yet," Tina yelled startling the two girls.

"Damn, girl we are inside," Allexis replied turning around to see what Doricka and Tina had on. "Ewe where are your clothes. Summer isn't here yet."

Tina had on a pair of blue Baby Phat shorts with practically half her ass hanging out, a white bikini top that left little to the imagination and a pair of four inch heels.

"Chick please, you know how I do," Tina said smacking her ass. She was in the mood to make some hoes jealous tonight.

Allexis was especially glad to see that Doricka had opted to change her clothes. She now sported a pair of Versace jean shorts, a cherry red halter-top and four inch heels as well. The red halter-top complemented her caramel skin and soft brown eyes. She now

Twisted Obsession

wore her natural hair styled in a wrap and it flowed freely down her back.

"Go ahead and start gettin' dressed," Tina said.

"I don't think I'm going," Allexis replied.

"Come on, Allexis," Passion begged. "It will be fun. I'll help you find something to wear," she offered. "Besides I don't want to be left alone with those two."

Placing her hand on her hip, Allexis thought about it. She hadn't been out in a while. She sighed, "Alright you go and look in my closet and I'm going to go jump in the shower real quick," Allexis said as both she and Passion headed upstairs.

While Allexis was in the shower Passion thumbed through her closet. Passion could only shake her head, because Allexis had tons of clothes; some still had the price tags on them. She picked out three outfits that she thought Allexis would look nice in. The first was a Gucci jean skirt, a pink spaghetti strap top, with a pair of laced up pink sandals to match. The second outfit was a pair of Apple Bottom jean capris, a winter green top, a white belt and a pair of four inch Jimmy Choo heels to match. And the third outfit Passion picked out she thought Allexis would look really nice in. It was a red, strapless form fitting Ralph Lauren dress that stopped just above the knee. She added black diamond ear rings, a black belt and black Jimmy Choo heels.

"Dang, I'm good," Passion said out loud, before leaving Allexis room to head back downstairs.

Allexis walked out of her bathroom with a towel wrapped around her body. She glanced at the outfits Passion had laid out on her bed. She had to admit that she was impressed, but she decided that she would keep it simple for tonight. She strolled to her closet and picked out a pair of sky blue Coogi Jeans, a red t-shirt and a pair of red flip flops.

"Girl, my services are not cheap," Passion said rolling her eyes, once Allexis was finally dressed and met them downstairs in the living room. But Passion had to admit was a very beautiful girl; she could wear anything and still look good.

Tina smacked her lips when her eyes landed on Allexis. "That's the white in you. Come here." Allexis did as she was told. Tina turned Allexis around and tied her shirt up in the back exposing her belly ring.

Twisted Obsession

"Tina," Allexis squealed.

"What? You got the darn thing, what you tryin' to hide it for?"

"She's right," Passion chimed in.

"Stop being so damned self-conscious, Allexis. Men are going to look regardless of what you have on. Because my baby sis got it goin' on," Tina said nodding her head.

"Okay," Allexis whispered feeling a tad bit shy.

"You'll be alright once you get some drinks up in you," Tina stated ruffling Allexis hair a little to give it more of a wild edgy look.

To look at Allexis and Tina standing side by side no one would guess they were sisters. Allexis was bright yellow with, with beautiful long wavy black hair that flowed down her back. She looked more like their father with her soft light brown eyes, oval face, and petite frame. Tina looked more like their mother, with the skin the color of honey, stunning green eyes and curly shoulder length hair. Since the age of twelve Tina was built like a brick house sporting a 32-24-40 frame.

"Okay hoes y'all ready?" Tina asked.

* * *

The club was packed from wall to wall. If Doricka hadn't known the bouncer there was no way they would have been able to get in. The club was nice, but smoke loomed in the air which was a turn off. Allexis and Passion chilled by the bar while Doricka and Tina hit the dance floor, rockin' to the ole school jam Crime Mobbs *Rock yo Hips*.

"Yo' shawty you wanna dance?"

Allexis turned her head to where the horrible stench stemmed from, only to be greeted by one of the most hideous creatures she has ever seen in her nineteen years of living. He stood there looking like Biggy Smalls twin on some serious crack, snot running down his nose wearing a knock off Armani suit. It had to be at least 80 degrees in the club. And she could tell dude was hot because the sweat spots beneath his armpits were clearly visible. Not to mention his breath had to be at least 10 degrees hotter than the club giving Allexis the urge to hurl.

Is he for real?

"Hmmmm . . . no thank you," she politely replied. At first she wasn't sure if he had heard her or not because he continued to stand there and stare at her.

But dude wasn't ready to give up just yet. He wasn't having any luck with the ladies tonight and was constantly getting no's left and right. So he figured he would give it another try. "Come on ma. You know you wanna shake that ass for me," he smirked, licking his cracked lips, and giving his chin a stroke.

Trying her best not to hurl from his bad breath and horrible body odor Allexis swallowed really hard and sweetly replied, "Not tonight, maybe some other time."

He looked at her like she was crazy. "Man, fuck you. I ain't wanna dance wit' ya ole stuck up ass anyway."

"Nigga please! How the hell you gone come over here and *mack* with your breath smellin' like stale chitlins. Not to mention that *fake* ass Armani suit and with *converses*," Passion stated rolling her eyes and neck, while Allexis could only stare on in disbelief.

But Passion wasn't quite finished giving homeboy the business. Putting up her index finger and curling up her lips she proceeded.

"You ain't foolin' nobody. You ain't cute, not even slightly attractive so that's one strike against you. Then you're a non dressin' nigga, so that's two. And three to come up in here hot as hell smelling like garbage, you deserve a medal," she said clapping her hands before taking a sip from her drink.

"Okay, Passion. I think we need to go now," Allexis said grabbing Passion by the arm.

Dude stood there speechless. He couldn't believe shorty carried him like that. He lifted up his armpit, taking a whiff. He damned near gagged himself, he had to admit he did kind of smell like hot garbage.

"Here wipe your nose," Passion said throwing him a napkin.

"Girl, are you crazy?" Allexis asked once she figured they were out of harm's way. She looked around cautiously to make sure the dude hadn't followed them.

"What? It was the truth."

"But he could have been a gangster or something. What if he had pulled out his nine?"

"Girl please! That wack ass Negro."

Allexis shook her head. "You deserve a medal."

They both shared a laugh before stepping out on the dance floor to groove to Ciara's

Twisted Obsession

Ride.

* * *

"Yo' this shit is whack," Dame stated emphatically placing his drink on the bar. He wanted to go New York. He was in the mood for some out of town pussy. The round bodacious ass of the cute Puerto Rican bar tender caught his attention, momentarily distracting him from his thoughts.

Is that who I think it is? He asked himself as another fine lady caught his eye. He took a sip of his drink as he watched her walk.

"Yo' have you seen her before?" Kane asked ignoring his friend's remark.

If you didn't know, Kane is the nigga to be and the nigga to see. He entered the game fearless at the tender age of fifteen, but his street smarts and don't give a fuck attitude allowed him to quickly rise in status. His name was well known in the streets because he wouldn't hesitate to put holes in anyone that crossed him. He owned and operated one the largest chop shops in Virginia Beach.

"Who?" Dame asked turning around to look to where Dame was pointing. "Which one the light one or the dark one? Damn, the dark one kinda thick." Dame didn't say it but the light one kind of looked like Kane's little sister Alex. He knew bringing up Alex was a sensitive subject for Kane and he didn't want to put a damper on the mood.

"The light one, shorty is fine as hell," Kane said more to himself than to Dame. He was memorized by her beauty. It was something about her that had him instantly intrigued. He continued to watch her dance, caught up in his own daydream.

"Man, she look like a stuck up bitch. Probably ain't even had her cherry popped yet," Dame joked. "Yo' you aight?" Dame asked taking note of the dazed look in Kane's eyes. He never seen Kane look like that over a woman.

"Yea nigga," Kane said shaking his head a little bringing himself back to reality. "Oh, we goin' to B-More tomorrow to handle some business," Kane said to Dame but his eyes still continued to stay glued to the beautiful creature before him. Thousands of women went out of their way to catch his eye but he only had eyes for one woman tonight.

"I hate fuckin' wit them B-More cats. They are some shiesty individuals for real. I don't see why we can't start the business in New York? I'm tellin' you Kane they be

Twisted Obsession

havin some high quality vehicles. Do you know how many celebrities visit New York on a daily basis?" Dame asked doing his best to convince Kane to start the business in New York, versus Baltimore.

But Kane wasn't hearing it. He already had his mind made up. "Quit bitchin'. You always talk that New York bullshit. If you ain't down to ride it's always another nigga who will," Kane stated finally giving his eyes a break, and giving Dame a sideways glance. Kane just felt his business would fare better in B-More because of the high drug use over there and besides it was closer. Once he got things settled over in Baltimore then he would consider expanding his operation to New York.

"*Damn!* It's like that? Ain't we boys?" Dame asked throwing his hands up in the air, a little turned off by Kane's arrogant attitude.

Although Dame still thought *their* business would fare better in New York he knew there was no talking with Kane, because what Kane wants Kane gets. But he was tired of Kane just up and making decisions without consulting with him first. They were supposed to be business partners. But lately Kane was acting like he was the HNIC, and over the years people started referring to Kane as such, and he was all of a sudden given the title of being the *second* in command. But what could he do whatever Kane said went and there was nothing Dame could do about it because he didn't have the heart to disagree with Kane. Not openly anyway.

"*Some say the ex, make the sex, spectacular"* Biggie Smalls blasted through the speakers causing the crowd to go wild.

Kane couldn't stand it any longer. He wasn't one to dance but tonight he was going to make an exception. "I'ma go holla at lil mama," he said standing up as he adjusted his collar to his short sleeve Armani shirt.

* * *

Allexis's groove was temporarily interrupted when she felt someone grab her from behind. She wasn't like most girls who needed a partner to dance with; she was fine all by herself. Her first thought was to politely walk away, but when she turned around and her eyes landed on one of the finest men she seen since being in Virginia, she quickly changed her mind. She felt her heart flutter when their eyes locked. The smooth chocolate

Twisted Obsession

skin brotha with the chinky eyes had her hypnotized. She glanced over at Passion to find her working the hell out of a fine Puerto Rican looking brotha. So Allexis followed suite and bent over what little ass she had, trying to give the dude behind her a run for his money.

Kane couldn't help but be memorized by the beautiful creature before him. He held onto her hips for dear life, while she continued to work. Just watching her made him want to fuck right then and there.

"Damn, baby you tryin' to make a nigga bust up in here," he whispered in her ear.

"Oh really," Allexis replied slyly cocking her head to the side, while continuing to work her hips in overdrive. The absolute and cranberry had her feeling sexy.

The duo continued to dance until the song was over. Allexis was suddenly hot and decided to go and get a drink with Kane in tow. He gently grabbed her arm not about to let her get too far.

"Damn! It's like that? You gone leave a nigga hangin'?" he asked trying to suppress the throbbing in his pants.

"Sorry about that," she giggled obviously feeling a little buzz.

She nervously looked around the club anything to avoid the fine man standing before her. She spotted Passion talking the dude she had danced with earlier, but she appeared to be a tad bit annoyed him. But unfortunately for Allexis she wasn't getting rid of Kane that easy.

"So, what's your name?" Kane asked his lips touching her ear.

Allexis felt a jolt of electricity run through her body. "Allexis," she answered shyly.

Kane was instantly turned on by her innocence. He could tell she was nothing like the chicken head bitches he was used to dealing with. She was special and he knew he would have to treat her a *little* different than the rest.

"Allexis, I like that," he replied giving her body the once over. Liking every bit of what he saw. "Let me buy you a drink?"

"Ummm. Sure. Let me see if my girl wants a drink too, if that's okay with you?"

"Yeah it's cool. Any friend of yours is a friend of mine."

"Okay, be right back," she said moving past him, thankful to be getting away from those beautiful eyes. Already he was sending her body in frenzy.

"I have a man," Allexis heard Passion say as she approached.

"Come on ma! Why you actin' all stuck up and shit?" Dame asked eyeing her ass through the tight dress she was wearing. He was in the mood to bang something tonight and home girl definitely fit the bill. He had a thing for asses, and hers' sat out just right.

Luckily for Dame Allexis came just in time. "Hey, Passion you want another drink?"

"Gladly," Passion stated rolling her eyes at Dame and waving her hand in his face. She wanted know parts of him. It was just a dance, he must have taken her for the groupie type because he was talking that hotel bullshit and she was not on that kind of time.

Allexis was relieved, she was nervous about being alone with . . . "Dayum, I didn't get his name," she said out loud to herself.

"Who's that?" Passion asked as she followed Allexis to the bar. She watched as Kane posted up against the bar with his drink in his hand. It was dark in the club but she could see that his eyes never left Allexis, not even for a second.

"I don't know but he is fine, isn't he?" Allexis asked feeling her stomach doing little flops.

"Girl indeed," Passion agreed. But she already knew what Kane was about. He was arrogant a blind person could see that. But they were there to have fun, no strings attached.

The quartet sat around the bar, Allexis sipped on her rum and coke while Passion opted for a Long Island Ice Tea. Allexis and Kane gradually made small talk while Passion continuously gave Dame the hand or her middle finger. Kane never took his eyes off of Allexis, he was honestly intrigued by her, and she reminded him a lot of someone.

And despite her own shady past and her lack of trust for men, for some reason Allexis was able to relax a little with Kane. He made her laugh and she could tell by his appearance that his cash flow was steady. He was well known too, it was ridiculous how so many people went out of their way just to greet Kane or to be in his presence. But he never offered more than two minutes of his time because he was more worried about getting to know Allexis.

But once again they were interrupted when Kane felt a tap on his shoulder.

Twisted Obsession

"Nigga, where the fuck you been?"

Kane turned around only find his baby mama Rasheeda, standing behind him with a cream colored dress that appeared to be painted on practically hugging her like a second skin. She stood there prancing with her hand on her hip.

"Damn," he muttered.

Does this bitch ever learn? Kane was already had to stomp a hole in her ass for running her mouth earlier that week for trying to show her ass in front of his crew. It looks like she still hadn't learned.

"Nigga you heard me," Rasheeda snapped sucking her teeth and rolling her eyes.

Rasheeda knew she was entering dangerous territory, fuckin' with Kane's crazy ass. But she reaped a lot of benefits being Kane's baby mama, but it also came with a lot of consequences. Since their last falling out he hadn't returned any of her phone calls and she was low on money and desperate. Besides he could do whatever he wanted to her as long as she got her money.

"What the fuck you want Rasheeda?" Kane asked obviously irritated, and forgetting Allexis was standing behind him.

"Well, you would know if you would just answer your damn phone," she said showing her ass. "But if you must know, I need some money," she said tapping her foot impatiently.

Kane was often generous with his money as long as you stayed in place and did as you were told. But sometimes Rasheeda had to be reminded that Kane was in charge and what he said goes.

Allexis stood up on the bar stool and peered over Kane's shoulder to get better view of what was going on. She wasn't sure if the young lady was Kane's current or ex – girlfriend. Either way she really wasn't in the mood for drama.

But Rasheeda wasn't the least bit thrilled about the yellow bitch parading in on her business. "Bitch what the *fuck* is you lookin' at?" Rasheeda snapped sidestepping Kane and putting a finger in Allexis's face. She and Kane weren't together but she should be number one because she was his one and only baby mama. But she knew Kane didn't see things that way.

Twisted Obsession

Oh, no she didn't!

"Bitch who do you thinks you are talking too?" Allexis asked as she smacked Rasheeda's finger out of her face.

Before the situation between Allexis and Rasheeda escalated any further, Kane gently pulled Allexis by the arm, while Dame quickly escorted a frantic Rasheeda to the back of the club. But unfortunately for Rasheeda there was no escaping Dame's firm death grip. Her rude ass mouth had gotten her in trouble once again. Dame thought he had seen Rasheeda in the club earlier, but opted not to tell Kane. He was hoping he would get lucky tonight and it seems he might have.

After the little episode that just transpired, Allexis and Passion decided to call it a night. Rasheeda had killed their buzz and neither girl was any longer in the mood to party. But before Allexis and Passion could make their exit Kane stopped them.

"I know this is kind of a bad way to start things off, but in the little time that we were able to share together I can honestly say that I am definitely feelin' you. Is there any way that I could your number?"

Allexis had to admit to herself that she was feeling Kane too, and she honestly didn't want things to end, but she wasn't down for all the drama either. Kane sensed her hesitation and did his best to reassure her that it wouldn't happen again. Thinking a moment longer Allexis opted to take his number instead. She couldn't resist it was something about Kane, that had her curiosity on full alert.

"Make sure you holla atcha boy," Kane hollered as he watched them exit the club. *Damn the dark one is kind of thick*, he thought to himself as he made his way to the back of the club to handle Rasheeda.

* * *

Rasheeda paced nervously back and forth in the dimly lit room awaiting Kane's arrival. She knew she had messed up and was about to pay dearly."Fuck!" she said out loud as she ran her perfectly manicured fingernails through her one thousand20 dollar weave. Knots formed in her stomach as she continued to steal glances at the door, contemplating on making a run for it, but she knew there was no way she was going to get through Dame.

Twisted Obsession

Maybe he won't be so mad, she thought. *I'm pretty sure it was just some groupie hoe trying to cash in on his ass.*

But she was dead wrong. She jumped as soon as she heard the door knob jiggle. Her heart pounded heavily and she lost her breath once Kane entered the room full force with the look of death in his eyes.

"Kane I-," she attempted putting her arms in the air in hopes of shielding herself from the blow she knew was more than bound to come.

But her words were cut short when he grabbed her by the throat, temporarily cutting off her oxygen supply, followed by his fist connecting with her nose. The sound of her nose breaking was eerie. She hollered out in excruciating pain, and immediately placed her hands to her nose and blood oozed between the cracks of her fingers, dripping onto her cream colored Christian Dior dress. Dame looked on as if he were watching a ghetto version of *The young and the Restless.* The only thing he was missing was a bag of popcorn.

"*Bitch*, what I tell you 'bout that shit? Always runnin' yo' fuckin' mouth!" Kane yelled. He was seething with anger. He couldn't stand a mouthy ass female and he always felt women should be tame and submissive. He often felt if someone would have tamed his mama his life would have turned out differently.

"Kane, I'm-,"

"Shut the fuck up," he roared raising his fist threatening to hit her again. Rasheeda cringed in fear, still reeling from the first powerful blow she had received. "I told ya black ass *I* would call you when *I'm* ready," he said adding emphasis by pointing to his chest. "But I guess what *Daddy* says goes in one ear an out the other. Well," he sighed running his fingers through his locks. "Daddy just gone have to *show* you what happens when you don't listen! Dame gone give this bitch what she wants," he said fixing his clothes as he headed out the door.

Rasheeda desperately makes a beeline for the door as Dame hungrily grabs her. She attempted to fight him off, but it was no use. He skillfully pinned her down and flipped her over on her stomach. She should be used to this by now.

"Dame, please," she muttered as blood mixed with snot traveled down her face and mouth, before dripping off her chin.

But Dame was too blinded by his own lust. He ripped off her dress along with her black *Victoria Secret* thong. "*Damn!*" he muttered as he fumbled with his pants before pulling out his dick. Quickly he spit on his hand before spreading her ass cheeks he rammed nine and a half inches of massive thickness in Rasheeda's ass without mercy.

"Ahhh," Rasheeda hollered in pain. She knew it would take him a long time to come.

* * *

Twisted Obsession

Chapter 2

The constant beeping of the alarm clock rocked Allexis out of an uncomfortable sleep. She peeked from underneath her covers before angrily giving her clock a tap. "Damn it!" she said out loud as she rolled over on her side. She was supposed to be to work at 10, but she called her boss and told him she was running late, but that was over 3 hours ago. But she knew that was her own fault for going out last night knowing that she had to work in the morning. She assured herself that would be the last time.

"Humph, I might as well stay home now," she yawned. She sat up and rubbed the sleep out of her eyes.

"Yo' Allexis. You up?" Tina yelled banging on the door as if she were the police.

"Damn, I am now," Allexis joked.

"My bad, anyway I fixed us something to eat so if you want some you better hurry up. You know how greedy Doricka's ass is."

"Alright here I come," Allexis said pulling back her covers.

She stood up and stretched a little, before calling her boss and informing him that she wouldn't be in, Which she was sure that he already had an pretty good idea by now that she wasn't coming in, but she still felt she should let him know. Then she used the bathroom and washed her face before heading downstairs.

The delicious aroma Allexis encountered once she entered the kitchen caused her stomach to grumble. Tina had cooked bacon, eggs, fried potatoes, sausage, and scrapple. Both Doricka and Passion were already seated at the table. Doricka was stuffing her face by the spoon full as if she were afraid her food was going to jump up at any second and run away from her.

"Damn! You can slow down, it isn't going anywhere," Allexis said frowning up her face in disgust.

"Good Morning to you too," Doricka sang giving Allexis the finger.

Allexis gave Doricka a sarcastic smile before sitting down at the table and pouring herself a cup of juice. Tina placed a plate of food in front of Allexis, but despite how hungry she was she said her grace first. Her mother always taught her to bless her food before eating.

"I saw you two puttin' in some work last night," Doricka said pointing to Allexis and Passion while advertising a mouth full of food in the process.

"Ugh, swallow then speak," Passion said.

These some seriously stuck up bitches!

Swallowing before speaking this time she continued, "I saw you and Passion wit Kane and his boy Dame last night."

"How do you know Kane?" Allexis asked before sticking a piece of bacon in her mouth.

Doricka jerked her neck as if she had asked her a stupid question. She waved her hand in the air. "Girl, please who don't know Kane? He is one of the biggest hustler's around here. Please believe honey that nigga is rolling in dough. Matter of fact I think him and Dom do business together. Ain't that right Tina?"

"Yea something likes that," Tina replied.

Well, I've never heard of him, Allexis thought.

"We were just dancing anyway," Allexis stated shrugging her shoulders.

"Humph, I was glad when the song was over. I think that nigga was seriously trying to bust a nut," Passion said as she thought back to the previous night's events. She had to admit that dude looked good but it was something about him that turned her off and his horrible breath really sealed the deal.

"You looked like you were helping him," Doricka smirked. Passion *flipped* her finger.

Finally Tina joined them at the table. "What happened between you and Rasheeda?" Tina asked as she took a sip of her coffee.

"Who is she?" Allexis asked.

"His baby mama."

"Well are they together?"

"Girl, please! Rasheeda is a hoe. She just happened to get Kane's ass caught up, who happened to have her eyes on Dom at one point and time. So of course I quickly shut that ass down. But nah I don't think it's more to their relationship than that. But on the other hand, that nigga does have money so you know women be on his dick."

"Oh," Allexis said a little relieved that Kane and Rasheeda weren't an item but her being his baby mama did put a damper on her parade. "Well, Kane and I were just talking and she interrupted our conversation," Allexis said answering Tina's question. "That chick tried to come at me wrong, so I had to let her ass fucking know," Allexis stated folding her arms across her chest. Just thinking about the incident between her and Rasheeda made her blood boil.

Doricka, Passion, and Tina all burst out laughing. Allexis jerked her neck wondering what was so funny.

"What? I did let that chick know," Allexis said with an attitude, a frown creasing her for head.

"You know ya ass cannot cuss," Doricka laughed while holding her stomach. "You talk so proper, it's not even funny. Are you sure you and her related?" Doricka asked Tina.

Allexis rolled her eyes, "Whatever y'all."

"No, seriously," Tina said getting ahold of herself. "I've never meant him personally, but I've heard some interesting things about that dude. So just be careful okay," Tina said looking Allexis directly in the eyes.

"Geez Louise. It was just a dance, but I promise to be careful okay," Allexis sighed. But she was feeling some kind of way about Kane. He lived on the wild side and it kind of excited her. Her life was pretty much boring, and she felt she was down for the ride.

The quartet continued to sit around and gossip while enjoying the meal Tina prepared for them.

* * *

Despite what Doricka and Tina said about Kane and the incident at the club, Allexis found herself constantly thinking about him. She continued to fold and unfold the business card he had given her with his name and number as she lay sprawled out on her queen sized bed. She picked up her phone and dialed the first three numbers before hanging up.

"Allexis, are you crazy?" she asked herself out loud. "He is a hustler, and hustler's mean trouble."

But she couldn't get over how fine Kane was. She memorized by his smooth chocolate skin which looked good enough to eat, the chinky eyes and those dreads. She could definitely see herself running her fingers through his locks. She couldn't help but to reminisce about the night before when they danced, the way he pressed his body against hers. She screamed into her pillow in order to control the throbbing between her legs.

"Okay, Allexis get ahold of yourself," she said as she sat up. "And besides one phone call won't hurt. Will it?"

After five minutes of doing nothing but holding the phone in her hand she mustered up enough courage to give Kane a call. He answered on the third ring.

"Whad up?" he asked.

Silence!

"Whad up?" Kane asked again ready to end the call. He figured it was one of his many chicken heads prank calling him. He was about to hang up until he heard an angelic voice.

"Hello, can I speak to Kane please?" Allexis asked after finally finding her voice. She could feel the butterflies fluttering in her stomach.

"Who dis?"

"My name is Allexis. We met at the club the other night."

"Oh, yeah I ain't think you was gone call me," he said knowing exactly who it was because she was the only female he ever directly gave his number to. And he could remember that voice anywhere. It was perfect and distinct, very proper and he kind of liked it.

"I didn't either after what happened," she replied rolling her eyes.

He snickered, "Sorry about that. My baby mom gets a little crazy at times. It's my fault because she well taken care of. But I can assure you that it won't happen again."

"Humph, I hope not."

"So let's cut the small talk. When you gone let me take you out?" Kane asked getting right to the point."

"You sure are confident. What makes you think I would let you take me out?"

"Awe come on ma. You know you want a chance to run your fingers through these locks."

Twisted Obsession

Allexis felt herself blush. "Okay player, so where are we going?"

Dang Allexis what's the rush?

"I don't know. Where you tryin' to go?"

"Surprise me," Allexis smiled to herself.

"Okay, be ready around seven."

They conversed a little longer she gave him her address before they said their good byes.

By the time Allexis made it back home from getting her hair and nails done it was going on 5:00. She quickly ran upstairs and ran herself a nice bath. While her water ran she thumbed through her cd collection, *X-scapes Traces of my Lipstick* caught her attention. She popped in the cd, wrapped her hair and relaxed to the smooth sounds of *Softest Place on Earth.* For some reason that song had her on over drive, as she had naughty fantasies about Kane.

After soaking close to forty-five minutes, she dried off and oiled her body. She walked out into her bedroom and stood in front of her closet, unsure of what she should wear. After trying on various outfits she opted to wear a pair of Gucci boy shorts, a white belly shirt and white gladiator sandals. She added diamond studded earrings, and a gold locket her mother had given her for her birthday when she was younger. It held a picture of both she and Tina, and their mother.

After giving herself a look over in the mirror she was satisfied with her appearance. That's when she finally came to the realization that she was actually going on her first date at the tender age of nineteen. She kind of laughed at the thought since most girls have their first date between the ages of fourteen and seventeen. But she's been through a lot in her young life and going out on a date was the furthest thing from her mind. Now that her life has somewhat settled down, she thought she could find the time to allow someone special in her life.

This was definitely something new for her. She still has a little phobia of men and her biggest fear was actually being alone with Kane, but he seemed harmless. But the one man she ever truly loved made her view all men as untrustworthy. Tonight she decided to put her past behind her and move forward with her future. And she knew in order to

Twisted Obsession

she couldn't let what her father did to her control every aspect of her life.

"Where do you think you are going?" Tina asked startling Allexis out of her thoughts.

"Sorry," she apologized, taking note of the bewildered look on Allexis's face.

Tina stood back in awe of her baby sister all grown up. She thought back to the days when they were little and Allexis would tell every little thing that Tina and her friends did. And how Allexis use to whine when she couldn't tag along with Tina and her friends to the mall, and to the movies. The fond memories bought a happy smile to her face. She wished she could go back to those happier times.

"You could knock," Allexis said with her hand on her hip.

"Girl, please," Tina said giving her a sideways glance, and flopping down on Allexis bed.

Allexis laughed. "I'm going on a date with Kane. How do I look?" she asked twirling around.

"Awe look at Allexis tryin' to get an ass," Tina teased. Her sister was looking sharp and the hairstyle she chose accentuated her delicate face. It was styled in bun with a swoop on the side. It was a simple style but Allexis wore it well.

"Tina!" Allexis squealed immediately walking up to her full length mirror to inspect her backside.

"Sike, you know I have to mess with you. But you look really nice."

"Do I really have a booty though?"

"Yea, a little pinch."

"That will do."

They both laughed.

"But you do look beautiful. I just can't believe your actually wearing those shorts," Tina said which she was, because Allexis has always been so self-conscious and rarely ever wore anything revealing. "But you look good in them though."

"You know what I can't either, but thanks sis," Allexis said inspecting her backside once more.

"Allexis, please be careful tonight okay. I've heard some things about Kane and I don't want to see you get hurt. Make sure your phone is on and don't hesitate to call. And if

Twisted Obsession

something were to pop off and he's not taking no for an answer, here I got this," she said holding up a mini mase spray container. "It goes on your key chain."

What in the world?

"And either go for the balls or the nose."

"Okkkkay I don't think it's that serious," Allexis said rolling her eyes towards the ceiling. "Besides you're the one always talking about how much clout Dom has in the streets? But anyway what are you doing tonight?" Allexis asked attempting to take some of the attention away from herself.

"Don't be tryin' to change the subject. Even though you nineteen I still have to give you the ropes okay missy." Allexis rolled her eyes. "But Dom and I are hangin' out tonight."

"So you're not mad at him anymore?" Allexis asked Referring to the incident where one of Dom's exes came up to him and said hi while Dom and Tina were out together. Allexis thought Tina went a little over board when she busted home girls behind, nearly sending her to the hospital.

"No," Tina replied rolling her eyes.

"You just need to stop being so damn jealous. You know that boy isn't going anywhere. He pays all your bills, keeps you and me for that matter in the finest clothing and not to mention he just bought you a brand new car. You need to hold onto him because he is definitely a good man if he is putting up with your crazy ass."

"Whatever chick and besides it wasn't exactly new it had 4,500 miles on it."

They both shared a laugh.

But Tina knew Allexis was right. Dom had saved her from a hell she had created for herself. She didn't know where she would be if it weren't for him.

* * *

Kane pulled up to Allexis house around 7:30 in his chromed out Escalade. He hopped out of the truck allowing the cool breeze to caress his skin. He admired Allexis cozy looking home. It stood out to him; it was a two story brick colonial, with a white picket fence and rose garden. It reminded him of something he would see in a magazine. It was impressive by far one of the biggest if not the biggest house in the area. He straightened

Twisted Obsession

his collar before knocking on the door.

Dayum! Kane thought to himself taking note of the beautiful creature before him. He had to stop himself from staring.

"You must be Kane? Hi, I'm Tina Allexis sister," Tina said extending her hand.

I don't remember him being this fine.

"It's a pleasure to meet you," Kane replied kissing her hand.

"You can come on in, Allexis will be ready in a few," Tina said stepping aside to allow him to enter her home. She slyly wiped the back of her hand on her pants.

Kane couldn't help but to watch as her round bodacious ass moved from side to side with each step she made, tantalizing him just a bit. He felt his dick jump. *Easy boy!*

"You can have a seat," she said directing him to the love seat.

Kane plopped down getting comfy hoping she would join him. If she sat down beside him he knew he had a chance.

"Would you like a drink?" Tina offered.

"No, I'm good."

"Okay cool," she said opting to remain standing.

Damn, he thought to himself as he admired the way her jeans hugged her thick thighs.

He looked around admiring their home. It was nice and very family oriented. Different forms of afro centric art decorated the walls. Maya Angelou's *Phenomenal Woman* hung above the fire place. He noticed a piano in the far corner of the living room and wondered if Allexis played.

"You play?" He asked nodding towards the piano.

"*No* that's just for decoration," she laughed.

There was a long uncomfortable silence before Tina finally decided to clear the air. "On a more serious note make sure you take care of my baby sis okay. She's all I got. And no funny stuff," she said with a serious look on her face. "This is-," she started to tell him this was Allexis first date but opted not to.

"Look, it's not the type of party," Kane calmly said although he was two seconds from putting her in her place. "I just want to take her out and chill and see how things go from there. Sounds good?"

Twisted Obsession

"Sounds good," Tina said skeptically. She was thrilled about Allexis going on her first date, but she wasn't thrilled about her *date*.

"Okay, I'm ready," Allexis announced descending the stairs. "How do I look?" she asked Kane shyly.

"Beautiful," Kane answered stuck in a daze. *She looks just like her.*

"*Kane!*" Allexis said waving her hand in front of him, a little creeped out by the look in his eyes.

"My bad ma," he said shaking his head a little. "You got a nigga in a daze," he stated giving her body a once over.

"Thank you," Allexis blushed.

"So where are you guys going?" Tina interrupted. She wasn't feeling the way Kane was eyeing her sister, it gave her the chills.

"I haven't decided yet. But I'll think of something," Kane replied never taking his eyes off of Allexis.

"You have your cell phone right?" Tina asked.

"Yes, Tina," Allexis said as she irritably waved her blackberry in the air.

Tina followed Allexis and Kane to the door. She was far from ecstatic about her sister going on a date with Kane. There was something about him, something in his eyes that made her feel uncomfortable. She stood and watched with her arms folded across her chest as Kane helped Allexis in his truck.

Nice wheels, she thought. She noticed Dom pulling in beside Kane and remained in the door way so that she could let him in.

"Hey babes," Dom said giving Tina a kiss on the cheek. "What you doin'?"

"Watchin' Allexis leave with her *date*," she answered dryly.

"Awe shit. Not Ms. Goody ass finally lettin' a nigga take her out," he joked. "Tubby been tryin' to holla at Allexis for a minute and she wouldn't even shell out the digits. Hold up a minute," he said as he thought back to who was pulling out of the driveway when he pulled up. "Who's her date?' he asked instantly dreading the answer.

Tina rolled her eyes. "Well unfortunately she's going out with the infamous Kane," she said trying to shake the funny feeling in the pit of her stomach.

Twisted Obsession

"Damn! How the hell did I know you were going to say that," he said shaking his head. Dom knew Kane's rap sheet all too well, and he was known to be abusive and very controlling with his women. He couldn't understand how Allexis could have gotten in with the likes of him.

"Yeah, tell me 'bout it," Tina said as she followed Dom out into the living room. He could only shake his head. "Yo' tell sis to be careful cause dude ain't nothin' nice."

Dom peeked outside once again watching as Kane pulled off. He had a funny feeling and was thinking about relocating Tina now that Kane knew where she lived. He did most of his business in New York just to keep his enemies from harming both Allexis and Tina. But unknowingly Allexis just may have invited the *enemy* in.

* * *

Damn this is a nice ride, Allexis thought as she and Kane traveled towards Virginia Beach. She couldn't help but to glance over at him periodically. He was fine, and she was already becoming infatuated with him. He wore a short sleeve Sean Jean dress shirt, with a beater peeking underneath, a pair of Sean Jean slacks, and a pair of Stacy Adams decorated his feet. His long dreads were pulled neatly back into a ponytail. The statement Kane made earlier about Allexis wanting to run her fingers through his locks only made Allexis blush, because if given the chance she knew she would.

She was a tad bit nervous her heart beating a mile a minute. He seemed nice enough but she couldn't help but to be a little shocked that she had actually gone out on a date so quickly with a guy she barely even knew. But she couldn't help herself, for some strange reason she was drawn to Kane.

The silence was making Allexis feel a little standoffish so she decided to do her best to relax and have some fun. "So do you know Dom?" she asked recalling what Tina said earlier about Dom and Kane doing business together.

"Yeah, you can say that," he answered not really wanting to get into detail.

"Oh, okay. So you think you all that?"

Her question caught him off guard. "What?"

"Do you think your all that?"

"Nah, why you say that?"

Twisted Obsession

"Just asking, but I figure a fly guy like you would say yes."

He gave her a sideways glance and shook his head. "You somthin."

She laughed. "I know right. Anyway so do you like sports?" Allexis asked starting to feel comfortable already.

"Definitely, I like football, basketball, and I watch a little tennis. You?' he asked not really taking her for a sports fanatic.

"I love basketball, and I like tennis. I watch a little football, and oh I like volleyball too."

"Oh, okay so who's your favorite basketball team?" he asked arching his eyebrow, and a little surprised to hear she liked basketball. To him she seemed more like a girly girl.

"I really don't have a particular team, I like players."

Kane curled his lips, "Bandwagon jumper."

"Whatever!" Allexis smiled because Dom often called her the same thing. "No I just like players. I like Steve Nash, Dwight Howard, Raja Bell, D wade, Chris Bosh, Chris Paul, Kevin Durant, Rajon Rondo, Kevin Garnett, but my number one is *Lebron James*."

"*What*! Awe you like that panzi ass nigga, he always whinin'."

"And he can afford too. Man he be balling. There isn't really anybody out there that can touch him right now. Well Kevin Durant is getting pretty close, but other than that I don't think so." Kane gave her another sideways glance. "Okay so who do you like?"

"I'm a Lakers fan. Kobe Bryant is one of the best, if not the best in the game right now. Lebron James, *please*," he smirked.

"What? The way Kobe is always crying and complaining," she said rolling her neck a little. Allexis had to admit that Kobe was one of the best in the game but she wasn't about to tell Kane that.

"Girl, Lebron can't even begin to touch Kobe. Besides it doesn't even matter because Kobe has four rings," he said while holding up four fingers. "And Lebron has how many?" he asked leaning over waiting for Allexis to answer, but she could only smile. "That's what I thought."

"Anyway that's okay Lebron is going to get his. He didn't have the help that Kobe had; because Shaq helped Kobe to get all of those rings. It wasn't just the Kobe Bryant show it was always Shaq and Kobe. That's why Shaq ended up leaving because Kobe was mad that Shaq was getting a lot of hype and he wasn't," she stated matter of factly crossing her arms across her chest.

Kane only laughed giving her a wave. He was actually enjoying himself. It reminded him of old times with someone special. And he actually allowed himself to relax. He wasn't usually uptight but it was something about this girl that made him feel indifferent and unlike his natural self.

"So you have children?" Allexis asked. But that only made her think if Rasheeda and she wondered if she and Kane made this work could she actually tolerate her. She hoped he didn't have a whole lot of baby mamas. *Okay Allexis you are definitely getting ahead of yourself*, she thought.

"Just one," he said holding up his index finger.

"Okay," Allexis said nodding her head liking the sound of that. "How old?" She looked over at Kane who seriously looked like he was in deep thought. "You really don't know how old your one and only child is?" she asked with a surprised expression on her face.

He laughed, "Come on ma you know men don't keep up and down with that shit," he answered seriously. But the truth was he didn't spend enough time with his child to know. He couldn't bring himself to bond, with the child because of its mother. To him Rasheeda was far from classy and he cursed himself for the day, she set him up and he has learned his lesson since then. Since Sean was his seed he felt obligated to at least be there for him financially. "But I'm going to say around two or three. Come on cut me some slack," he said taking note of the puzzled look in Allexis eyes.

"Okay," she said rolling her eyes towards the ceiling. "Is it a boy or girl?"

"A boy, you sure are asking a lot of questions. What are you ready to play step mom?" he asked jokingly. But from the expression on Allexis face he thought she was actually considering it.

"Maybe," she joked, honestly feeling some kind of way because he asked her that. "Okay one last question about the little one and then I'm done."

"Shoot."

"What's his name?"

Twisted Obsession

"Sean."

"Sean, I like that," Allexis smiled already in a daydream of her own about playing mommy. She thought she would make a good mother, nothing like her own mother, because she vowed to protect and keep her child from harm. She smiled when she noticed Kane staring at her. "So where are we going?"

"You know what I still don't know yet. So why you just set don't back and enjoy the ride."

He fiddled with some buttons on his stereo before R. Kelly's *Twelve Play* filled the truck. So Allexis did just what Kane told her, allowing R Kelly's smooth voice to penetrate her mind.

Kane took Allexis to a nice exclusive restaurant in Virginia Beach. For both there was an instant connection. Their conversation flowed gracefully as if they had known each other for years. They learned a little more each other. Kane discussed his current business, his chop shop how he rebuilds and details cars. While Allexis discussed she was currently in her first year of college and how she worked part time. Despite their feelings towards each other neither one was quite ready to discuss their dark past.

After dinner they went for a walk on the beach holding hands. Allexis loved the feel of the sand in between her toes it made her giggle, and the sound of the ocean's waves was soothing. And the moon overshadowed the ocean just right giving it a romantic feel. And to her own surprise she wasn't nervous, but very relaxed.

"You okay ma?" Kane asked. "You kinda quiet."

"Yeah I'm alright," Allexis replied feeling a slight buzz from the wine she had earlier with dinner. "Just enjoying the moment," sighed, looking up at him and giving him a smile.

"Yeah, me too," he said feeling a warm spot in his heart.

"What's your full name?"

"Why?"

Allexis shrugged her shoulders. "Just making casual conversation, if that's alright with you?" she asked smiling.

He couldn't help but to smile before answering. "My full name is Dion Kane Stanley."

Twisted Obsession

"Kane is an odd name."

"Yeah, but that's why I like it. Dion is average and Kane was my mother's maiden name."

"Was? Did something happen to her?" Allexis asked. Immediately Allexis felt the tension in Kane's hand, and she knew she must have hit a sore spot. "I'm sorry," Allexis whispered taking note of the distant look in his eyes.

Kane snapped out of his trance like state and looked down at Allexis, "It's cool, and it's just that my mother died a long time ago. And I'm really not comfortable talking about it."

"I'm sorry," Allexis apologized again.

"It's cool ma. You didn't know."

As they continued to walk along the beach, Kane couldn't help himself as he constantly stole glances at Allexis. She was so beautiful, he admired her innocence. He was used to chicken head bitches that showed they asses and played a lot of games, but she was different. He liked what he saw and the way she conducted herself like a lady. That's what he was looking for in a woman. And she reminded him so much of his baby sister. The resemblance between the two was uncanny, even almost creepy.

Kane could no longer control his urges. He needed to feel her, and taste her. He grabbed an unsuspecting Allexis who was still caught up in the moment of being on her first date and pulled her closely to him. Cupping her beautiful baby doll face in his strong hands, he kissed her ever so gently. Allexis panicked a little and attempted to pull away.

"Relax baby," he said looking deeply into her eyes. "I will never hurt," he said with sincerity.

As Allexis gazed into his beautiful brown eyes a sense of safety came over her, and once again she felt her muscles relax. To her at that moment Kane could do no wrong, and just like that she was putty in his hands. He kissed her again, but this time more passionately, pulling her even closer to him. A slight moan escaped her lips once Kane's lips traveled to her neck, while his hands became disobedient and traveled down her body.

He pulled her hair causing her bun to unravel, and it fell loosely around her shoulders and down her back. While Kane held both of Allexis breast in his hands they continued to explore each other's mouths with their tongues. Another moan escaped Allexis lips once Kane began to make circular motions around her nipples with the tips of his fingers, causing her knees to buckle. At that point Kane knew he had her exactly where he wanted her as he eased her down into the sand.

"Kane I-," Allexis attempted to protest. There was no doubt her body was jumping for joy but she couldn't help but to feel a tad bit guilty about what was taking place.

"Shhhh," he said placing a finger to his lips. "Its okay baby tonight is all about you. I want you to feel good. Let someone take care of you. I'll show what it's like to feel good."

Slowly and gently he pushed her back until she was lying flat on her back. And in one swift motion, Allexis shirt was over her head with her bra in toe before she could even blink. Not giving Allexis a chance to protest he placed one of her pink nipples into his awaiting mouth.

"Hmmmm," Allexis moaned and arched her back once Kane's warm tongue began to make circular motions. While Allexis enjoyed what was going with the top half of her body, Kane fiddled with her shorts so that he could explore the bottom half as well. "Kane," Allexis said feeling the need to protest but not really wanting to.

"Don't worry baby," he whispered once her shorts and thong were completely off. But he could go any further he had to sit back and admire the beauty before him. She looked so exotic lying naked with her hair spread out in the sand. And the look of pure pleasure in Allexis eyes turned him on.

This time he started from the top, kissing her eyes to her toes before traveling back up towards her center to her neatly shaved pussy. Spreading her legs apart he came eye to eye with his victim before taking it hostage in his warm mouth. Kane worked his tongue quickly and sharply around Allexis clit. Instantly Allexis jumped from the feeling her pussy was experiencing. Jolts of pleasure sprang through her body causing it to tingle, and shutter in pleasure.

"Kane," she moaned as he continued to go up, down and around and around. She grabbed his head and began to hump his face intensifying her pleasure. "Oh, please don't stop," she begged opening her eyes temporarily looking down at him.

They locked eyes for a brief moment. Allexis seductively licking her lips before closing her eyes again to enjoy this new found bliss. Kane obeyed her commands as his tongue continued to do tricks around her clit, while he continued palming and fingering her tight wet pussy.

"I think I'm about to cum," Allexis squealed her eyes rolling in the back of her head, as she dug her fingers in the sand as she screamed out *"Kane!"* Allexis body jerked and she began to convulse. Never had she experienced anything like that before.

Kane continued to slurp her juices never taking his eyes off her, while Allexis continued tremble from the pleasure she had just received. After her convulsions ended Kane wiped his mouth and kissed Allexis softly on the lips. He then picked her up and carried her to his house. He somehow managed to his house, up a flight of stairs and into his master bedroom where he laid her gently on the bed. He stood and watched her sleep for a moment before heading to the bathroom to take a shower. Kane wasn't upset that he didn't get the pussy, he knew eventually he would and vowed to be the only man that did.

* * *

Twisted Obsession

Chapter 3

"I don't know 'bout this yo'," Neko said as he pulled hard on his blunt, coughing a little once the smoke invaded his lungs.

"Come on nigga. I told you Deuce is ridin' solo tonight. Just follow the plan like I said," Dame stated.

But Neko was feeling uneasy about the whole situation. What Dame wanted him to do was suicide if Kane ever found out. He was already scheming Kane's product now but according to Dame Kane would be long gone by the time he found out. On the other hand Neko wasn't so sure and this heist Dame cooked up was just too risky.

Kane supplies drugs for a couple of cats in New York and about once a month he would send a car that way. And Dame had come up with the idea of robbing Deuce before he made the drop in New York. Neko thought it would make better sense to wait until the deal was done, then Rob Deuce once he makes it back to Virginia. That way they will already have the money. But Dame thought because he has been in the game longer that his thinking and planning was more logical.

After a long moment of silence Neko decided to go ahead and do the job, but he was going to do it his way without Dame knowing. And besides he had to show his loyalty and that he was down to ride. Dame promised to make his second in command once Kane was out of the picture. Which he hoped would be sooner rather than later. The fear of Kane finding out about him being unloyal caused him to lose a lot of sleep at night.

"Aight, I'll call you once everything has been done."

"Make sure you go alone," Dame warned. "We don't need a whole lotta niggas in on this. You know niggas gossip more than bitches do."

"What you take me for?" Neko asked feeling a little giddy from the weed. But it was already too late for that pep talk.

"Good, I'll check you lata."

Dame stuffed his phone in his back pant pocket of his Sergio jeans. He scanned his surroundings as he sat in his Ford Expedition. Being in the game he knew he had to be on guard. He stroked his goatee as he checked for anything out of place, everything seemed kosher. He checked the mirror and admired his reflection.

Twisted Obsession

"Damn, I look good," he said out loud before hoping out his truck. He swiftly walked towards the hotel ready to carry out plan B just in case plan A didn't work.

"Where the fuck is he anyway?" Detective Coleman asked, loosening his tie.

"He out wit some bitch," Dame replied taking a pull from his blunt.

Detective Coleman frowned his nose as he watched Dame blow smoke in the air. He turned around and peeped out the window. He wanted to make sure no one knew his whereabouts. Detective Coleman was a short stubby man with a round stomach and plump ass. For the last five years he has been after Kane, but each time Kane somehow managed to slip through his fingers. He knew Kane had officers on his payroll, so he decided to work exclusively on this case only certain personnel could accompany him. And thanks to the new addition on Dame, Kane's second in command he may finally have the means to put Kane's black ass behind bars. Where Detective Coleman felt Kane so rightfully belonged. Just the thought of the victory put a smile on his face.

"So do you have enough evidence to arrest him yet?" Dame asked growing impatient as he tapped the armrest of the chair he casually lounged in.

"Well, not exactly," Detective Coleman lied. "I need his connect to seal the deal."

"What the fuck you need all that for? You got all his spots," Dame said aggravated. He was ready to get this done and over with. All the sneaking around made him irritable, causing him to lose a lot of sleep at night. "Besides I don't even know who the nigga is. He will only do business with Kane."

"You mean to tell me you are Kane's number one lieutenant and you never met his fucking connect?" Detective Coleman smirked.

"*Man fuck you!*"

"Oh, you will," Coleman said smiling. "It's just that I know the DA thinks we will have a stronger case if we bring down Kane's connect as well."

The truth was Detective Coleman had enough evidence to bring down Kane and his entire operation. He was just being greedy. He wanted to secure another bonus, by catching Kane's connect. Doing so would be huge for his career; it would look good on his resume when he ran for Mayor.

"Yo' I have to ask is there more to this story?" Dame asked. He knew Detective

Twisted Obsession

Coleman wanted to get Kane, but it seemed he was really obsessed with the case.

"Huh?" Detective Coleman asked caught off guard by the question.

"I mean you just really seem to be after Kane. It's something in your eyes that makes me think it's more to the story that you're not telling me," Dame said as he smashed the last of the blunt into the astray.

There was no doubt it definitely was more to the story, but Detective Coleman had no plans of telling Dame that. "Nah, it's strictly business."

"Good, are we done talkin'? So that *we* can get down to *business*." Dame asked grabbing his crotch as he stared hard at Detective Coleman.

Coleman walked over to the King sized bed and unbuckled his pants allowing them to drop to his ankles. "Yeah, I 'm ready daddy," Coleman giggled as bended over.

Dame stood up with his dick in hand and stood directly behind Coleman. He then separated both of Coleman's ass cheeks before licking the crack of Coleman's ass from top to bottom before easing his dick into Coleman's donut hole. Coleman squealed in delight squeezing his butt cheeks as tight as he could around Dame's dick. Basking in the pleasure of each thrust Dame sent his way

Detective Coleman and undercover homosexual never minded the weight he gained over the years, taking pride in the shapely form of his backside. Meanwhile Dame grew up without a father and has always secretly longed for the affections of a man.

* * *

After his late night rendezvous with Dame, Detective Coleman decided to meet up with his own bunch of recruits. They all met at the comfortable home of his mistress Shelley. They sat around her dining room table sipping on cheap wine, and homemade cheese with crackers.

There was Detective Cruise his best friend and partner for fifteen years. And then he had Officer Dionte a rookie but Detective Coleman could easily tell that he was young and eager to get criminals off the street. So Detective Coleman figured he would try to get him while the young kid was in his prime. And then there was D.A Marissa Coleman the best in the state whose never lost a case. With this team Detective Coleman was more than confident they could bring Kane down.

Twisted Obsession

"Okay so far we really don't have any concrete evidence against Kane," Marissa stood up and said after a long moment of silence.

"What?" Detective Coleman said with a frown on his face. "What about the video surveillance tapes. And we have plenty of witnesses that are willing to testify against him."

"I know but it's all circumstantial. We have him going in and out of his houses but he's never been caught handling any product. We need something concrete just in case some of our witnesses do come up with a sudden case of amniesia."

"What about his phone? It's been tapped right?" Officer Dionte asked taking a bite of his stale cracker.

"Kane's too smart for that. He has a house phone but rarely uses it. Believe me I checked," Detective Cruise added.

Detective Coleman nodded his head in understanding of Marissa's point of view. "Okay as you all know that I am working with Dame Kane's second in command in hopes of getting new evidence. I've asked him to see if he can find out who Kane's connect is. Maybe in the process of doing so we can finally catch Kane's ass in the act of doing something, and he's willing to testify," Detective Coleman added, but he seriously doubted he just hoped it would give the team a little more umph.

"You think this will work? And can this Dame character be trusted? I don't think so from the looks of him he looks like a snake," Marissa stated.

"He's fine and yea I assume it will work," Detective Coleman said or at least he hoped. "This could be our only way of actually catching Kane doing something in the act." Detective Coleman stated shifting in his seat. His case appeared to be full of loop holes and it didn't sit too well with him.

"Can I get you guys anything else?" Shelly asked barging in on the crew once again. She was a rather tall woman, taller than Detective Coleman matter of fact. Stood about 6'2 with broad shoulders and a very thin frame, she wasn't curvaceous like the average woman but a straight up and down figure which suited Detective Coleman well. Despite her hard shape she had a very pretty face with delicate features, soft brown eyes, a very thin nose, and square jaw line. Her blonde shoulder length hair often sat in a messy bun at the top of her head. She wasn't what you would call beautiful but average looking, enough to catch a couple of eyes.

"Can I have a glass of water if it's not too much trouble," Marissa smiled taking note of her large hands as she picked up the empty wine bottle and replaced it with a new one.

"Dammit Shelly," Detective yelled banging his hand on the frail wooden table. "Were trying to fuckin' work here. We don't need any more of this fuckin' shit. Now go upstairs and if I need something I will call you."

Shelly jumped placing her hands to her throat. Her lip quivered so she bit down on it hard to prevent herself from crying. Disowned by her family, very few friends and with Detective Coleman being married she was very lonesome and often left alone. It felt good to have company and here he was messing it up for her.

Detective Coleman looked away, he couldn't stand the pained look in her eyes. He knew he was wrong but he was frustrated and she was his outlet. Without another word Shelly spun on her heels and quickly left the room. Officer Dionte shot daggers Detective Colemans way, he liked Shelly and felt he was too harsh on her. He felt no woman should be talked to or treated that way, especially Shelly he had a real soft spot for her.

"Okay it boils down to this. We have to get closer find a way to bug his house, his businesses. Whatever it takes," Detective Coleman said standing up. "Now if were done here this meeting is adjourned."

"I'll see if I can possibly do about the bugs," Detective Cruise offered. "Hey were going to get him," he said taking note of the distant look in his partner's eyes even though deep down inside he wasn't so sure himself. He knew Coleman had put a lot of hard work and effort into this but he felt they clearly didn't have the man power to go against Kane. But his partner was dead set on bringing Kane down himself and refused to seek anymore outside help because it was to risky.

Detective Coleman simply nodded his head nowhere near as confident as he was when the meeting first began.

* * *

Twisted Obsession

Chapter 4

Allexis jumped up unsure of her surroundings. She looked around before replaying the events of the night before. She dug her fists in her eyes in hopes of removing the sleep out of her eyes. Once she was finished she was greeted by her knight in shining armor.

"Good morning sexy," Kane smiled carrying a breakfast tray he had his cook Carmelita prepare. He was already dressed and looking fresh. "Look I have to make a run, but I left you the keys to the Nav over there," he said nodding with his head towards the dresser before sitting down on the bed placing the tray in front of Allexis. "You do know how to get back to Norfolk right?" Allexis nodded her head yes. "Just checkin'."

Did he just say Navi, Allexis realized?

"Where are you going this early looking this fly?" she teased as positioned herself comfortably on the bed, covering her upper half with the covers since she was still naked. She used the wet ones on the tray to wipe her hands before sticking a piece of bacon in her mouth.

He ignored her question. "I had fun last night."

Allexis felt herself blush.

He smiled, "Look I know we just met but I'm really feelin' you. I don't know it's just something about you. So why don't you let me take care of you? All of this could be yours."

The sincerity in his eyes was evident and she had to admit she was feeling him too, and wouldn't mind having Kane take care of her but she felt they were moving too fast. Last night was the furthest she has ever gone with a guy, of her own free will anyway. But she wanted to take things just a tad bit slower and she didn't want to give him the wrong impression of her.

"I do think we are moving a tad bit too fast," she answered shyly as she twirled a strand of hair around her finger.

Damn, he thought. He loved her innocence. He wasn't pleased with her answer but he had a feeling that was the answer he was going to get. But he wasn't going to force the issue, he had a feeling she would change her mind soon.

"Okay ma when you're ready I will be here," he said kissing her on the cheek, already claiming his prize. He continued to stare, causing Allexis to blush again.

"Kane, why are you staring at me like that?"

He caressed her cheek ever so gently. "You remind me of someone from my past, a sweet, innocent girl that had my heart."

Allexis was unsure of how she should respond to that comment. She didn't know if she should be jealous or not. After a few moments in thought she finally asked, "What happened to her?"

He looked deep into her eyes before answering. "I let her down," he said putting his head down. He squeezed his eyes tightly shut, doing his best to suppress memories from his painful past.

"Oh, I'm sorry to hear that," Allexis said placing her hand over his, sensing his pain.

He moved his hand and ran it through his dreads. He shook his head a little in order to relieve himself from his own thoughts. He hated getting caught up in his emotions. To him showing emotion like that was a sign of weakness.

He smiled at Allexis, "Make sure your back here around eight. I have a surprise for you. Aight baby girl I'm out," he said giving her a peck on the cheek. "Oh there's the bathroom and the Nav has GPS just in case you get lost." With that he left, not the least bit concerned about his worldly possessions, this was a little test and besides he knew where to find her if anything came up missing.

Allexis laid back down getting lost in the comfort of his enormous bed. She couldn't believe he trusted her in his home and left the keys to his Nav. She felt like she hit the jack pot. She grabbed the pillow Kane had slept on and held it close to her body. His Sean Jean cologne lingered, filling Allexis nostrils with the sweet scent.

Laying there she thought back to what happened to the night before. The way he looked at her and the way he touched her. She often wondered how she would react when the time came for her to be intimate with a guy. And to her it honestly went well. It felt good to see that what her father did to her didn't tarnish her completely. Kane said he would make her feel good and he did just that.

Allexis hated to admit it but she was completely infatuated with Kane, and felt she was falling in love already. As much as she wanted to lay there and daydream she knew she

Twisted Obsession

had to get up. So slowly she dragged herself out of bed and walked over to the dresser where Kane laid the keys to the Navi. She picked up the envelope with her name written on it and opened it. She counted its contents and the total was five grand.

Humph, he's trying to give Dom a run for his money.

Being nosey she rummaged through Kane's dressers and closet, trying to find a hint of any other woman, when she found nothing she was a little relieved. But that was only a small part of his house; since he allowed her to stay she was going to take advantage. That was as soon as she took a shower. She grabbed a pair of Kane's basketball shorts and a beater before heading to his bathroom.

After taking a bath in Kane's Jacuzzi she found herself wandering from room to room. And to say his house was huge would be an understatement. Fifteen rooms in total she believed she had counted. Allexis couldn't help but in awe of her surroundings her family wasn't poor but they weren't this rich either.

She descended the stairs once again. "I can't believe he carried me up all these steps," she said out loud.

Allexis admired the chandelier that hung from the ceiling in the center of his living room. The crystals were shaped just like diamonds and his plush carpet was so thick she felt her feet would get lost in it. On the mantle above the fireplace held a picture of a beautiful black woman she assumed was Kane's mother. She had those same chinky eyes. And Allexis immediately fell in love with his kitchen despite the fact that she couldn't cook. The floor was made of marble, a grayish color with a hint of green. And the cupboards were made of stone with glass doors.

Your love is a one in a million Allexis sang. She looked down at her caller ID.

"Damn I forgot to call her. Hello."

"Where are you at?" Tina asked.

"None of your business, I'm grown."

"Girl, quit playin'."

Allexis laughed, "I'm at Kane's house."

"Ewe Allexis you nasty, and already I didn't think you would get down like that. And you could have called and let your sis know. I called you like three thousand times already," Tina said sounding more like her mother than her sister.

"Sorry," Allexis said rolling her eyes towards the ceiling.

"So, when you comin' home?"

"I'm on my way now."

"Well how you gettin' here?"

"Damn, none of your business. Goodbye," Allexis said pressing the end button on her cell phone. She pulled her hair back into a ponytail, taking one last look at her surrounding she headed out the door.

Damn, I could really get use to this.

* * *

"Who the hell is that?" Doricka asked Tina taking note of the lavish Navigator that pulled up in front of Tina's driveway.

Tina peeped outside. "I don't know," she said shrugging her shoulders. "But that's a tight ride though," she said admiring the deeply tinted windows. She strained to see who it was, with Dom 9 mm.

"Awe no this chick didn't. That's Allexis," Passion said excitedly.

"Not huh, you lyin," Tina said.

"Ooooh, you know what, it sure is. You should be able to recognize her pale ass anywhere," Doricka said.

Tina laughed before mushing Doricka in the back of the head.

Allexis bounced from around the side of the truck with her keys in hand as she made her way up the sidewalk with a little more pep in her step. The weather described her mood, sunny and bright. She smiled as she thought about her date with Kane. Before she could even stick her key in the key hole the door swung wide open, with Doricka, Passion and Tina damn near ready to fall out.

"Damn, you got it like that?' Tina asked rolling her neck.

Allexis rolled her eyes. "Girl, *whatever!*" she said as she squeezed past the trio and moved into the living room with them all following closely behind.

"You let him hit it *already*? Damn he ain't waste any time did he?" Doricka asked with a hint of sarcasm in her voice.

Twisted Obsession

"Allexis did you at least use protection? Girl, I just don't know what to think," Tina said shaking her head with her arms folded underneath her chest.

"Uggg!" Allexis said as she flopped down on the couch. "The last time I checked I could have sworn I was grown. Besides he didn't touch the pussy. Not with his dick anyway," she smiled devishly.

"Ewe, I want details," Passion said joining Allexis on the couch.

Tina wasn't the least bit thrilled and wasn't even sure if she wanted to stick around and hear the details. She couldn't believe Allexis was getting down in nasty and with Kane of all people. She wanted so more for Allexis, but what could she say when she was dating a hustler herself.

"Okay," Allexis said smiling. She was in a zone as she kindly interrupted Tina's thoughts, by giving explicit details of the events that took place the night before.

"Are you sure you know what you are doing?" Passion asked, thinking Allexis was way in over her head and way out of her league. Passion really liked Allexis a lot but thought she was very naïve. She felt it was okay for Allexis to have fun, but the look on Allexis face let Passion know that girl was falling and hard.

"Girl, I hope so," Allexis answered truthfully. "Anybody want to go shopping? He dropped me five g's," she boasted.

"I'm down," Passion said eagerly.

"Awe that's just chump change," Tina said waving her hand in the air trying to put a damper on Allexis mood. "But y'all two go ahead I'm not up to shopping today," Tina said as she was deep in thought.

"Well you might as well enjoy it while it last," Doricka smirked, raining in on Allexis parade as well. "Next month he'll be looking for the next new pussy in town."

"*Damn,* y'all really hating on a chick right now!" Allexis said before standing up and stomping out of the room.

"Doricka you really need to stop hatin'," Passion said with an attitude.

"What it's the truth," Doricka replied.

The truth was Doricka was jealous. She been after Kane since he got in the game, but he never tried to get down with her like that. So one night at the club they both happened to chillin in vip and he asked her to go back to the hotel with him which she quickly obliged. That night Doricka whipped her best ever head game just for Kane, but that didn't make that nigga stick around. He dropped her five hundred and bounced. She hadn't heard from him since. She had to wonder what made that high yellow bitch so special.

* * *

Doricka opted not to go shopping with Allexis and Passion. She wasn't in the mood to be around a bunch of females. She knew she was being sheisty, but she couldn't help it. The way she looked at it, Tina had Dom and now Allexis had Kane. When would she be able to find a nice rich dick to call her own?

"Man, fuck them high yellow bitches," she whispered to herself, even though deep down she was feeling guilty but didn't want to admit it to herself.

"You say something?" Toi the Asian nail stylist asked her as she put the finishing touches on Doricka's nails.

"No, me talking to me self," Doricka smiled.

Toi shook her head, "Not good."

"Your right," Doricka laughed as she admired the money tips on her fingernails. She decided to treat herself to a fresh manicure and pedicure. The tips on her fingers had her feeling a tad bit better.

She felt her phone vibrate in her jeans, careful not to smudge her nails she dug in her jean pocket and retrieved her cell phone. But immediately rolled her eyes and sucked het teeth once she realized it was Tina. She didn't even bother to answer it but simply stuffed it back in her pocket. Guilt found its way in her heart. She was being selfish and she knew it. But she couldn't help it; she couldn't help but to feel as if she were living in Tina's shadow.

Doricka stepped out of the nail salon, but not before putting on her Fendi shades. She was dressed down with just a red tee shirt which she had tied in the back to show her belly ring, a pair of blue skinny jeans and her timberland boots despite the fact that she just got a manicure.

"What's up ma?"

Twisted Obsession

Doricka turned her head only to face a suave looking brother. He looked familiar but she couldn't quite place him yet.

"What's yo' name?" he asked as he stroked his goatee allowing his eyes to travel up and down her small yet curvaceous frame.

"Doricka," she replied seductively. "And yours?"

"Dame," he replied. But he could care less about her name. Shorty was fine as hell. *Damn that's right*, she thought to herself. *How could I not recognize his fine ass? Humph second best ain't bad.*

"You think I could get your number so I can holla at you lata? Maybe do dinner and a movie," he said pleased because from the look in her eyes once he said his name he knew it was a rap and believed he would be banging shorty by the end of the night.

Sure," Doricka said shelling out her digits while he entered them into his phone.

The two agreed to meet up later that night. She gave him a quick peck on the cheek before hopping into her Chevy Tahoe. She didn't pull off right away; she wanted to see what type of whip he was pushing. Pretending to fiddle around in her purse she watched him as he walked over to a cream colored Mercedes Benz.

"Dayum, this might be my lucky day," she said kissing her nails.

* * *

Twisted Obsession

Chapter 5

Passion handed Allexis the mirror. "Girl, you know your *ish* don't you?" Allexis said as she admired her new look. Passion had styled and cut it in three layers adding honey blonde high lights.

"I told you, but you were sleeping on my skills," Passion replied brushing hair off her apron.

"I'm next," Tina said sitting down in the chair once Allexis got up. "So where are you and Kane going tonight?" Tina asked admiring her sister's new look.

"I don't know," Allexis said unable to remove her eyes from the mirror. "He said it was a surprise."

"Just be careful sis."

"Okay geesh. How many times are you going to keep saying that?" Allexis asked a little irritated. "Damn you act like he's killed somebody or something," she shrugged her shoulders nonchalantly.

Both Passion and Tina looked at each other in shock. They couldn't believe how naïve Allexis was.

"What?" Allexis asked taking note of the uneasy glances and Passion and Tina were giving each other.

"Allexis you are so caught up in the glitz and glam and the thrill of being with a hustler. But Kane is a hustler that's waist *deep* in the game. Racking up bodies comes with the territory," Passion said.

"Yea girl, Dom has racked up plenty of bodies under his belt too. On top of that he is constantly dodging stray bullets. Being with a hustler is no easy task. So yes I'm going to say it and keep saying it. *Be careful*."

"Okay," Allexis said as she left the kitchen and headed towards her bedroom.

She never *really* took the time out to go over all Kane's aspects of Kane's life of a hustler. The reality of what he stood for finally set in and it left a funny feeling in the pit of her stomach. But on the second hand she couldn't deny her strong feelings for Kane and they had such a strong connection. But she wondered could their strong connection outweigh his heavy flaws?

* * *

"And you now just tellin' me this?" Kane questioned gripping the wheel so tight his knuckles began to hurt, but he ignored the pain as he allowed the news that Dimples just told him to penetrate his mind.

"Nigga I called you last night and all this mornin'. What the fuck is I pose to do?" Dimples asked as he sat on the passenger side of Kane's escalade.

"I can't believe this bitch ass nigga. Dude's was tellin' me bout his ass yo'," Kane growled as he punished his steering wheel banging it with his fist. He was past the point of anger. It was his own fault because Dimples had tried to get in contact with him, he cursed himself for sleeping. He had to get focused, because he knew a lot of young cats out here were thirsty.

"Yeah, he called me up and asked me if I wanted to get in on it too."

"I just wonder how the hell it all it went down? I told Duece no stops, period. Somebody else has to be in on this. Not everybody knows about this drop, and plus I don't think that lil nigga has enough balls to cross me on his own. Somebody's definitely been talking in his ear."

"That's easy, *Dame.*"

Kane didn't reply because Dame was the first person that popped into his mind. He didn't want to believe it but that's what his gut was telling him. Both he and Dame had grown up in the game together. Dame was the closest thing he had to a brother. But over the years they have grown apart. Kane was more business minded, while Dame was caught up in the money and living the flashy life style.

"You know that nigga despises you," Dimples started again once Kane didn't respond to his comment. "You can see it in his eyes when he looks at you. He wants to be you. He can't stand how nigga's go outta dey way to get up close and personal with you, while he's left back in the cut," Dimples said as he burst out laughing. Dimples couldn't stand Dame anyway. He felt Dame was wink link and only feared and respected because of his affiliations with Kane.

After a long uncomfortable silence, Kane sent Dino on his way. He needed some time to think, close to half a million dollars stolen. He had to figure out a way he was going to

Twisted Obsession

get it back, and bury the nigga that was behind it. He picked up his cell phone and dialed Dame's number and once again no answer. He dialed Neko and got no answer.

This time Kane took his frustrations out on his dash board. He knew this had to be handled, so that the next nigga lookin' for a come up would think twice. He decided to cancel his plans with Allexis, his business had to come first. It was because of her that he got side tracked in the first place. But she was special, and he felt she would be worth it in the end. He dialed her number.

"Hello."

"Whad up ma?" he said. The sound of her voice put him at ease.

"Hey babe what's up?" Allexis asked, instantly forgetting her previous thoughts about Kane's current lifestyle.

"Yo' somethin' came up and were gone have to reschedule," Kane said a little disappointed, not really wanting to because he really was hoping to see shorty tonight. But he knew his business came first.

"Is everything okay?" she asked genuinely concerned.

"Yeah just business, nothin' I can't handle. But how was your day?"

"Fine just had you on my mind."

"You've been on my mind too. I enjoyed my time with you last night," Kane said thinking back to how sweet Allexis taste.

"Me too," she smiled.

"But I'ma have to get at you lata. Promise I will make it up to you."

"Okay," Allexis said a little disappointed. She was really looking forward to seeing him. "Be careful."

"I will ma, take care," he said before ending the call.

"Fuck," Kane yelled his mind immediately drifting back to Neko. "This lil nigga really ain't tryin' to see me." He dialed Dames number again and still no answer. "Fuck him too. His days are numbered. I can handle this bitch ass nigga myself," he growled as he started off his truck and sped off, having little regard for pedestrians or other motorist.

* * *

Twisted Obsession

Chapter 6

"Yes, daddy," Doricka yelled as Dame pounded her pussy.

"You like that?" Dame asked her as her grabbed a handful of her hair and gave her ass a smack. Loving the way it jiggled he smacked it again, this time a little harder.

"Oooooo, shit yes," she moaned. "Bang this pussy like you own that shit," she said throwing it back at him, doing her best to match his strokes.

Dame obeyed continuing to pound away until he felt himself nearing a climax. He pulled out his sticky dick and splattered his cum all over Doricka's back and ass.

"Damn, that was good," he said as he stretched out on the king sized bed, while Doricka walked to the bathroom.

Since their meeting outside of the nail salon, Dame and Doricka have been inseparable. They went and got something to eat as planned, then headed straight to the nearest hotel.

Dame grabbed his phone from off the lamp stand. It had been vibrating non-stop he been too occupied with his time. "Shit," he cursed himself once he realized it was Kane. He knew he should been on alert so that Kane wouldn't get suspicious once the robbery took place, but he let his little head outthink his big head. He immediately dialed Kane's number.

"Yo' nigga where the fuck you at?" Kane barked into the phone.

"Damn, nigga. Calm the fuck down," Dame replied.

"Yo' what is up with that nigga Neko?" Kane asked getting down to business.

Dame immediately sat up. *Oh, shit,* he thought to himself. But he decided to play it cool.

"What about him? He's cool with me."

"You see that's where you fucked up," Kane barked. "Dat nigga been stealin' from me. Skimmin shit off the top, then sellin' that shit on the side. Then the nigga robbed and killed Duece. I can't believe dat nigga. Where you find his punk ass? And you talkin' bout he good peoples."

"Yo' do you even know if this bull shit is even true?" Dame asked not really feeling the fact that Kane was talkin' all that me shit, but he kept his peace.

"Hell yeah it's true. Especially the way shit playin' out right about now. And the nigga

Twisted Obsession

not answerin' his phone and I've been blowin' them the fuck up. And he ain't even home, and nobody has seen this mutha fucka. Damn right somethin' goin' wit dat dude."

Dammit I knew that nigga couldn't hold his tongue.

He wondered who Neko could have told and how much this person actually knew and how much Kane knew for that matter. But he felt confident that Kane was still in the dark about his involvement or at least he hoped so.

"Well, who is this dude you been talkin to?" Dame asked. "And how do you even know if you dat nigga can be trusted?"

"You know what don't even worry about all that right now, we already been on the phone too long. What you need to do is meet me at my crib now," Kane snapped, ending the call before Dame could respond.

Dame quickly jumped up and threw on his clothes a little agitated about the way Kane ended the call but he would worry about that later, he knew he had to get to Neko before Kane did.

"Where you goin'?" Doricka asked strolling out of the bathroom with a towel around her waist.

"Yo' look I got some business to take care of," Dame said as he dug in his pockets and pulled a thick wad of money. He hurriedly peeled off four one hundred dollar bills. With pussy like that he thought she deserved something. He tossed the money on the bed, roughly kissed her on the cheek and then bounced.

"Dammit," Doricka sulked as she watched the hotel door slam in her face. Well at least she had her own whip this time.

Once inside his car Dame dialed Neko's number.

"Yo' what's up nigga?" Neko yelled into the phone. "Bout time I thought you forgot about me."

"What's good lil nigga?"

"Same, what's good wit you?"

"Can't anybody get ahold of you? Why you ain't been answerin' yo phone?"

Neko sighed. "I don't know yo'. Kane been callin' all damn day. Made me think some shit is up."

"See lil nigga. That's where you fuckin' up at. When you don't answer your phone after shit goes down that makes you look suspect."

"So what should I do?"

"Just tell em you lost it and had to go outta town for some family business. But he was callin' because he wanted to give you another corner. Damn, yo' could of added a little more gold to yo' pot," Dame smirked to himself.

Bitch ass nigga

"Man, Kane ain't goin' for that shit yo'. Damn he was gone give me another corner though yo'."

"Yelp," Dame said trying hard not to laugh.

"Damn," Neko said shaking his head. Yeah it was a little extra work but the money would look extra good in his pocket. Especially with all the expenses he has accumulated over the years, trying to keep his girl happy. He smiled as he thought about her. He glanced at the clock knowing she would be there any minute.

"Where you at?" Dame asked as he retrieved the glock he had tucked safely underneath his seat.

"At the crib chillin, why what's up?"

"Which one?"

"My hide out," Neko said regretting the words as soon as he they left his mouth. Only few people knew where his secret spot was. That was himself, Dame and his girl. A funny bone was telling him he should have never told Dame and he couldn't seem to shake the funny feeling in the pit of his stomach.

But he couldn't help but to smile to himself as he looked around at his elegant middle class home. It was a two story, split level, with four bedrooms, one and half baths, an outdoor swimming pool and Jacuzzi. So he was proud that all of his hard work in the streets hadn't gone in vain. It would be a shame if he had to leave it all behind.

"Good I'm on my way so that we can discuss some more business. Plus I need you to put some money at the spot. Think you can meet me outside? Oh you did take the product to the spot right?"

"Yea I did," he lied. "What you thinkin' bout doin' another job already?" Neko was all

Twisted Obsession

about getting his money but if Dame was talking about doing another job, then Dame wasn't thinking rationally.

"Yeah nigga that was just chump change. That was a test. You ready for big dough now," Dame said laying it on thick. "What you scared nigga? Its money to be had playa and It's our time to shine, Kane has reigned long enough. Don't you think?"

Neko thought about it for a moment, Dame was right he did think it was *his* time to shine after putting in years of work and still only pushing pennies while Kane pumped out quarters.

"Aight yo' I'm on my way outside now," Neko said.

"Good I'm right down the street," Dame said before hanging up.

Neko got a funny feeling in the pit of his stomach. This was highly unusual for Dame to come out his way. And he still had the feeling that Kane was calling about more than giving him a promotion. Everything was telling him that something was wrong. But he knew the consequences when he first got into the game. It all boiled down to, could be Dame be trusted?

The answer was no, so after this meeting tonight he was gone get ghost. Whatever money Dame gave him tonight he planned on using that along with his other score to get gone.

All this shit wasn't worth his life. He grabbed his bullet proof vest off the chair, contemplating on whether or not he should even go outside because he was planning on leaving after tonight. He glanced at the clock and decided to proceed, he knew his girl should be strolling through any minute and he didn't want her to get caught up in any of his bullshit.

Look at this bitch ass nigga?

Dame hid in the bushes outside of Neko's house. Dame watched as Neko paced back and forth in front of his house, puffing hard on his cigarette. He had to admit the little nigga put in work, but he was a liability. He didn't think straight and couldn't keep his mouth shut. For a second Dame felt guilty about sending Neko to meet his maker, but only for a second. Besides his plan was to only keep Neko around for one purpose and that was to kill Kane. He never planned on making Neko a part of his operation. Dame

Twisted Obsession

knew that if he so easily turned on Kane it was a possibility Neko would do the same to him.

Neko continued to pace back and forth, oblivious to what was about to go down. Two minutes after the bullet exited his skull he was dead.

Before leaving his hiding place Dame looked around to make sure the coast was clear. The night was eerily silent, and the only sound that could be heard was his own heavy breathing. He quickly got up and walked briskly to get to his truck which was parked two blocks away. The area was pretty secluded he hoped nobody would find the body until morning. He knew this was a fast and hasty kill, he just hoped Kane wouldn't suspect otherwise.

"Almost there," Natalia breathed as she jogged lightly, her long ponytail swinging from side to side.

The night was warm but a light breeze was sent her way every so often to cool her body. Her mother and sister often chastised her for jogging alone so late at night. But Natalia wasn't the least bit afraid. She was very fit, not muscular but lean. And besides she had her stun gun if a nigga wanted to get funny.

"What do we have here?" she said as her jogging came to a sudden halt, a little disappointed about the object that interrupted her flow.

Blood flirted with the tip of her tennis shoes as she stood over his body with her hands on her hips. Staring down at him she cocked her head sideways. He was clearly dead, but she decided to check his pulse anyway, while allowing her other hand to check his pockets. She took the rolled up twenties that were in his pocket and stuffed it in her pouch that sat tightly around her waist, along with his diamond studded pinky ring.

"Poor Neko," she muttered shaking her head as she dialed 911. "I'm going to miss his pretty dick." She wanted to hit it one last time before taking him out herself but it seems as if someone had already beat her to it. *Oh, well* she thought. She already took the goods from the *money spot* now all she had to do was get the money from his little score last night.

Neko gossiped worse than a bitch and couldn't hold water. He couldn't hold water. He more than eagerly told Natalia about his little planned heist with Dame. She planned on telling Kane, but figured Dame being on the opposite side could be to her advantage.

"911 what is your emergency?"

"Hello I would like to report a homicide"

* * *

Frustration turned to anger once Dame realized that Neko never put the money in the spot like he said he did. This was an little idea Dame concocted convincing Neko to start a money pot to go towards their operation. But he figured Neko must have been one up on him because the pot was completely empty. The money pot was just a couple of safety deposit box buried underneath and abandoned house.

"Damn," he said out loud as he swatted at spider webs that loomed in the air. He checked his watch. He knew he had to get going, before Kane became suspicious.

"Damn nigga. Where the fuck you been?" Kane yelled.

"Chill man damn. I was in Richmond," Dame lied entering the kitchen taking a seat next to Kane.

Kane looked at him suspiciously while taking a puff on his Cuban cigar. He blew out a puff smoke before proceeding with the conversation. "Yo' I've been trying to get ahold of bitch ass Neko all day, but he ain't answerin' his phone. You know what? What is up wit you nigga's not answerin' yall phone. What the fuck is that shit about?" he asked angrily punishing his cigar by jamming it in the ashtray until it broke in two. "I'm getting' real paranoid Dame and you know how the fuck I get when I get paranoid. I'll blast a nigga just for lookin' at me wrong."

"Come on Kane," Dame said knowing all too well the statement he just made was true. "Are you sure that lil nigga would have enough heart to cross you?" Dame asked trying to direct the conversation back to Neko.

"Yeah, one of his homeboys put a bug in my ear about him, said Neko tried to put him on too. That's why business slacked off and a few of the regulars refused to do business. I knew somethin' was up wit that nigga. Matter of fact I asked yo' ass to look into," Kane said to Dame giving him a sideways glare.

Dame swallowed hard, his words stuck in his throat. Just then Kane's cell phone rang. Dame's thoughts were all over the place. He wondered who Neko ran his mouth to and how this person actually knew. And how much Kane knew for that matter. But he had a feeling Kane still didn't suspect him because he would have been dead as soon he stepped foot in the door. Dame's thoughts were rudely interrupted when he felt the butt of Kane's gun connect with his forehead.

"Yo' what the fuck?" Dame asked in complete shock, as the deep gash in his forehead began to bleed profusely. Blood trickled down his face onto his shirt.

But Kane wasn't finished just yet. He then proceeded to wrap his strong arm around Dame's neck, and pressed his gun against the side of Dame's head. Dame nearly shit himself once he felt the cold hard steel pressed firmly against his temple.

"I'm going to ask you one more time, where the fuck have you been?" Kane asked in a calm yet scary tone.

"Yo' Kane . . . I told you man. I was in Richmond shoppin' and fuckin' with some bitch. Come on dawg, what . . . the fuck?" Dame asked barely able to breath.

"Somebody just mirked Neko. Nigga you don't think that's *suspect*?" he barked maintaining the strong grip around Dame's neck.

"Kane," Dame managed gasping for air as he squirmed around in the chair. "I swear . . . on my life," he begged. Kane released his grip. Dame grabbed his neck, doing his best to take in huge gulps of air.

Something in Kanes gut was telling him not to let Dame leave his house alive, but Dame must have sensed this also.

"Come on nigga, we . . . go . . . way back," Dame coughed. "We like brothers. I would never disrespect you."

Kane gave him the look of death. "You betta not be fuckin' wit me. Or it's ya ass nigga. Now get the fuck outta my house."

Dame left without saying a word. He ignored his throbbing headache. His anger wouldn't allow him to feel any pain. He cursed himself for allowing Kane to punk him like some bitch. This fueled his hate for Kane even more.

* * *

Twisted Obsession

Chapter 7

Allexis sat on her bed and tried to concentrate on the book she was reading, but she couldn't because she had Kane on her mind. It's been almost three weeks since she last seen him, and it was tearing her insides apart. He's been so busy working on opening his new shop in Baltimore that he hasn't had the chance to come and see her. Yea she talked to him almost every day but it wasn't the same as being able to have him near. So the only thing she could do was savor the first night they shared together and cherish their late night phone calls. But she so yearned to see his face, it was sickening.

Just then Tina burst through her bedroom door and snatched the book out of her hand. "Sis you still have the keys to the Navi, so you might as well use it."

"I don't feel like it," Allexis said rolling her eyes.

Tina sucked her teeth. "Come on, we can call up the girls and go out. Just chill and do somethin' to get your mind off Kane.

She knew her sister was love sick over Kane, and upset because she hasn't been able to see him for a while. She was actually hoping that was the last she has seen of Kane, but she knew it wasn't because Allexis still has his truck. She knew Kane's type all too well; he only wanted to claim Allexis cherry. And from the looks of it he might have, because Allexis has done nothing but mope the last couple of weeks.

"No I have to study for my exams," Allexis said attempting to take the book back from Tina.

But Tina snatched the books back out of Allexis reach. "Girl, whatever you know this shit like the back of your hand. Besides if you're not going to take if it for a spin, I will," Tina threatened dangling the keys.

Allexis shrugged her shoulders, "Go ahead I don't care."

"Maybe this is a sign about you and Kane Allexis. Maybe-,"

"Tina," Allexis said giving her a sideways glance. "You don't even know him. Give him a chance so far he's been nothing but a gentleman," Allexis said with dreamy eyes.

The look in Allexis eyes made Tina want to throw up. Her sister was seriously tripping over Kane. *How is she going to tell me that I barely know him, when don't even know him her damn self* is what she wanted to say.

Twisted Obsession

But instead Tina flopped on the bed beside her sister and sighed. "Allexis you do realize that all this comes with being the lady of a hustler. He works all kinds of hours and doesn't have a regular schedule like we do. All of this comes with the territory. I thought you knew that. You should, look at Dom and me. How often do you see him over here?"

"I know but I can't help it," Allexis whined. She replayed her first date with Kane in her head. "Our first night together he made me feel so sexy, yet beautiful at the same time. Every time I close my eyes, I see his beautiful face. I know I should leave it alone because he has got me feeling some kind of way and I haven't even had the dick yet."

The two girls shared a laugh.

"But honestly Allexis you don't think your only feeling this way because Kane is the only man you ever allowed to take you out. I mean you have never given anyone else a chance. Just don't up and fall in love with the first guy you go out with Allexis. Sometimes you have to keep fishing until you catch the big fish."

"Your right Tina, but for some reason Kane is the only man I have ever came close to trusting." Tina looked at her funny. "I know, what you're thinking I barely know him. But I just think this is right Tina. I think he is the one for me."

Tina felt her heart drop to the pit of her stomach, because the look Allexis eyes let her know that her baby sis was serious. She could only think back to that night they went out and now she wished she would have let her baby sis stay home and continues to study. But after some convincing she was able to get Allexis out of the house. They rode around their housing development blasting Shop Boyz *Party Like a Rockstar* while their white neighbors gave them disapproving looks.

"Let's swing around and pick up Doricka and Passion," Tina said bobbing her head to the music.

Allexis honked the horn several times before making a complete stop outside of Doricka's house.

"Y'all chicks crazy," Doricka yelled coming out of the house in a pair of boy shorts and a beater with Passion in tow. "What y'all want anyway?" she asked with a bit of an attitude.

Twisted Obsession

Tina snapped her neck back. *"Excuse you!"* she said rolling her neck a tad turned off by Doricka's stank attitude. "Well we just wanted to know if you and Passion wanted to ride to the mall wit us?"

"Nah, we aight."

"Speak for yourself," Passion said as she climbed in the back of the truck.

"Yo' what has been up with you? You have been on some foul shit lately and you know what, it really stinks," Tina said rolling her eyes.

"Well maybe I just don't want to be stuck up y'all asses all the damn time," Doricka snapped.

"Well fuck you too, ya *sideways hoe*."

Passion looked at Allexis and mouthed *Sideways Hoe*. Allexis shook her head and shrugged her shoulders.

"Come on Doricka don't be like that. Besides y'all have been friends to long be on some ole bull shit like this," Passion said.

Doricka knew Passion was right. She had been throwing a lot of Shade Tina's way which Tina truly didn't deserve.

"Give me a minute, Punta," Doricka said mushing Tina in the head and darting out of Tina's reach before Tina could get her back.

"And get my purse it's in the bedroom under my bed," Passion yelled.

On the way to the mall Doricka and Tina decided to spark up a blunt and by the time they arrived at the mall they were feeling nice. They all walked around aimlessly just to do something. None of the stores seemed to pique their interest.

"Damn, I hungry," Doricka shouted, as she scratched her head and patted her stomach. She hadn't eaten anything since early that morning.

"Dang Doricka," Tina said covering her ear. "We right here."

"Oops my bad," Doricka giggled.

While they were eating in the food court Allexis noticed a dark-skinned girl was staring at her. Allexis assumed they might have crossed paths so out of kindness she waved but the girl looked at her as if she were crazy and folded her arms underneath her breast. Allexis shrugged her shoulders and stuffed some french-fries in her mouth.

Twisted Obsession

"Yo' that chick is really clockin' you," Passion said taking note of the dark skinned girl starring Allexis down as well.

"Yeah why the fuck is that black chinese bag starring at you?" Tina asked.

Black Chinese bag, that's different, Passion thought to herself.

"I don't know," Allexis replied shrugging her shoulders, completely un-phased by the chick who was staring at her.

"It's probably one of Kane's chicks," Tina said. "I told you Allexis about Kane having females on his dick. This type of shit you have to worry about. Now I have to mess my knuckles up to bust a bitch ass."

Allexis smacked her lips. "Whatever Tina you don't even know what the scoop is yet. It probably doesn't even have anything to do with Kane."

They watched as the dark-skinned girl whispered something to her friends, and the trio proceeded to make their way over to where Allexis and the crew were sitting.

"Hi, my name is Tamela," she said with a fake smile plastered on her face. "I just want to know who you are and why do you have Kane's truck?" she asked maintaining her fake fascade.

Tamela peeped Allexis hopping out of Kane's truck when she first pulled up to the mall. She could spot Kane's truck anywhere because the license plates spelled out his name. She wanted to know who the yellow bitch was and why she was privileged to drive the Navi. Tamela and Kane had been kicking it for a while, and she thought they had something special. Granted the nigga could get a little crazy sometimes, but the nigga was paid, and that made up for a lot of things in her mind. She wondered what had been up with Kane when he didn't return her calls. Now she knows why? But what did this yellow bitch have that she didn't. Tamela stood with her arms folded across her chest, waiting for the yellow bitch to answer.

"See Allexis I told you yo'," Tina said with a knowing look on her face. She knew the game all too. She had to break a couple of bitches down because of Dom's ass.

"Whatever Tina," Allexis said rolling her eyes not appreciating the chick coming over to her table. "Are you and him dating or something?"

Twisted Obsession

"I'm not going to lie me and him are *really* good friends," she stated with an attitude.

"Well if you and Kane are such good friends, than you can ask him yourself," Allexis said sweetly.

"Bitch don't get cute," Tamela said not liking Allexis sarcastic attitude. She was already feeling some kind of way because she bent over backwards for that nigga and he brushed her off like a fly on a wall.

Out of nowhere Doricka jumped up yelling, "Bitch, *can you see? Can you see?*"

Tina jumped up knocking her tray over. "Yeah, ya *wow*," she yelled. She was raring to go, cause she hadn't tagged nothin' in a minute. Plus she was use to dealin' with chicken heads like this, because of Dom.

Allexis and Passion looked at each other. *These chicks must have been smoking some hard shit*, Passion thought to herself.

"I was talking to her," Tamela said pointing to Allexis, not really wanting it with Tina, but she knew it was too late.

But it was too bad for Tamela she picked the wrong chick to mess with, because Tina didn't play when it came to her baby sister. Tina stood up and spit in Tamela's face. Tamela attempted to retaliate by grabbing Tina's shirt but before Tamela could get a lick off Doricka jumped in and cocked home girl with right hook knocking the girl down taking Tina with her. Passion peeped one of Tamela's friends about to hit Doricka, she took her tray and banged the girl over the head with it. Apparently the other friend didn't want to get involved; she quickly turned around and walked away from the ordeal.

And just like that Doricka and Tina took off running. Allexis and Passion quickly followed suit. The four girls ran to the nearest exit. They didn't stop until they reached the Navi. They quickly hopped into the truck and peeled out of the mall parking lot and onto the highway.

Twisted Obsession

"Y'all . . . two . . . some crazy . . . bitches," Passion said gasping for air, pointing at Doricka and Tina. "Tina, of all the names in the book, and the only thing you could come up with. *YA WOW!*" Passion said laughing.

"Now that I think about it, it was kind of corny," Tina said laughing.

"And Doricka. What in the world did you mean, can you see? Can you see?" Allexis asked mocking Doricka.

"Okay," Doricka said rolling her eyes. "I messed up. I meant to say, can we see you? That bitch was black as hell!"

"She was pretty though. Besides it still wouldn't matter, considering we were in the mall, and it's still daylight outside," Allexis said doubling over as she thought back to the incident.

"Shut-up, damn. You sure do know how to make a hard dick go limp," Doricka yelled.

They all shared a laugh at Doricka's expense.

"See Allexis you ready to deal with type of drama?" Tina asked. "This what you would have to deal with on a daily basis. Not to mention his crazy ass baby mama, Rasheeda. And I heard that chick could go," Tina lied, knowing damn well Rasheeda couldn't fight to save her life and was only good at running her mouth.

"I'm a down ass bitch," Allexis replied, nodding her head feeling herself a little.

Tina could only shake her head and look up at the ceiling. *She can't even cuss right, talkin' bout she down. I give up she is just going to have to learn on her own.*

Allexis checked her cell phone as she unlocked the door.

"Damn, move. I gotta pee," Tina said, pushing past her. "You up here trying to play love connection.

Twisted Obsession

"Whatever," she laughed. She noticed Kane called seven times. *Damn, is it that serious*, she asked herself, dialing his number. He answered on the third ring.

"Yo' where you been? I've been trying to get ahold of you all day?" Kane asked in a bit of a harsh tone.

Allexis looked at her phone, and then put it back to her ear. "Hello, to you too," she replied with an attitude. She couldn't believe the nerve of him when he's been missing in action all this time.

"Come on ma, don't play with me," he said in a joking manner, but he was dead serious. "Where you been?"

"To the mall," she replied, as she walked into the living room placing her pocket book on the couch.

"You get anything?"

"No."

"If you needed some money all you had to do was ask."

"Is that the only way I will get to see you?" she asked, just the sound of his voice was making her vulnerable although she didn't want to. She still wanted to stay mad at him, but she was finding it hard to.

"Awe, come on ma. Don't be like that. Besides if you're not nice to me, I'm a take somebody else wit' me."

Allexis sucked her teeth. "Where are you going?" she asked.

"Paris."

"Where?" Allexis asked, unsure if she heard him correctly. "Did you just say-?"

"Paris," he said finishing her sentence. Kane really didn't want to travel all the way to

Paris at the moment. He just opened up his shop in B-More, and he wasn't too keen on the idea of leaving Dame in charge. He just had a feeling Dame was up to some shady shit. Kane's connect was on vacation in Paris, and he wanted Kane to fly out to discuss business. Like it couldn't wait until he got back, but he couldn't complain because he didn't have to pay for anything. Allexis's screaming rocked him out of his thoughts.

"Oh my gosh Kane! Are you serious?" she asked panting and jumping up and down unable to contain her excitement, while Passion smiled and looked at her like she was crazy. She had never been to Paris and has always wanted to go.

Kane smiled already knowing the answer to his next question. "So, you gon' ride wit' me, or stay mad at me?"

"Both," she teased.

"Well, get ready. I'm outside."

"But what about my clothes, do I have time to pack?" she asked checking her watch. She figured if he at least gave her twenty minutes she could throw something in her suit case.

"Don't worry about that ma, I got you."

"You sure?" she asked.

Kane didn't like the fact that she was questioning him; he would definitely have to talk to her about that later. "Yes, so come on so we can go."

"Alright, give me a second."

"Hurry up we don't want to miss our flight," he said flipping his phone shut.

The thought of flying never entered into her mind. She quickly ran into Tina's room and told her where she was going. Of course Tina wanted details, so Allexis promised her she would fill her in later. She gave Tina a quick peck on the check, grabbed her Chanel purse and headed for the door.

"Have fun," Passion stopped her before she left giving her hug. "And be careful," she stated seriously.

"Okay, I will," Allexis smiled.

When she stepped outside she was greeted by a stretch limo. She smiled as Kane stepped out of the limo and greeted her with a sexy smile. "Hey ma. You ready?" he asked embracing her in his arms.

She savored him for a moment, closing her eyes and inhaling his scent. 'As ready as I'll ever be," she replied, smiling up at him He opened the door for her, she hopped in.

They both got cozy with each other. She lay comfortable wrapped in his strong arms, each with a glass on moet in hand. Kane hugged her tight and nestled his nose in her hair. He hadn't realized how much he actually missed her until now. Allexis enjoyed the quiet ride for the moment but she felt it would be the perfect time to ask.

"Kane?"

"Yes babes," he asked in his own fantasy with Allexis snuggled next to him.

"Whose Tamela?"

* * *

Twisted Obsession

Chapter 8

They arrived at the hotel around 12 o'clock in the afternoon. The plane was a long, but a smooth ride thanks to flying first class. Allexis set her purse down, jumped in the huge bed and drifted off into a peaceful sleep. Kane removed her shoes, lay down beside her and was knocked out within seconds.

When Kane woke up it was around 1:30 only to find that Allexis still asleep. He gently kissed her on the cheek and headed in the bathroom to take a shower. After showering he dried off and put on a white G-unit tank top, G-unit jean shorts, and a pair of all white Flavs. He looked at Allexis one more time before leaving.

Allexis stretched her arms, rolled over onto her side only to find Kane had left already. She scanned the room, *Damn, he could have at least of left a note.* She was so tired when they first arrived at the hotel room she never even bothered to check out her surroundings. Their suite was huge. It had a living room area with an entertainment system, kitchen area, and a mini bar. After thoroughly checking out her suite she decided to call Tina, to let her know she arrived in Paris safe and sound. After that was done she headed toward the bathroom, she wanted to take a nice long hot bath once she saw the enormous tub. She soaked in the tub relaxing her muscles for at least an hour. With no clean clothes, she decided to let her body air-dry. Having eaten anything since the plane ride she was extremely hungry. Allexis walked out into the kitchen unaware of Kane and his company.

"Allexis, what the fuck are you doing?" Kane yelled.

Allexis was startled by Kane's harsh tone; she blushed and did her best to cover herself up when she noticed Kane's companion eyeing her body. He admired her with those intense blue eyes allowing his eyes to travel from head to toe. It seemed like minutes she had been standing when in fact it was only a couple of seconds. Embarrassed and humiliated, Allexis quickly ran into the bedroom.

"Oh, my gosh this is so embarrassing," she mumbled to herself as she darted for the clothes she had on yesterday which were sprawled across the bed.

"I'll be back," Kane mumbled to his guest.

Kane headed to the bedroom nostrils flaring. He found Allexis in the bedroom half dressed, still attempting to put her clothes on. He quickly grabbed her by her shoulder and spun her around so that she was now facing him.

The tight grip Kane had on her shoulder caused her to wince just a little. "Kane you are hurting me," she cried with a bewildered look in her eyes. The look in his eyes was scary and it caused Allexis to shiver just a little.

"Do you realize how you *embarrassed* me?" he yelled, as his spit decorated her face. "Go out there and apologize."

"Are you *serious*?" Allexis asked in disbelief. "Kane it was an honest mistake. How was I supposed to know you had company out there?" she asked pleading her case, even though she didn't know why the topic was even up for discussion. It was an accident.

Without saying a word he grabbed a handful of her hair, and escorted her out into the living room. "Yo' Joey my girl needs to apologize for the way she acted earlier."

Joey stood, "No, it was just a misunderstanding." He said holding up his hands. "There's no need for her to apologize," Joey said a little hypnotized by Allexis beauty.

Allexis tried to loosen the hold he had on her hair. He smacked her hand as if she were a child. "*No*! She has to learn how to respect my company." He looked Allexis dead in the eyes, "Allexis, *apologize*."

Allexis felt her heart jumped into her throat. She swallowed really hard, closed her eyes, and took a deep breath. "I'm really sorry," she heard herself say. This was beyond humiliation, and she couldn't believe she actually apologized.

"Good girl," Kane said, releasing her hair and smacking her on the butt. "Go get those bags by the door. I went shopping for you today."

Without saying a word Allexis did as she was told; all the while fighting back the tears that so badly wanted to escape her eyelids. She carried the bags back into the bedroom, keeping her head down the entire time.

Joey couldn't help but to feel sorry for the young beauty. It was an obvious misunderstanding, and he couldn't understand why Kane didn't see that. He shook his head at Kane in disgust; she didn't deserve to be treated that way.

Joseph D'Amico b.k.a. Joey, Kane's connect. Joey is the son of Jonathan D'Amico, and the grandson of the notorious Joseph D'Amico. The D'Amico family is one of the most powerful mafia families in the United States. Their primary rackets include narcotics, loan-sharking, extortion rackets, pornography, car theft, restaurants, gambling, and waste management. To say Joey was rolling in dough would be an understatement. He knew he would never be as powerful as father or grandfather, because he wasn't a full blooded Italian. But he still reaped the many benefits of being the grandson of Joseph D'Amico, who wouldn't mind burying anyone who had ill feelings towards his grandson.

He was a smooth, tall, fine drink of water. Muscular build and beautiful blue eyes that always set the women in frenzy (black women that is). The ladies often tell him that looks just like Prison Breaks Michael Scofield. Joey discreetly dealt with Kane simply because he didn't like the idea of keeping to many business partners. He figured men could be like women sometimes, jealous and petty. Besides, he didn't like Dame; there was something about Dame that didn't sit too well with Joey.

"So you took care of that problem?" Joey asked, referring to Neko.

"Yeah," Kane lied. "His home boy Dimples put me onto the foul shit he was doing." Kane was pissed that Joey already knew about his little problem. But he should have known Joey would eventually find out, considering it was fucking with his candy.

"Are you sure he was working alone?" Joey asked. "You think he had the balls to cross

Twisted Obsession

you on his own? Or did somebody put a bug in his ear?"

"What is up wit everybody asking me that question?" Kane asked a little annoyed. "Look, its taken care of," Kane snapped.

Joey put his hands up in the air, "Okay, just making sure. "Oh, Thursday, Frankie and Dom, are going to meet you in New York, Brooklyn Ave around 11:15. We just got a new product in, pink candy. Dom tells me it sells just like candy too." Kane nodded; even though he was pissed Dom had already reaped the benefits of this new product before he did. "Look, I have to run, let's meet for dinner. The La' Tour d' Argent, around eight, my limo will pick you up. Oh, and bring your friend. She wouldn't happen to have a playmate for me?" he asked wickedly.

"Nah," Kane answered, seeing Joey out. Kane secretly despised all the black pussy Joey tagged on a daily basis. He wasn't jealous or anything, he just couldn't stand the idea of a *white man* fucking a *black woman*. Despite the fact Joey was a third black, and that Joey often told him that he was part Italian not white, Kane *still* considered him to be white? He felt Joey lived the life of a white man. He didn't have to struggle like he did.

* * *

Allexis never even so much as glance inside of the bags she retrieved from Kane. She had cast them aside, jumped on the bed and cried like a baby. She couldn't believe he embarrassed her like that, and over nothing. Granted, his companion saw her naked, but it was an accident. She heard him enter the room. And she felt when he plopped down on the bed beside her but she refused to turn around and acknowledge him. She was hurt and can't believe he had treated her so rudely. As if she was a child instead of an adult.

He turned her over on her side, facing him. He wiped her eyes, "Look, ma I know you mad at me, but that was my business partner. I can't have my girl walking around butt ass naked in front of my company like that. You have to conduct yourself like a lady."

"I' am a lady," Allexis sniffled. "And it was an accident," she cried looking up at him with sad puppy dog eyes.

"I know ma, but next time think," he said pointing toward his head not phased by her tears he honestly felt he was teaching her a lesson. "So that we won't have to go through this again," he said, kissing her softly on the cheek. "I don't like seeing you hurt Allexis. But it's just some things I won't tolerate. I like classy women, who conduct themselves accordingly."

"But-," Allexis attempted to protest.

"Shhh- we will discuss all of that later," he said placing a finger to her lips, but he had already decided that was the end of the discussion. "I want you to see what I've bought you."

Allexis sat up as Kane retrieved the bags she cast aside. He reached in one of the bags and pulled out a long black, form fitting, Versace dress. Allexis couldn't stop the smile that spread across her face, as she imagined how good she would look in the dress. "This is Versace baby," holding the fabric against her skin. "I know you will look good in this ma."

She peeped the price tag, *damn 2900 dollars*, she thought to herself. Oh, well she thought she deserved it after the he had treated her earlier. Kane purchased several more designer garments for her, Gucci, Prada, Valentino, Dolce and Gabana. Allexis was in love all over again, despite the fact she realized he never apologized.

Kane hired a hairstylist to do Allexis hair. The stylist decided on an updo to showcase her beautiful face. They were both ready and dressed by 7:30. She admired herself in the Versace dress, with a long split on the side advertising her thigh. Kane purchased a diamond-encrusted necklace with a matching bracelet for her. She couldn't stay out of the mirror; she loved the way the necklace felt against her skin.

"You ready baby?" Kane asked. He was decked out in his all black Versace suit, and black pimp hat, that rested on top of his dreads. He checked out his reflection in the

Twisted Obsession

mirror and couldn't help but to thank his momma for his features. That's about the only good thing she had ever done for him.

Allexis smiled, admiring how good he looked. "Yes, I'm ready."

He walked up to Allexis and pulled her close to him, inhaling her Glo perfume. He cupped her face in his hands, and kissed her gently yet passionately. Allexis felt her knees get weak.

"You look beautiful," he whispered into her ear, and she truly did.

Allexis blushed, "You don't look to bad yourself," she teased.

Kane hooked his arms in hers and escorted her to their awaiting limo. All eyes were on them as they walked through the hotel lobby, and Allexis couldn't help but feel like a superstar from the attention she was receiving from the hotel guests and patrons. She smiled as woman looked on envious, while men gave her flirtaous and inviting glances. But none of them stood a chance, because she already found her prince charming or so she thought, as she smiled over at Kane.

* * *

Joey did his best to control the throbbing in his pants as he stole glances at Allexis. He thought she was beautiful when he saw her at the hotel earlier, but she was even more stunning now. He couldn't understand how a delicate creature got involved with an animal like Kane. He watched her as she gracefully made her way back over to their table, catching glimpses of her thigh, with each step she made. *Damn*, he thought to himself. Flashes of Allexis naked in the hotel room tormented his mind. One way or another he was going to have Allexis.

"Were you able to find the bathroom okay?" Joey asked. Joey thought his companion Natalia, would be more attentive to the young girl, but for some reason she seemed a little standoffish tonight.

"Yes, no trouble at all," she replied. Allexis rejoined them at the table and sipped her wine. It was kind of odd being around Joey because of what happened earlier. But the wine helped her loosen up a little. The restaurant was an elegant, yet formal place to dine. The main entree was duck. She never had duck before, and she had to admit she enjoyed it.

"So Kane, business is doing well I assume?" Joey asked trying to make small talk.

"Yeah, business is boomin'," Kane, replied taking a sip of his wine. "I just opened up a shop in Maryland."

"In Maryland, huh? You're brave. The crime rate is extremely high over there. Make sure security is tight."

"It's cool I got things on lock," Kane said, taking another sip of his wine.

Joey smirked, "I have to come by and check it out sometime." He already had tabs of where the business was located. With the constant drug use on the rise in the area, Joey figured it might not be such a bad move for Kane.

"Yeah, you do that," he replied, as he watched Joey repeatedly steal glances at Allexis. He caressed her arm. "You okay baby?" he asked affectionately.

Allexis nodded her head, "Yes."

He smiled and kissed her passionately enjoying the fact that Joey was watching. He knew Joey had a hard on for Allexis it was evident in his eye. But it was obvious to Kane that Allexis had no eyes for Joey, but for him stroking his ego.

"When we get back to the states I need a car. A family car a station wagon or maybe even a minivan," Joey said unable to take his eyes off Allexis. He couldn't help himself and couldn't get over how beautiful she was. He had been so wrapped up in Allexis he wasn't really paying his own date any mind.

"Okay," Kane mumbled continuing to nibble on Allexis neck.

"Baby," Allexis squealed, a little embarrassed.

"So, Allexis, is it?" Natalia asked, interrupting Allexis and Kane's love session.

"Yes," Allexis nodded her head.

"What is it that you do?"

"Well, right now I'm enrolled in college. And I work part-time for a telemarketing firm."

Natalia smirked, "How *cute*," she replied, not really liking Allexis already.

Allexis looked at her for a moment, unsure of how to handle her comment. Natalia smiled and so Allexis smiled back, also taking note of the way Natalia eyed Kane every so often. She was a little intimidated by Natalia, because she was strikingly beautiful. Coal black hair with piercing blue eyes that seemed to just cut right through you and her deeply tanned skin was flawless. The deep red lipstick she wore stood out perfectly against her skin tone.

From the looks of things Natalia could tell that Allexis was different. The way Kane watched her every move let her know that he had a soft spot for the yellow bitch which didn't sit very well with Natalia. He was hers in her mind and wanted him so badly all to herself. Every other black dick she has fucked was like putty in her hands, but Kane was different. Her nostrils began to flare as a look of rage began to consume her body as she continued to watch Kane stroke Allexis arm.

"Are you okay darling?" Joey asked taking note Natalia's sudden mood change. She had a scary look in her eye, as if she were about to do something crazy.

Touching base with reality Natalia flashed her million dollar smile. "Of course, you know me. Work is always on my mind," she laughed a little before taking a sip of her

Twisted Obsession

wine. She slyly sent Allexis and evil glare.

Natalia secretly seethed inside as she watched Kane play house with his little trophy wishing she was on his arm instead. Sure she was with Joey for the night, but Joey was soft in bed, and Kane's entire persona screamed power and authority, just the sight of him made her panties wet. Even though he made it clear, they weren't together and there was no chance in hell they ever would be, whenever she saw him with another chick she became insanely jealous. Some kind of way before the trip was over she vowed to make sure Kane's dick somehow landed in her mouth.

The quartet drank some more, all of them loosening up a little a bit more with the exception of Natalia who secretly continued to send evil glances Allexis way. But after getting pissy drunk they all decided to call it a night.

* * *

Allexis and Kane stumbled into their suite around 1 o' clock that morning. Kane hungrily groped Allexis from behind, planting wet kisses on her neck. Jus the taste of her flesh had him excited. A slight moan escaped Allexis lips as they somehow managed to find their way to the bedroom.

Once in the bedroom Allexis became the aggressor, pushing Kane down on the bed. She slowly and seductively slipped out of her dress, and Victoria Secrets matching bra and thong. Kane massaged his dick as he watched Allexis wind seductively. She attempted to place her manicured toes into his mouth, but fell during the process.

Oh, shit, Kane thought to himself, suppressing a laugh as he watched Allexis struggle to stand to her feet.

He scooped up her naked body and gently laid her on the bed. His hands freely explored her body. He placed his face between her legs, and made love to her pussy with his tongue. She humped his face, until her juices made his lips shine. She grabbed his face and kissed him passionately. "Make love to me Kane," she whispered. He quickly undressed and placed himself between her legs. He placed the tip of his dick inside her warm wet opening. She winced a little from the pain. He kissed her passionately as eased the rest of his nine-inch dick into her warmth.

Allexis buried her face in Kane's chest as he stroked her pussy. The pain was unbearable at first, but soon her pain turned into pleasure. She did her best to match his strokes as he rocked her to ecstasy. Kane couldn't believe how good her pussy felt; he thought he was in heaven.

"Allexis. . . Allexis, damn, baby," he whispered in her ear. He kissed her aggressively to smother the scream that was rising in his throat. The warmth of her pure pussy was driving him insane. In and out he rocked of her mist as he planted soft kisses around her neck.

"Mmmmm . . . Kane," Allexis moaned deeply as she looked deep into Kane's eyes. And at that moment she knew. "I love you Kane," she whispered.

Kane memorized by the pure innocence in her eyes went deep well within her walls. "I love you too Allex . . . is," he moaned. Kane's mind ran wild as he continued to go deep into Allexis ;pussy, it reminded him of the first time he'd ever had sex.

They both moaned and grinded into each other until both of their bodies trembled from pure satisfaction. They drifted off to sleep entangled in each other's arms.

* * *

Allexis woke up with a pounding headache. She looked over at Kane who was still sleeping. She kissed him lightly on the cheek before stretching and rubbing the sleep out of her eyes. In between her legs she felt a little sore, that's when the realization of losing her virginity hit her. She shrugged her shoulders; she felt it was worth it as she stared down at her prince charming. Careful not to wake Kane she made her way to the bathroom, washed her face, and brushed her teeth. She hopped into the shower, letting the hot water soothe her body. After a couple of minutes she heard the shower door slide open. She turned around to a grinning Kane.

"Can I join you?" Kane asked easing in beside her. The scent of strawberries and crème filled his nostrils. He eyed her body intently the sight of her with her hair hanging, and the mixture soap dripping off her body made his mouth water.

"Sure," Allexis said smiling.

Kane grabbed her, pulling her body close to his. He sucked her bottom lip and palmed her ass. He swiftly picked up her, pressing her back against the shower wall. Allexis squealed in delight and fear, she didn't want him to slip and fall. He planted kisses on her neck as he slipped his dick into her awaiting pussy.

She held on for dear life with each stroke. Kane licked and teased her nipples. Allexis arched her back in ecstasy, "Kane," she moaned enjoying every moment of their love making.

He admired how sexy she looked, hair and body dripping wet and the expression on her face let him know that he was hitting it just right. And the sound of his balls slapping her ass turned him on even more, causing him to pump even harder. They grinded their bodies into one another chest to chest while gazing into each other's eyes. "Allexis," he whispered as he came long and hard well within her womb.

The remainder of their stay in Paris went smoothly. Kane took Allexis shopping, Gucci, Prada, Donna Karen, Christian Dior, any shop or boutique Allexis wanted to enter. After shopping they went to eat at a local restaurant. Kane even took Allexis to see the William's sister play tennis at the French Open. Allexis was ecstatic when she got Serena William's autograph.

* * *

They had so much fun Allexis didn't want to leave Paris, but she knew she had finals to prepare for. As soon as she got home, she made it upstairs to her bedroom and went to sleep forgetting about all her worldly possessions. Her body was tired on only wanted to sleep.

The limo driver helped Kane carry Allexis's bags upstairs. Kane kissed Allexis softly and headed out the door. After a couple hours of sleep, Allexis felt refresh enough to do some last minute studying.

"Damn, you did it up didn't you?" Tina asked referring to the shopping bags that decorated Allexis bedroom. "I want details!" Tina said excitedly.

"Okay," Allexis replied setting her book aside. Allexis described her trip to Paris, minus the incident involving Joey. She revealed that she finally lost her virginity. She was in the middle of telling her about how she met Serena Williams but Tina cut her off.

"So hold up, you're not a virgin anymore?" Tina asked praying that she heard Allexis wrong or that Allexis was playing some sort of practical joke.

"Nope," Allexis said shaking her head.

"But why so soon Allexis?"

"You were the one always joking about how I should be getting my pus wet. Well I finally have and you're acting all funny about it."

"I mean I just didn't think. Well you don't. I don't know what the hell I'm trying to say," Tina finally said, still in shock about her sister losing her virginity to Kane. She knew that it would eventually happen, but not this soon. Her sister barely knew Kane, and she knew Kane was jumping for joy for claiming his prize.

"Tina, please don't make a big deal about this okay. I can tell from the look in your eyes that you're not too fond of Kane, but I'm a grown woman now and I feel I can handle myself. So can you please just be happy for me?"

The look in her sister's eyes said it all. No matter what Tina said Allexis was going to

Twisted Obsession

do what she wanted to do anyway. The only thing she could do was go with the flow even if she didn't like the way the current was going.

"I'm jealous. Dom doesn't take me anywhere," she said changing the subject.

"Whatever! You have been to Jamaica, Hawii and Honduras. Need I say more?" Allexis asked, glad that she and her sister had an understanding. She knew her sister only wanted the best for her, but Allexis felt she could handle herself.

Tina rolled her eyes. "Yeah but I ain't never been to Paris." Allexis laughed. "Well, I'm just glad you had fun." Tina smiled. She was glad her sister had a good time.

* * *

Twisted Obsession

Chapter 9

"What?" Kane yelled into the phone for the umpteenth time. This was the third time he had received a call from a private caller and it was starting to irate the hell out of him. He started not to answer but whoever the hell it was, was persistent. "Hello," he said again and still no answer.

But this time he hung up. He figured it was probably Rasheeda playin' dumb ass games. Well, he figured he would just have to teach her another lesson. He honestly believed she liked it when he had to fuck her up. But he would be lyin' if he said he didn't get some kind of thrill out of her putting her or any woman for that matter in their place, it gave him an adrenaline rush. He lay there in his king size bed with nothing but his boxers on thinking about Allexis and how he wished she occupied the space next to him. The smell of her hair still lingered on his pillow. He lifted it up and held it to his nose savoring her scent. That girl definitely had him twisted.

His phone rang again, so caught up in Allexis he didn't even bother to look at his cell phone or he would have seen that he it was another blocked call. "Whad up?" But the caller hung up. Kane looked hard at the ceiling as he angrily placed the pillow back down on the bed.

Ring! Ring!

"Yo' what the fuck is up?" he roared into the phone, not liking this shit. This has gone on to many times and was starting to put him on edge. He sat up and checked his drawer to make sure is glock was in place which it was.

"Aye papi," Natilia teased. "Just hearing his aggression through the phone made her kitty wet. What's wrong?" she asked, faking like she was concerned.

"Natalia?" Kane asked.

"The one and only baby."

Twisted Obsession

"What you want?" Kane asked not even bothering to hide his irritation. He now knew it was her crazy ass calling all him them times and hanging up. She did petty shit like that whenever she couldn't get her way. And to think he was all set and ready to stomp a hole in Rasheeda's ass.

"I'm lonely Kane and my kitty misses you," she cooed all the while stroking herself as she lay spread eagle in her king sized bed.

The offer was definitely tempting to Kane, but he wasn't about to drive all the way across town to get some pussy, there was only one woman he would do that for and that was Allexis. And he damn sure wasn't about to let her crazy ass up in his house. He could honestly say that Natalia was one bitch he might not be able tame. He could beat the shit out of her and he honestly felt that she would get off on it.

"Kane," Natalia said getting a little impatient. She was hoping he wouldn't turn her down. She honestly missed him.

"Nah, not tonight Natalia," he yawned.

"Kane," she whined. "I am so fucking horny. And I really need for you to get your black ass over here and ram that pretty dick of yours up this tight ass," she said sweetly.

"Natalia, I ain't got time for these bull shit ass games. I already told you not tonight," he said hanging up the phone, not even giving her chance to respond. He would eventually call her later on that week because he did have some work for her to do. He really didn't want to but opted to turn off his phone, because he wouldn't get any rest if he didn't. After wards he turned over on his stomach and drifted off to sleep dreaming about Allexis.

"That son of a bitch," she mumbled after literally calling him thirty times only to be directed to his voice mail.

Angry she tossed her cell phone which landed against the wall before shattering to pieces on her bedroom floor. Mumbling obscenities to herself she turned on the lamp that

Twisted Obsession

sat on her nightstand and roughly opened the drawer. She still couldn't believe he turned her down. She grabbed a cell phone, one of several that she purchased when she had incidents such as these and ripped open the package before getting up and retrieving the battery out of her old phone and putting it in her new one. She then placed it on the charger.

She was still beyond anger. All week she has been trying to get Kane to come over and each time he claimed he was too busy. She knew he was busy alright too busy with his new found bitch. Just the thought of Allexis made her queasy. There was doubt in her mind that she wanted Kane, all to herself but he didn't return the feelings, all because he said she was white. He told her that he couldn't see himself settling down with a white woman.

"But Kane, I'm not white," she remembered telling him. "I'm Italian."

He only shrugged his shoulders and said, "What's the difference."

No matter how Kane treated her she knew, that if he asked her to do something she would. That's how deeply in love she was with him, but that didn't mean that he was going to get away with hanging up on her either. Sitting down on her bed, she laid back, her black negligee hugging her in all the right places and she began to think.

"Good one Natalia," she said to herself as an idea popped in her head. Her blue eyes lighting up, as she grabbed her cell phone.

"Talk to me," Dame said into the phone, wide awake and unable to fall asleep after snorting a couple lines of coke.

"Hello, daddy," she purred.

"Who is this?" Dame asked, but liking the sound of the voice on the other end of the receiver.

"It's me silly, Natalia."

Twisted Obsession

A puzzled look took over Dame's face. He knew Natalia but they never really got down like that. She was always stuck up under Kane's dick to ever give him the time of day, and here it was one o'clock in the morning and she was calling him. It kind of put him on edge, and made him think something was up. But none the less he decided to entertain himself.

"Dame," Natalia said, after a couple minutes of silence.

"I'm sorry ma. What's good?" he asked as she sat up in bed.

"I was thinking about you," Natalia said laying it on a little thick. She knew Dame always wanted a piece but he wasn't her type. He tried to play hard, but underneath it all she knew his heart was soft.

"Oh really?" Dame asked licking his pink lips, as he looked down at the bulge that was starting to form in his boxers. "What you got on your mind?"

"Well to be honest I have a lot on my mind actually. But I have a proposition for you. That's if your interested?"

"And what might that be?" Dame asked with thoughts of sex on his mind. He began to imagine what it would feel like to have Natalia's lips wrapped around his dick. The first thing that came to his mind was the red lip stick she always wore.

"I'd rather not discuss things over the phone, but in private."

"Would you like me to come over now?" he asked all set and ready to take that trip across town.

Natalia covered her mouth trying not to laugh. It was funny to hear how desperate Dame was. But she really hadn't planned on fucking him, if she had to she would. But she felt she wouldn't have too once she dropped the bug in his ear about what she planned on doing.

"No, not right now I have the early shift tomorrow. Matter of fact I'm leaving here in half an hour," she lied.

"Oh," Dame said a little disappointed as he looked down at his bulging erection that was full of cum begging to be released. "So when would be a good day time for you?"

"I'm shooting for tomorrow once I get some things in order. I will give you a call with more details. How does that sound?"

"I'm game," Dame said trying not to let his disappointment be heard.

"Good talk to ya later," she said cheerfully as she hung up the phone. "I'm a bad bitch," she said out loud to herself. She already knew that Dame was going to be down because of his hatred for Kane, a blind person could see the animosity he held towards Kane. All she had to do was get Marco's code to the shop down in B-More and things would go smoothly. She knew this wouldn't hurt Kane too much, but she felt he had to be taught a lesson for treating her so harshly. But she knew she had to carry this plan out carefully, she didn't' want Kane getting to suspicious.

That nigga has to be out of his mind if he doesn't want all of this, she thought to herself as she looked down at her shapely thighs and perfectly manicured toes. With thoughts of Kane and revenge on her mind, she allowed her legs to explore her body before allowing her fingers to sensate her pulsating gem. She squealed and panted loudly as she rotated her fingers bringing herself to a powerful climax before drifting off to sleep.

* * *

Twisted Obsession

Chapter 10

"*Shit*," Allexis cursed as she walked into the campus library. She knew Kane was going to piss a bitch; he has become very possessive lately. Making her quit her job, wanting to know her whereabouts at all times, constantly calling to check up on her. Treating her more like his daughter than his girlfriend. But he always told her it was for her own good, simply because of his position in the game.

"Is everything okay?" Passion asked.

"Yeah," Allexis answered nervously. She quickly dialed Kane's number. He answered on the third ring.

"Yo' where the fuck you been?" he asked in a calm, yet serious tone. He called her three times already. He was ready to stomp a hole in her ass when she didn't answer the first time.

"I'm at the college returning my books," she answered nervously. Passion was staring at her with a look of concern on her face.

"Right, so you can't call a nigga and let him know?"

"I'm sorry, Kane it must have slipped my mind," she as she nervously bit her bottom lip. She hated when Kane got angry he it was like Dr. Jekyll and Hyde. He became a completely different person. She always tried to do her best not get on his bad side.

"Don't worry about it ma. Just don't let it happen again. Hey make sure you are home by 8 o' clock. I have a surprise for you."

"Okay," she replied, relieved Kane didn't chew her out.

"I love you beautiful."

That bought a smile to Allexis face. The way he said it made her heart melt. "I love you too, Kane."

Twisted Obsession

"What was that all about?" Passion asked, as soon as Allexis got off of the phone. The entire scenario was all too familiar to her.

"Huh? Oh, nothing," Allexis said still in a little bit of a dream.

"That didn't sound like nothing to me."

Allexis sighed, "He just wanted to know where I was at. He has a surprise for me and wants me to be home around 7:30. Is that a problem?"

"No, that's not a problem. But him controlling every aspect of your life is."

"He's just really protective of me. You know he has a lot of enemies out here in these streets and he truly has my best interest at heart."

"Oh, really," Passion said not truly convinced. "Be careful Allexis, I know how these controlling type men can be." Passion turned around and lifted up her shirt, exposing a large and ugly scar. Allexis gasped. "An ex of mine. Use to beat the crap out of me, and over little shit. He got mad because a male cousin of mines picked me up from school one day. *Battery acid*," she said, pulling her shirt down and turning to face Allexis. "It burned the shit out of me and after that I just couldn't take it anymore."

"Look, it's nothing like that with me and Kane. We have an understanding."

"Okay, Allexis. Just remember, it never stops."

"So, what happened to the guy?"

"*Dom.*"

* * *

Kane pressed Rasheeda's head further down on his shaft. He thrust his dick into her throat, ignoring the coughing and gagging sounds she was making. "Allexis," he screamed as he shot his sperm down her throat.

Twisted Obsession

An irritated look claimed Rasheeda's face as she sat up and wiped her mouth. She wasn't too keen on him calling out another chicks name especially since she knew it was that bitch at the club, but what could she say. Kane did what he wanted, when he wanted, she was starting to get use to his little spats by now. Kane handed her a duffel bag full of money. "Aren't you gon' come up and see Sean?" she asked.

"Nah, I'll get a little man later," he replied.

Rasheeda wanted to protest but what could she do. It was her own fault that she made a daddy out of a no good son of a bitch like Kane. The only bad part was that her son had to suffer. She could count on her hand's how many times Kane comes to see his son each month. It just pained her to have to make up excuses to her son as to why his dad can't come see him. She gave him credit for being their financially but there was more to being a father.

"What you still in here for?" Kane asked a little annoyed. "I said I would get at little man later."

"Nothing," Rasheeda mumbled as she hopped out of the truck. She just didn't have the energy to fight with Kane today. Maybe Sean was better off, because he truly had a psychopath for a daddy. She watched the headlights from his Mercedes fade in the distance before turning around to walk into her extravagant home. It was a lovely but very lonely home. Kane wouldn't allow any other males around his son even though he was barely there himself. But what could she do, this was the life she had chosen for her and her child. With the heavy duffel bag in hand and her head held low she made the long walk up her driveway.

Kane needed to relive some stress, and Rasheeda's head game was just what he needed to relieve some of the tension he has been experiencing lately. Apparently some local niggas broke into his shop in B-More stealing over 100 grand in cash. Not to mention the damage they did to the place. Marco one of his workers had conveniently forgotten to set the alarm. And Kane had to conveniently stomped his ass dead into the ground. But that wasn't all that was troubling him. He still had no clues as to who put the hit out on Duece and took his money. He still had his suspicions about Dame but he hasn't done any out of the ordinary spending, so that kind of made him put Dame in the back seat. But the fact that a nigga would cross him like that still lingered in his mind.

"Maybe Dame was right about starting the business in New York," he said to himself starting up the Mercedes, screeching onto the highway.

Kane pulled up to Allexis house around 9:30. He dialed her number.

"Hello," she answered.

"Yo' come outside," was all he said before hanging up the phone.

Kane got out and stood outside of the car to greet Allexis. She came outside her eyes immediately landing on the new whip Kane was pushing. He couldn't help but to stare at her delicate face each and every time he saw her. She was enchanting.

"Hey baby," she said giving him a big hug. And as usual the smell of his cologne teased her nostrils. She stood back for a second and admired how fine her man was. She couldn't help but to feel like the luckiest woman in the world.

"Hi beautiful," he said kissing her on the forehead. "You miss me?"

"Always," she answered truthfully gazing into his eyes. He smiled which made Allexis knees buckle. It was something about that man that set her ablaze. Despite his little flaws she was *dangerously* in love with him.

"You like these wheels baby?" he asked referring to the all-black Mercedes Benz S600 he drove up in. He didn't even have to ask the expression on her face said it all.

"Yeah, baby it's nice."

He handed her the keys, "It's yours. If you want it?"

Twisted Obsession

Allexis snatched the keys out of his hands and jumped into his arms. "Oh, my gosh. Thank you baby!" she said excitedly planting kisses on his face. She stood there for a moment and admired the all black Benz trimmed in silver.

"Come on, take me for a ride," he said snapping out of her daze.

The keys jiggled in her hands as she excitedly got in the driver's seat and waited for Kane to get in. It was so nice inside with leather interior and wood grain. She was almost in tears. She could only sit there a moment in awe. Of course she pushed Tina's old Lexus once Dom bought her a new car, which Tina's Lexus was nice, but this was a newer model and it was tight.

Kane got cozy and laid his back on the head rest. He was glad with the choice he made himself. "You aight ma?' he asked once he noticed Allexis hadn't started up the car yet/

But Allexis could only nod her head. "Thank you baby," she finally managed to say. "You don't know how much this means to me." She was extremely touched by the gesture. Of course Kane practically gave her the Navi, but this was her very own. And she could still smell the newness of it.

"You're welcome baby," he said giving her a kiss on the forehead. "Besides you deserve something to call your own. Now come on so you can see how smooth this baby rides."

Doing as she was told started her baby up and put it in drive. It rode smoothly and Allexis was in love with her new ride already. She looked over at her lover man and smiled. He was everything she wanted in a man and more, but at the same time she felt she knew nothing about him. He was secretive at times about a lot of things and Allexis couldn't understand why. "Kane?"

"Yeah baby," Kane said as he head lay against the head rest, with his eyes closed.

"How come you never talk about your mother or anyone in your family?"

Twisted Obsession

Kane sat quietly for a moment. Because the truth was he didn't have any family. Not any that he cared about anyway. He never knew his daddy and his mother and sister died a long time ago. "I don't have any family Allexis. Your all I got."

"Oh," Allexis said out loud. "So you don't have any aunts or uncles?"

"Nah, my mother was an only child. And like most black kids out here, I didn't know my daddy, only thing I know of him is my last name. Why you wanna know?"

"I just figured I'd ask because you never talk about anyone in your family, and I want to know everything about you."

"Be careful what you wish for." Allexis smiled. "But you don't really talk much about your family either ma. Why is it that you never talk about your mother and father?" Kane already knew he just wanted to hear it from her.

Allexis took a deep breath. She knew this conversation would eventually surface. She had been putting it off for a while simply because she felt extremely dirty whenever she discussed her father. Sometimes talking about it only made it worse, and she would have nightmares at night. But she felt that Kane told her some of his past, so it was only fair that he knew about hers. She looked over at Kane who was starring directly back at her, waiting for her to respond. Gripping the wheel tightly she took a deep breath.

"Well, to make a long story short," she began. She inhaled and exhaled three times before starting again. "My father molested me for almost a year. And . . . ," her voice trailed off before she pulled over and began to cry, it was still hard for her to talk about. And for some reason she felt Kane would look at her differently. Unfortunately the scars of her abuse were still there but somehow she managed to keep them well hidden.

"It's okay baby you don't have to say anymore," Kane said rubbing her back and stroking her hair. "You don't have to worry about dat nigga no more. I'm here and I'll be the father that he could never be." Allexis was a tad bit disturbed by that comment but didn't say anything. "But," he said cupping her chin in his hand, "I don't play games

Allexis. Don't cross me or hurt me. I'll protect you and mold you into the classy young lady that you are meant to be."

Okay where is he going with this?

"Don't ever leave me Alex, promise," Kane said losing touch with reality for just a second.

Allexis nodded her head. She never heard him call her that name before. And the look in his eyes made her feel queasy. It was as if he were talking at her and not to her. But none the less she continued to listen and nod at all his commands like a good girl.

"Good girl," he smiled and patted her head. "Come on let's go."

Allexis started up the car. Their conversation left her wondering. She wondered what he meant when he said he would be the father that her real father couldn't be. And she wondered what he meant when he told her don't *play games*. She didn't play games; at least she thought she didn't. She continued to drive wrapped up in her own thoughts wondering what she had just gotten herself into.

"Oh," Kane said interrupting her thoughts. "I think it's time you met my son, don't you think?" Kane asked with a sly grin on his face already knowing the answer.

The smile that spread across Allexis face couldn't be prevented even if she tried. "Yes, Kane I do." Allexis had her heart on having a real family, something she always wished she had.

* * *

Twisted Obsession

Chapter 11

"Where are you taking my baby?" Rasheeda asked as her voice trembled in fear of asking, because she knew how hostile Kane could get whenever she questioned him. But this situation was different; this was her child at stake. Sean was all she had and she would never forgive herself if something were to ever happen to him. She continued to hold on to her baby for dear life while Kane continued to rummage through Seans room.

"Write down a list of things that he likes to eat?" Kane said ignoring Rasheeda's question as he continued to stuff a few of Sean's belongings into the bag.

Rasheeda trembled with fear. Kane has never taken Sean out, and she has never spent a day apart from him. Tears welled in her eyes before falling down her cheeks. Sean sensed something was wrong with his mommy and attempted to wipe her tears away. Rasheeda smiled at the gesture. She looked at her beautiful son; he was the apple of her eye, her shining star. And the fact that Kane wanted to play house all of a sudden didn't sit too well with her.

"Put his ass down, and write down a list of things that he likes to eat," Kane roared taking a break from what he was doing to yell at Rasheeda. He put down the bag and walked towards her with a menacing gaze. "Don't make me tell you again."

"Mommy," Sean whispered hugging onto his mother's neck tightly.

Rasheeda trembled in fear as she put Sean down on the floor. Immediately he began to cry. Her heart ached as he attempted to follow out the room but Kane blocked his path with a sinister grin plastered on his face. She watched in awe as Kane roughly grabbed Sean and practically threw him on the bed.

"You betta shut all that noise up," he yelled causing both Rasheeda and Sean to flinch. "Please, he doesn't know you Kane. That's why he's acting like that," Rasheeda said attempting to console her frightened son. But a hard blow to the face sent her flying backwards. "Ahhh," she screamed in agony as she hit her head on the wall with a loud thud. The hard blow caused her to see stars. She grabbed the back of her head, she feel that a knot was beginning to form.

He calmly walked over and kneeled down in front of Rasheeda's face. "I'm sick and tired of you questioning me. Why do you continue to do that shit when you know I don't like it? Ain't nothin' wrong wit that lil nigga you just got his ass spoiled? Now go write down a list of what he likes to eat, so I can spend some time with my son. Isn't that what you always wanted?" he asked with a sarcastic grin.

Rasheeda glanced over at her son, who appeared to be terrified and at that moment she felt so helpless. She admitted that she always wanted them all to be a family and Kane be a part of his child's life, but not like this. Slowly she stood to her feet, wiping the blood that dribbled from her lip. She looked at her son one last time before leaving the room. Once again her child began to cry. Rasheeda had to balance herself on the walls to keep from breaking down and crying.

"Okay so we will be back in a couple of days," Kane said as he watched Rasheeda strap a sleeping Sean in his car seat.

Rasheeda kissed him on the cheek doing her best to hold back tears. She knew he was going to cry and look for her when he woke up. She knew tonight she wouldn't get any sleep, Sean has always been with her, they never been more than a day apart. Slowly she willed herself to step away from the truck.

"Kane, please tell me where you guys are going?" Rasheeda said as a river of tears flowed freely down her face. This was a first and she was honestly terrified of Kane being alone with her son.

"Relax," Kane said not fazed by Rasheeda's tears. He fiddled on his cell phone a little more before he finally decided to address Rasheeda. "I'm taking him to meet Allexis. She is the new woman in my life I felt it was time for Sean to meet her."

Twisted Obsession

Hearing that eased some of her tension, she relaxed a little. "When will he be back?"

"Soon," he simply replied closing the door.

"Okay, Kane his favorite stuffed animal is Mr. Bubbles a stuffed teddy bear. Sometimes he wakes up in the middle of the night I always give him half a chocolate chip cookie and some warm milk. And also sometimes it helps when you sing to him when he's scared," Raseeda rattled off while nervously wringing her hands. Perspiration clung to her body and clothes, while her heart practically beat her chest.

"Alright Rasheeda. Did you write all that down?"

She nodded her head yes. "Please take care of my baby Kane. He's all I got," she sniffled.

He gave her the once over before hopping into his truck. "Bitch I got this. Get yo' ass on in the house." He threw her a couple of deuces as he drove by.

Rasheeda took a deep breath, sucking in as much air her lungs would allow. It took everything within her being not to run after that truck. She watched as the truck disappeared out of her driveway. She contemplated on calling the police but she knew it wouldn't do any good because they had joint custody. Defeat consumed her body as she slowly walked back to her house with hunched shoulders and her head held low.

* * *

"This is a nice ass house," Passion said as she towel dried Allexis hair. "You sure Kane doesn't mind me being here?"

"No why would you say that? He knows you and I are good friends," Allexis said as she continued to thumb through *Essence* magazine.

Passion made a face that Allexis couldn't see. Allexis never took note but Passion always peeped the disapproving looks Kane sent her way whenever she was around. But she decided not to bring it to Allexis attention because she knew Allexis would only make excuses. She continued to towel dry Allexis hair, momentarily stopping once a dark blemish on the back of Allexis neck caught her eye. A frown interrupted her smooth forehead. She was all set and ready to question Allexis about it, but Allexis interrupted her thought process.

"Guess what?" Allexis said excitedly, as she placed her magazine on the stand.

"What?" Passion asked.

"I'm going to meet Kane's son today." Allexis said excitedly clapping her hands. She turned around to face Passion with a goofy expression on her face which immediately turned into a frown when she noticed Passion didn't share the same enthusiasm. "What? Aren't you happy for me?"

A huge sigh escaped Passion's lips. She really didn't want to put a damper on her friend's happy mood, but she felt Allexis should tap the brakes on her relationship with Kane. She felt Allexis was moving too fast, and Allexis was too naïve to see what she saw in Kane. While Allexis was infatuated with Kane's fantasy cars, and his floss, Passion could see that he was dominating and controlling.

Kane reminded her of her past situation with a controlling boyfriend. She was young and naïve just like Allexis. He was her first crush, first date, first kiss, first dick, first love, and the first to break her heart. People tried to tell her about him, but she wouldn't listen because in person he was the complete opposite of the monster people said he was. But after a couple months of wooing her in, he did a complete one eighty. Gone were compliments, and the love notes replaced with endless phone calls, and constant beatings.

Months and months of emotional and physical abuse, but no matter what he did or how bad he beat her she couldn't break away. It was like he had some kind of hold on her. Until that one fateful night, the incident with the battery acid, something in her broke she didn't know if it was because it was the anniversary of Dom's mother's death or what, but whatever it was she knew that was the last straw. She didn't want to live in fear anymore, Dom sat her down and made her realize her worth and to him she was forever grateful. With her approval, her ex was never seen or heard from again. She smiled thinking that Dom always tried to be the knight and shining armor.

"Are you there?" Allexis asked waving her arms in front of Passion's face.

Passion let a slight chuckle escape her lips. "Yea I heard you. Allexis," Passion began shaking her head as she tried to conjure up the right words to express how she felt. But Kane's loud voice interrupted her.

"Allexis."

Passion took a step back as she watched Allexis jump up out of the chair practically knocking it over. And before she could blink Allexis was out the door her heavy footsteps could be heard as she ran down the stairs. Passion rolled her eyes and shook her head as she bent down and picked up the chair. She took off her apron and dried her hands before meeting Allexis down stairs.

"Shhhh," Kane said placing his hand to his lips as he watched Allexis eagerly approach him. He gently set the car seat with a sleeping Sean still tucked securely in it, beside his couch. The smile that claimed his face quickly evaporated once Passion came down the steps to join them.

"Oh, my gosh Kane. He is so cute," Allexis cooed. She stood there in awe placing her hands over her mouth. She stood next to Kane locking her arm in his as she stared down at a peaceful looking Sean. He was Rasheeda's complexion but one look at him and she could easily tell he was Kane's seed.

"Hi Kane," Passion said as she stood beside Allexis.

He grunted before giving her a slight nod. He checked his watch. "Okay I have to make a run. I won't be back until later on tonight. His mother wrote everything down, and it's in the bag."

Twisted Obsession

Allexis didn't bother to hide the bewildered look on her face. "Wha . . . What? Your leaving me by myself with him. He doesn't know me Kane. What if he starts crying? I don't know if you should leave so soon. At least wait until he gets use to me." When Kane shot her a look she immediately put her head down. Usually when he shot her a look he met business. She uneasily glanced at Passion and smiled, but Passion didn't return the favor.

"You'll be aight," he said wrapping her in a bear hug and kissing her passionately. He cupped her face and looked deep into her eyes. "You know I love you right?"

"Yes, and I love you too," she said standing on her tippy toes and planting a warm kiss on his lips.

He smiled, knowing he had caught a good girl. "Good," he said patting her on her butt. Turning around to leave but not before sending Passion a hard sideways glance which she didn't mind returning.

Passion stuck her tongue out at Kane's back as she stood beside Allexis with her arms folded tightly across her chest. She shook her head thinking that this was some way to start off a formal introduction. "So are you ready for mother hood?"

Allexis gave her a sideways glance. "As ready as I'll ever be."

Bam!

They both jumped when they heard the door slam. They looked down at Sean who was now wide awake starring at the both of them. Allexis stooped down in front of him and began to unhook him from his carseat.

"Hi little man," she cooed.

He looked from Passion to Allexis. Then from Allexis to Passion. "Whaaaaaaaaaaaaaaaaaaaaaaa!"

Twisted Obsession

* * *

Twisted Obsession

Chapter 12

Dame followed behind Kane Closely as he entered New York. He nervously put out his tenth cigarette and counting. He was so nervous thinking that Kane could find him out at any moment. He just hoped that he had lagged far enough behind for majority of the trip, so that Kane wouldn't get suspicious. He was already treading on thin ice with Kane. But he didn't care because he was desperate and desperate measures calls for desperate things. He was tired of being in Kane's shadow and was going to great lengths, including risking his life just to find out Kane's connect so that Coleman would finally put Kane's pussy ass in jail.

The day that it happened he would be more than happy he would be ecstatic. After recruiting Neko which he ultimately failed Dame decided to roll solo other than working with Detective Coleman. Neko had crossed also. Before he offed Neko, Neko told him that he put the money he robbed from Kane in the spot but when Dame went all of the money they had put in the spot was gone. So that met one of two things Neko planned on double crossing him along or two he ran his mouth to a bitch. Which Dame was leaning more to number two because Neko had a weakness for women? His baby mom hit up for money all the time, and like a bitch Neko would give it to her.

Dame could only shake his head, because all that hard work of trying to set up the heist and someone else reaped the benefits of it. But he still hoped that this worked. He was tired of living in Kane's shadow. And since the night Kane punked him Dame has held nothing but resentment and animosity towards Kane. As Kane pulled into what appeared a vacant lot Dame dialed Detecitve Colemans number.

"Detective Coleman, speaking."

"What's up boss man?"

"Dame?" Detective Coleman, asked unsure.

"The one and only, you wouldn't believe what I'm doing right now."

"I don't have time for the games, so let me have it."

"Damn! You must be missin' me. Anyway I'm on my way to New York right now. Kane's about to re-up and I'am tagging along unsuspected," Dame said keeping his eyes closely on Kane's every move.

"Are you sure he's going to meet his connect?"

"I'm pretty sure, the Baltimore niggas is putting in work and-," Dame trailed off. "You know what never mind all of that. I'm more than sure."

"Okay, I tell you what. Get a license plate number and try to get some more information out of Kane if you can. Do you have a camera phone?" Detective Coleman asked.

"Yeah, why?"

"Take plenty of pictures of his so called accomplice and if you can take a picture of Kane transporting the drugs from their car to his car, or however you want to do it. We just put a wire on Kane's truck anyway," Detective Coleman said remaining optimistic that he would eventually catch Kane.

Dame wondered which truck because Kane had so many but didn't bother to ask.

"You askin' a helluva lot. What if I get caught? Kane ain't the one to be played wit'. I've seen first-hand how ruthless he can be."

"You're already into deep. So quit whining you knew what you were getting yourself into before we started. It's too late to turn back now. Besides you're already up there, if you're going to play detective than at least do it right."

He had to admit Detective Coleman was right. There was no way he could turn back. "I'll see what I can do," Dame said hanging up the phone. He was beginning to regret he made the phone call in the first place.

Dame got comfortable as he took out his camera and hit record.

* * *

"Hello boys," Kane said as he exited his Ford F-150. The cool breeze felt good against his skin. He flicked the cigar that he held his mouth on the ground, before adjusting the cowboy hat that sat on the top of his head.

Frankie, Joey's right hand man didn't respond simply grunted. He wasn't in the mood for all the small talk. He wanted to hurry up and get this transaction over with. He hated coming to New York, but it was all a part of the job.

"You got the goods?" Carlos asked.

"I wouldn't make this trip all the way over here for nothing," Kane replied sarcastically not really liking doing business with Carlos. He knew Carlos was racist and often tried to be slick and let the n word slip out of his mouth.

"Slick talkin' ni-,"

"Hey, Carlos that's enough, were here to do business not bust balls," Frankie interjected wanting to get it done and over with. Frankie really didn't care too much for neither one of the sons of bitches and under any other circumstances would have gladly let them off one another.

Kane looked at Carlos sideways. Ready to bust a cap before the full words could escape his lips. "Where'sJoey and Dom?" Kane asked giving Carlos a disdainful look. If given the chance he would blow Carlos brains out. The mere thought brought a sinister smile to his lips.

"Out," Frankie replied shortly. "Let's cut the bull shit. Step into my office."

Which Kane did so as Frankie showed him the candy? Kane ripped open one of the bags and tested the product. He nodded his he approving. He walked back to his truck, getting the money. For some reason he had the strange feeling he was being watched, he scanned the area his eyes landing on a Black Toyota Camry, it stood out to him. He squinted getting a better look at it and knew something was up.

"Are you going to stand here and sight see, or do business?' Carlos asked growing agitated. He hated doing business with niggers. They were trying to take over what the Italians started and thought what they so rightfully deserved. He cracked his knuckles in anticipation.

"Here," Kane said forcibly throwing the bag Carlos way, which Carlos caught the impact sent him stumbling backwards. "We definitely should change the location again we've been here twice already," Kane said still getting the funny feeling he was being watched. He adjusted his hat so that it was well over his eyes.

"Agreed," Frankie said getting a little paranoid by the way Kane was acting, because he had the same feeling, he looked around cautiously the black Toyota Camry catching his eye as well. "Alright let's get the fuck outta here," he said grabbing the other bag Kane had in his hand.

Kane retrieved both bags filled with product and quickly escorted it to his truck. He looked around one last time his eyes landing on the black Toyota Camry. He continued to stare at it, for some reason it seemed out of place in the vacant lot.

* * *

"Yes," Dame said excitedly as his Toyota Scion rocked from side to side. No words could express what he was feeling at that moment. Well couple came to mind, *boss*. His only hope was that this would finally be enough for Detective Coleman.

He texted Detective Coleman to tell him the good news. He was so caught up that he never even noticed the black Toyota Camry following him.

* * *

"So what did he say?" Marissa asked.

Once again the "Bring down Kane" task force sat around Detective Coleman's dining room table. And from the looks of things they were no closer to bringing Kane down then they were the first meeting. There were a lot of flaws in their case, and several witnesses were starting to get nervous about facing Kane in court. They knew all too what happened to the *others* in the past.

Detective Coleman rubbed his chin. "Dame is hopefully in the making of breaking this case for us. Right now he says Kane is in New York doing a drug deal with his connect."

"So that's why you told him to take several pictures?" Detective Cruise asked.

He nodded his head. "You think it will be enough?"

Marissa shook her head no. "You should have told him to call the police. I've seen Kane's attorney in action before. He will make some kind of an excuse, unless he can get up close and personal. He has to do this right. Are you sure he can pull this off?" Marissa asked drained from the tension in the room. Hours of meetings and still they haven't gotten any closer to catching Kane. For the first time in a long time she honestly felt like giving up.

"Will you calm the fuck down?" Detective Coleman stressed running his fingers through his thinning hair. "He's got this," he stated mad with himself for picking a clan of fucking rejects. When he first presented the project to them all, they were all for it. But lately it seemed like he was the only one thrilled about the quest to bring Kane down.

"Fuck you," Marissa spat. "I don't want to go to court with a half assed case. Your fucking this up not me. What kind of fucking organization is this anyway? You don't have shit, but the shit you forgot to clean outta your ass." Marissa had had enough she quickly began to gather up her things.

"Okay," Shelly said barging in once again, but this time at the right time. "I've brought the good stuff," she said placing two tall bottles of Jack Daniels in the middle of the table followed by shot glasses. "Your beer Officer Dionte," she said her smile a tad bit brighter than before. She placed the can in his hands allowing her fingertips to meet his.

"Thank you," he said with a flirty smile, staring at her rare form of a backside in the tight jeans she was wearing.

Detective Coleman ignored the flirting session between his mistress and the young officer. Instead he focused his attention on one of the bottles in the center of the table. He stood and took the cap off one of the bottles before placing it to his lips. He allowed the brown liquid to glide down his throat. It burned but in a good way. He took another long swig, set and ready to numb this feeling called defeat.

"Take it easy partner," Detective Cruise said giving his partner a pat on the back. "We will get him," he added doing his best to sound confident but not sure himself. He still felt they should get more outside help but Coleman was stubborn and refused to do so. He had to wonder what was the vendetta Coleman had against Kane ?

"Yea," he nodded not so enthused. But his mind was in another place. He could possibly have one more link. If that didn't work out he would be forced to resort to not so lawful measures.

"I'm out of here," Marissa said with her suit case in hand. She swung her long fire read hair over her shoulders. "Call me when you get something concrete. I'm tired of dealing pennies." With that she stormed out of the room, her stiletto heels clicking on the hard wood floors.

"Bitch," Coleman muttered before taking another long swig of his new found friend Jack Daniels. "This meeting is adjourned," he said standing up leaving the rest of his crew behind. He was going home to his wife, she always knew the right words to say when he was in need of uplifting spirits.

"I'm right behind you buddy," Detective Cruise said standing up and throwing on his jacket. "Hey you comin?" he asked taking note that Dionte hadn't budged from his seat.

Twisted Obsession

"Uh, yeah once I down this beer and take a piss. You go ahead I'm right behind you."

He looked from Shelly to Dionte and shrugged his shoulders. *Fuckin' fags* he thought to himself as he turned to leave.

Once they heard a second slam of the door Shelly walked over and sat on Dionte's lap. She engaged him in a passionate kiss. Since their last meeting they have been inseparable. They continued to kiss allowing their hands to grope one another. He placed her on the table before taking off his jacket. Shelly seductively licked her lips and grabbed the huge visible bulge in his pants.

Sounds of pleasure escaped both of their lips. Unable to stand the friction that held them apart any longer, Shelly dropped to her knees and quickly unbuckled his pants. She pulled out his huge anaconda. Her eyes quickly lit up, she had never seen one so big before. The first time they made love they used a whole bottle of lubrication, because of the thickness of his man tool. She thought back to Coleman and compared the two it was like the Papa snake and the baby snake.

"Hmmm," Dionte moaned once he felt the warmth of Shelly's mouth around his long pole. He gently glided her further and further down onto his rock solid shaft. Gyrating his hips in circular motion, as he squeezed his butt cheeks tighter ready to buckle at any moment from her immaculate head game. His eyes rolled in the back of his head as he felt the veins in his dick began to swell. "That's it Shelly," he encouraged feeling himself nearing an orgasm. "Yes," he panted his knees buckled once his semen exploded from his penis and into Shelly's awaiting mouth. His orgasm intensified as he watched her take all of him and his cum into her mouth. "Oh, Shelly baby," moaned.

He couldn't stand it any longer. Roughly he stood her to her feet and turned her around. Stroking her inner thigh ever so gently, she leaned her head back in pleasure. He sucked her neck doing his best to leave his mark of passion. Shelly reached behind her and stroked his love machine back to full attention. She was tired of the four play and was all but ready to feel his powerful thrust.

"Please," she begged. "I want to feel you," she moaned.

"Okay baby," he said unable to deny her request. He pulled down her pants and revealing her square bottom which turned him on. Just the sight of it made his dick grow and inch longer. He carefully removed the tape that held her own tool together. He dug in his back bent over and fished around in his pants pocket and pulled out lubrication.

"I can't wait for the operation so that we don't have to go through this anymore," Shelly pouted.

Dionte continued to moisten his monstrous snake. "Its fine baby, besides daddy likes you just the way you are." Without another word being he glided his dick into her opening. He thrust himself deep within her until he could go no more. Long and hard he stroked her, turned on by her powerful moans of pleasure. He rocked and rocked them both to ecstasy. Then they both collapsed on the dining room table.

* * *

Twisted Obsession

Chapter 13

Allexis sat cozily on the couch in the living room reading one of her favorite books, Sister Souljah *The Coldest Winter Ever*.

"Yo', sis is that your new whip? Dom asked, taking note of the all black Mercedes Benz that sat outside in the driveway. His little sis was growing up and trying to do big things. He just hoped she was careful. Kane had a long rap sheet when it came to his women.

"Yup," Allexis replied never looking up from her book.

He glanced over at a sleeping Sean that lay on the love seat with mounds of pillows in the floor waiting just in case he fell. Dom shook his head because Sean was bad as hell and had Allexis wrapped around his finger all ready. "You tryin' to get step mother of the year?"

Allexis looked up from her book and smiled. She looked over at Sean. "Whatever."

"Rasheeda doesn't mind you taking Sean?" he quizzed.

Allexis shook her head no. "Well at first but once she realized that Sean practically adored me she was at ease. But she's pretty cool sometimes."

He nodded his head as looked toward Tina's bedroom to make sure she was still out of earshot. "Yo' walk outside with me for a second?" Dom asked, heading toward the door. "He'll be alright with all those damn pillows," he said once he took note Allexis hadn't followed him.

"Okay but don' t be too long," she said reluctantly as she looked over at Sean who laid on his back with his mouth slightly ajar, a little dribble running down his chin.

He opened the door for her and they walked outside, they stood in front of his Range

Twisted Obsession

Rover. "Okay, you know Tina's birthday is coming up right?" Allexis nodded her head. "Aight, this year I want to do somethin' big. Maybe throw a party in New York. And . . . I'm thinkin' 'bout askin' Tina to be my wife," Dom said pulling a box out of his pant pocket.

When Allexis eyes landed on the rock Dom held in his hand her eyes nearly popped out of her head. "Oh my gosh," she screamed jumping up and down with excitement. Allexis was leaping for joy because she has always loved Dom like a big brother. And to her this would make it official; he would be stuck with the both of them for good.

He grabbed her and put his fingers to his lips. "Shhh . . ." Dom said looking towards the house making sure Tina hadn't overheard them. "I would like to keep it a secret," he joked.

"Oh, my gosh Dom. She is going to be so ecstatic!" Allexis said doing her best to contain her excitement.

"Yeah, I know. I just need you to bring her to the spot next Saturday. You think you can handle that for me?"

"Yes, I can handle that," she replied.

"Thanks sis," he said, planting a kiss on her cheek, before hopping into his Range Rover.

Allexis was at a loss for words. She truly loved Dom with all her heart, because when Allexis came knocking on Tina's door that fateful night he willingly accepted her with open arms. He never tried anything out of the way and she truly loved him for that. She could sit down and trip with him like she could the girls. And she could honestly say that other than Kane *he* was truly the only man that she ever trusted.

"Dom, thanks," she said sincerely.

"For what ma?" he asked with a confused look on his face.

Twisted Obsession

"Just for being there for her when she fell so low. For . . . giving me my sister back."

Dom wiped the tear that had escaped her eye and traveled freely down her cheek. "Girl, get on in the house before you have me cryin' like some punk," he joked. "But sis real talk, how are things going with you and Kane?"

"Pretty good so far. Why?" she asked cocking her head to the side and putting her hand on her hip. She had to wonder if Tina put him up to this.

As if he were reading her mind he replied, "I know what you're thinking and no your sister didn't ask me to talk to you. It's just that you're like my little sis and I only want to see the best for you. You don't need to be out here in these streets wit some hood nigga, and definitely not with a hood nigga like Kane."

"Dom, you know Kane doesn't live in the hood. He owns a beautiful home on the beach."

"Real cute Allexis, but I think you know what I mean. I know his type all too well. And I've heard some not so nice stories about him regarding his women. So Allexis I'm telling you, if that nigga even as so much look at you wrong don't hesitate to call."

Allexis eyes shifted a little. "Okay, Dom."

"I'm serious Allexis," Dom said attempting to engage her in eye contact but she refused to look at him. "You're too good for him anyway, the sooner you realize that the better off you will be."

"Okay Dominique," she smiled knowing that he hated to be called by his government name. "I assure you if anything were to happen I would let you know. But everything's fine. I promise."

"Aight baby sis," he said still skeptical. "And you know I don't like that name. So quit playin'," he smiled.

Twisted Obsession

Allexis smiled and waved as he backed out of the driveway. She found it hard to contain her excitement. Kane sat across the street in his Escalade, banging his dashboard. He was definitely going to have to teach her a lesson about letting other men their hands on her. Allexis never noticed him pull off into the opposite direction.

* * *

"Baby, I thought we were going out to eat?" Allexis asked, as she climbed up the steps cradling a sleeping Sean in her arms. It was struggle to walk up the steps with the little fella because he was so heavy but she didn't mind doing so, because she had grown completely fond of the little man.

Their first encounter started off kind of rocky. Allexis smiled because she was so in over her head, and thank God for Passion. She thought back to that night, when Sean first woke up he was beside himself because of the unfamiliar surroundings. Frantically she ran like a chicken with her head cut off. Finally Passion found the instructions Rasheeda had left for them, Mr. Bubbles worked for like five minutes before he went into rare form again. Passion suggested that she sang to him, so Allexis did so but he wanted no parts of her horrible singing and Passion didn't either. So Allexis let Passion try and once Passion opened up her mouth her smooth angelic voice put Sean at ease. And it was smooth sailing from there.

She walked into his room which Kane let her decorate herself. It was the red and blue the spider man theme. His bed was shaped like spider man even though he didn't sleep in yet because he was still sleeping in a crib. But Allexis thought it was so cute. Gently she placed him in his crib before planting a gentle kiss on his forehead. "Good night little man," she whispered.

Kane held his bedroom door open for her without even so much as a word. Allexis walked into the bedroom, dropping her Chanel bag on the dresser. She walked over and plopped down on the bed kicking off her Jimmy Choo sandals. Allexis nervously twirled a loose strand of her hair, she was a little uneasy. Kane's silence was putting her on edge.

Twisted Obsession

She shivered a little and rubbed her arms to relieve the chill.

Kane stood in front of Allexis and glared at her. He was beyond livid. He specifically told her that he didn't play games, and that's what he thought she was doing the other night. He definitely had to teach her a lesson about playing games and playing with his emotions. Just seeing another man touching her was enough to send him over the edge.

Allexis felt her heart jump into her throat. She swallowed hard. "Baby is there something wrong?" she asked as she looked up at Kane. His eyes were ice cold and it made her shiver. The funny feeling she had in the pit of her stomach told her something bad was about to happen and when she got that feeling she was usually right.

"Take off your clothes," he demanded, ignoring her question.

"Baby-," he cut her off with a smack to the face. Allexis grabbed her cheek. "Take . . . off . . . your . . . fuckin' clothes," he ordered through clenched teeth.

Allexis slowly slipped out of her Christian Dior pantsuit. "Kane," she whispered, her body trembling in fear of what he was about to do. "You promised never to *hurt* me again," she said as a single tear slid down her face.

"Take those off too," he said referring to her matching Victoria Secrets bra and panties.

She slowly removed her undergarments. And without warning he smacked her again, knocking her off the bed. Allexis cried out in pain, and quickly scurried across the room. She made a beeline for to the door only to find that it was locked. Holding up her hands she attempted to plead, "Kane, please. You promised," she said with her back flat against the door, standing on her tippy toes.

But her pleas fell on deaf ears. Kane grabbed a handful of her hair and effortlessly threw her on the bed. "I saw yo' fresh ass last night, talkin' to that no good nigga," Kane spat. Just the thought another man touching Allexis made him insanely jealous.

Allexis frantically thought back to the night before and the only person she could think of was Dom. "Who, Dom? It wasn't like that, Kane. Please," she begged. "You know I wouldn't do anything to hurt you Kane. You said conduct yourself like a lady and that's what I've been doing. Kane' please," she said as her bottom lip quivered.

"It didn't look like nothin' to me," he replied removing his leather belt. "Smilin' and laughin' in some nigga face. What the fuck do you take me for? Huh? "

Panic set in once she saw the belt in his hands. "Oh, God Kane. Please, don't," she begged. "Not with Sean in the other room. What if he wakes up? He might hear you," she cried hoping he would take in son in consideration but he didn't.

"I warned you Allexis. I told you don't play games with me," he said shaking his head. "I guess daddy is just gone have to show you what happens when you play games, because I always win. This gon' *hurt me*, more than its gon' *hurt you*," he said grabbing a handful of her hair.

Allexis yelped out in pain when she felt the warm leather slap her skin. " Don't . . . you . . . ever . . . let me . . . catch . . . you wit' . . . some . . . other . . . nigga," Kane said whipping her mercilessly. The more he thought about Dom touching her, the angrier he became and the harder he beat her. He felt little regard for her feelings as the leather tore into her delicate flesh.

When he was finished he released her hair. Allexis lay limp on the bed, unable to move. Her entire body ached from the beating she just received. She sobbed uncontrollably. Kane put his belt back on. Mad that no matter what he did, they never seemed to learn. But he vowed to make Allexis learn, she was special. He would instill moral values in her, even if it killed her.

"I have some business to attend to. And you best not leave this house," he threatened. "I love you Alex and only want the best for you. You have to learn." He kissed her on the cheek and left.

Allexis lay there on the floor too tired and sore to move. She cried herself to sleep.

Twisted Obsession

When Allexis woke up the next day her body was screaming in pain. She somehow managed to make it into the bathroom. She ran herself a nice hot bath. "Ahhh . . . ," she screamed, as she dipped her sore body into the steaming water. She was covered in black and blue bruises; there wasn't a part of her body Kane didn't somehow manage to touch. Allexis sat in the tub, and buried her face into her hands. She wondered how she could be so stupid. She knew Kane loved her and only wanted the best for her, but she always found a way to fuck it up.

Kane left Allexis secluded in the house for three days. The only company she had was Sean, Kane's maid Maina, and his cook Carmelita who cooked her two meals a day. She and Sean mostly stayed cramped up in the room watching re-runs of Eve. Tina called her Wednesday, wondering where she was, and to make sure she was okay. Of course Allexis told her what she wanted to hear. And Rasheeda called also to check on Sean.

"Hey girl where's my baby?" Rasheeda said while rudely smacking her gum.

"Hold on he's right here," Allexis said annoyed with the childish antics. She knew Rasheeda still felt some kind of way about her and Kane but she shrugged it off because she knew that Kane was the least bit interested in Rasheeda. "Here Sean," she said attempting to hand him the phone.

But he adamantly shook his head to preoccupied with the game his dad had brought him. He continued to push buttons as if he knew what he were doing, all the while intrigued by the movements on the screen.

Allexis tried again, "It's mommy on the phone," she said placing the phone to his ear.

"No," he said shortly inching away from Allexis.

She only laughed because any other time he would practically tear the phone out of her hands. "Girl, he is too wrapped up in this game," she sighed.

"Awe my baby doesn't miss me?" Rasheeda asked not hiding her disappointment. She couldn't help but to feel jealous especially on occasions when Allexis would drop him off after spending the weekend and he didn't want any parts of her and would actually cry after Allexis when she left. It was only natural for her feelings to be hurt. But other than that she was more than relieved to find that Sean was spending most of his time with Allexis rather than Kane.

"Yea, he swear he is doing something on this computer," Allexis chuckled.

"So how are things with you and Kane?" Rasheeda asked, being nosey. She wondered if things were different with her, because she could tell Kane really had it in for Allexis for some odd reason.

"Good," Allexis said shortly not really wanting to discuss her personal life with Rasheeda.

"Oh, that's nice. I know how Kane can get at times."

"What is that supposed to mean?"

"I think you know."

"No, I don't," Allexis pressed wondering if Kane's abusive habits were a pattern. She didn't know why it really mattered because she knew no matter the outcome of what Rasheeda said she was still going to be with Kane. Besides she didn't know if this was a ploy from Rasheeda to break up her and Kane.

Rasheeda sat there quietly for a moment, pondering the idea of spilling the abuse she herself had suffered at the hands of Kane. After a long moment in thought she figured what the hell it couldn't hurt anything, she was more than sure Allexis has met with Kane's evil twin.

"I'm just saying that Kane has a problem with his hands. On several occasion's I've been forced to walk around with a black eye or two for that matter. That nigga is plain straight bonified crazy," Rasheeda managed to say all in one breath.

Twisted Obsession

Allexis held the phone tightly in her hands. She didn't know why this was news for her, deep down she suspected. She decided not to air out her own dirty laundry. "Well, Kane has been nothing but a gentleman to me. He's never put his hands on me. *Never*," Allexis squeaked her hands shaking a little.

"Okay," Rasheeda smirked, smelling the bullshit Allexis was sending her way. "Just be careful, and whatever you do make sure that Dame is nowhere near when Kane get's mad."

"Dame, why?" Allexis asked a confused look overtaking her face. She rarely ever see's Dame.

"Because . . . because . . . ," Rasheeda took a deep breath not really believing she was actually about to reveal this source of information. The most humiliating form of abuse that she has suffered at the hands of Kane. She opened her mouth but couldn't form her mouth to say the words.

"Rasheeda," Allexis said after a long pause. Her heart skipped several beats as she waited in anticipation of the answer, but her mind was telling her that she really didn't want to know. Allexis hand began to shake, she gripped the phone tighter afraid she might drop it.

"Never mind," Rasheeda said in a voice barely above a whisper. She couldn't bring herself to do so. "I'm going to go. Give Sean a kiss for me."

"Okay I will," Allexis said a little relieved. A part of her was actually scared to find out what Rasheeda had to say, although she had a feeling what it was going to be, and the very thought made her sick to her stomach.

A wave of nausea washed over her body. She quickly got up and grabbed the glass of water that sat on her night stand and consumed it. It took her awhile to get herself together. She sat down on Kane's bed in deep thought. She heard Sean playing underneath the bed and quickly scolded him.

Twisted Obsession

"Ook," he said holding up a mahogany jewelry box barely able to hold onto it.

"Give me that," she said taking it out of his hands. "I'm sorry," she cooed caressing the side of his cheek once she realized she hurt his feelings. "Go get your game," she encouraged clapping her hands. He smiled and obliged.

Allexis cradled the box in her hands and continued to stare at it. It was so beautiful the smooth wood. At first she felt guilty about having the jewelry box in her possession; getting the feeling she was somehow betraying Kane. After what seemed like hours she started to put it back but curiosity got the best of her and she couldn't bring herself to do so.

When she finally opened it she nearly gasped. There starring right back her was a picture of herself. Allexis picked the picture up and flipped it over.

To K my big brother and confidone.

Love Alex

Age 15

"Alex . . . Alex . . .," she repeated to herself. Why did that name sound familiar. And that's when she remembered Kane called her Alex the night he bought her the Mercedes. Allexis quickly closed the jewelry box and put it back where she found it. For some odd reason seeing that picture gave her the creeps.

She didn't know much about Kane's sister except that she died at a young age. He really didn't like to talk much about his family so Allexis never bothered to ask. But seeing that picture of Kane's sister left a bad taste in her mouth.

* * *

Kane came home around 3:30 Thursday morning a little tipsy. Allexis was lying down on the bed wide-awake. She turned around to look at him. She watched him as he moved

Twisted Obsession

around in the dark. "Kane," Allexis whispered.

"Yeah," he replied, slipping out of his shorts and hopping in the bed beside her. He pulled her on top of him. Her body was still a little sore, but she ignored it. He pulled off the oversized tee shirt she wore, exposing her full breast. He placed her nipple in his mouth. She arched her back as he placed his hand inside her panties, playing in her wetness. He ripped off her panties. "Sit on my face," he commanded.

Allexis got on her knees and slid her body toward his face. She grabbed the headboard as Kane darted his tongue in and out of her wetness. He teased her clit with his tongue, causing Allexis to moan. "Kane," she whispered. She rode his tongue, "Oh shit," she yelled as she reached her climax.

Kane flipped her over on her back and entered her soaked pussy, stroking her walls, hitting her spot. "This is my pussy," he softly whispered in her ear.

"Yes, this is your pussy," she moaned.

"That's right. Don't . . . you . . . forget . . . it," he moaned.

The love making was passionate, their bodies enter twined tightly. He stroked her hair ever so gently while kissing her eyes, her nose, her neck, and breast. Just the gentleness of his touch made Allexis shiver. She wondered how one minute he could be ready to tear her head off and the next be as sweet as angel. Allexis couldn't help but to melt as he gazed upon her with the soft brown eyes while he rocked her to ecstasy.

"Kane," she cried into his chest as they reached their climax. "I love you," she whispered as a single tear rolled down her cheek. She didn't know what it was about this man, she couldn't let go.

The next day Kane dropped Allexis off to her house and she was glad to be home. Without saying a word to her sister she went upstairs and locked herself in her room. Tina noticed a drastic change in her sister's behavior. Allexis never hung out with them before, but she rarely ever came out with them anymore. And Allexis loved to go shopping with them but she would always make up and excuse about why she couldn't come out. Tina knew something was definitely up but she couldn't figure out what it was. She just hoped everything was okay between her and Kane.

* * *

Twisted Obsession

Chapter 14

Natalia watched with open eyes as Kane continued to talk on his cell phone. She wasn't the least bit concerned about it though, because it wasn't another female, because if it was she would have definitely showed her ass, not even flinching about the consequences. She loved Kane's aggressive nature. In all her twenty some years of living she has never endured a man like Kane. He had some kind of hold over her, and she couldn't shake it. Once he was off the phone she stood directly in front of him, looking exotic in her skin tight leggings that showed every curve in her body along with a black long sleeve shirt. She loved the color black.

"So it's true?" Kane asked, but deep down he knew the answer all along, but for some reason he was having a hard time acknowledging the fact that his boy turned against him.

"Yes, it's true daddy," Natalia said as she dropped the manila folder in Kane's lap, which he didn't try to catch and the papers and photos went flying everywhere.

Even though Kane had his suspicions he still felt some kind of way once his suspicions were confirmed. He glanced down at the photos of Dame and Detective Cole in some not so compromising situations. It was one thing for Dame to be a snitch but to be a fucking homosexual. He couldn't believe all these years and he never knew, that would look bad on him if anyone ever found out. He wondered if Dame had ever thought about coming him sideways. Just the thought had him on the verge of throwing up. He thought homosexuals were the most disgusting creatures on earth.

"So that was you in the Black Toyota Camry?" Kane asked, eyeing her deep red lipstick which stood out against her all black attire.

"Yelp," Natalia said wanting to get down to other business. "So how do you want this handled? You want Joey to handle it? Or I could handle it for you if you like?" she said seductively.

"Man I don't even want that nigga to know, but more than likely he already know.

Twisted Obsession

Damn to much shit has gone now. Do you know this looks?" Kane asked agitated. He felt the very operation that he put so much hard work into was falling apart. "Shit how much more fucked up can this shit get. I need to get back on my mutha fuckin grind. Nigga's is really sleepin' on me yo'. Fuck a second in command, I'm runnin' this shit by my damn self. Get on yo' fuckin' knees," Kane growled grabbing Natalia roughly by the hair.

She slapped his hand away in an attempt to protest, but he gave her a back hand for being disobedient. Blood dribbled down her lip, which she quickly licked seductively looking Kane in the eyes. He unbuckled his pants and allowed her lips to penetrate his dick.

"That's right spit on that shit," he said as his eyes traveled down to her, but his eyes unwillinging traveled to the picture of Detective Coleman and Dame and instantly he lost his erection. "Move I can't do this shit right now," he said pushing Natalia back.

"Ouch," she said as she landed on her elbow.

"Right now, I have to get down to business," Kane pulled up his pants. "Find out who else is working on the case with Detective Coleman. Anybody and everybody that is linked to this shit is dead."

Natalia watched pissed off that her chance to get fucked just the way she liked it walk out of the door. She looked down at the pictured of Detective Coleman and Dame. She shivered at the fact that she was two seconds from fucking Dame one night. She looked up at the ceiling and mouthed, *thank you Lord.*

Natalia gathered up the pictures that lay scattered across the floor. She knew it was time for her to get to work. She knew she had to prevent Kane and Joey from getting caught, because her ass was on the line too.

* * *

Dame was getting impatient with Detective Coleman. He wanted Kane out of the picture as of today. He wondered if Detective Coleman was able to find out who Kane's connect was? On the up side he had a little extra spending money thanks to scheme he pulled in Baltimore. "Kane should have listened to me," he said out loud pointing to his chest. He chuckled to himself as he snorted a few lines of coke. He was in the mood for some pussy to relieve some stress, so he decided to call up Doricka.

"Hey baby," she purred.

"What you doin'?" he asked grabbing himself, "Daddy, misses you."

"Oh, really. I can't tell," she replied sucking her teeth.

"Awe, don't be like that," he replied. "Look, how 'bout I go pick up a couple movies, swing over to pick you up and bring you back to my place. How does that sound?"

Doricka had to admit the dick was tight. She sighed "Aight," she replied, sucking her teeth. "I'll be ready around 9 o' clock. And be ready to give this pussy a black eye," she replied seductively before hanging up.

Dame smiled as he grabbed his car keys. He dialed Detective Coleman's number, only to get his voice mail. If Detective Coleman took too long Dame planned on finding other ways to put Kane away, for good.

* * *

Joey had been trying to get a hold of Kane for the last half an hour. He found out some disturbing news about his friend Dame. He recently discovered Dame has been working undercover with Detective Coleman, way undercover for that matter. He dialed Kane's number again.

"Whad up?" Kane bellowed into the phone, knowing it was Joey but only wanting to make him mad. He loved getting a rise out of Joey, because he always tried to act so calm and collected.

"It's about fuckin' time," Joey yelled pissed that so many of his phone calls had gone unanswered.

"Yo' what the fuck?" Kane asked a little agitated.

"Where are you?

"On my way back from Maryland, Distributing some of the product I picked up last night. Why?" Kane asked not really liking all the questions.

"Meet me at Taverna's at 9:15," he replied, hanging up before Kane could respond

Kane looked at his phone, "Who the fuck does this bitch, think he is? He'll be dining by his damn self" Kane said as hedialed Allexis number. He decided to let her attend her sister's party. He felt that she had definitely learned her lesson.

"Hello," she answered.

"Whad up?" he asked. "What you doing?"

"Nothing, just reading a book," Allexis smiled a little glad to hear the softness in Kane's voice.

"So, what you wearin' to your sister's party tomorrow night?"

"Really Kane. I can go?"

"Yeah, so what are you wearing?"

Allexis smiled. Up until that point she wasn't sure if he was going to let her attend Tina's party. And she knew Tina would be devastated if she didn't show up. "I don't know. I'm pretty sure I can find something to wear," she replied excitedly. "Thank you baby. Thank you for giving me another chance. I know I messed up, it won't happen again. I promise."

"I'll have a limo pick you and the girls up tomorrow morning," he said smiling, glad to hear that she was trying to be a better person.

"Thank you baby," Allexis said once more.

"Bye," he said hanging up. *I think she may be wifey material*, he thought to herself. He called up Dimples and told him to meet him at the shop in V.A to discuss the details on the Hummer, forgetting about his dinner date with Joey.

* * *

Twisted Obsession

Chapter 15

"Happy birthday," Allexis, Doricka, and Passion sang in unison.

Tina looked at her clock, it read 7:30. "Y'all bitches better go back to bed. It's too damn early in the morning for this shit. And Doricka, that hot pink sweater is *blindin'* a bitch," she said turning over on her side giving her back to the trio.

"She is right. That pink is a little loud," Passion, commented. Doricka gave Passion her middle finger.

"No," Allexis said, pulling the covers back. "Get up! We are going shopping!"

Tina sat up, "Where?" she asked rubbing the sleep out of her eyes.

"You are so predictable. Anyway, we are going to New York." Allexis said rolling her eyes.

"Awe that ain't nothin', Dom takes me there all the time," she said ready to lay down and go back to sleep. She wasn't all that excited about her birthday, to her she was just another year older. Although she must say she is thankful to see another day.

"But everything is on Kane today. And were riding in a stretch limo," Allexis replied doing her best to make the trip sound exciting. "We can get our hair and nails done. Go clubbin' afterwards. You know, big shit."

Tina laughed, "Clubbin' huh?"

"Come on bitch! Splash some water on the cat and let's roll," Doricka said, itching to spend the dime, Dame dropped on her.

"Aight, damn. Give me fifteen minutes."

"*All I got is fifteen minutes, and I wanna get up in it,*" Doricka sang as the trio left Tina's room.

Twisted Obsession

The quartet was dressed and ready to go by 10:30. Kane rented an all-black stretch Navigator, fully equipped. They had Cristal, Dom P, Moet, the works. He sealed the deal by giving her a duffel bag full of cash. They made a toast to Tina on her birthday, before heading off to New York.

The girls were a little tipsy by the time they reached New York, thanks to the Cristal. They all decided on doing a little shopping in Times Square.

"Damn, my ass looks phat in this dress," Tina giggled, bending over and shaking her ass for the security guard who stood by the store's entrance. He blushed and looked the other way. She admired the way her body filled out the black one shoulder belly dress.

"Girl, you need to stop," Passion said laughing.

"What! He was lookin' so I decided to give him a little show," Tina replied innocently placing a finger to her lip.

"Girl, you are *wearin'* the hell out of that dress!" Doricka exclaimed.

"I know right!" Tina said smiling, admiring her figure. "Look at my fatty," she teased, shaking her ass a little.

"Don't rub it in Tina," Allexis said a little jealous because her backside wasn't as curvaceous.

"What if you got it you got," she said kissing the tip of her finger before placing it on her bottom.

They all shared a laugh.

"So, you gon' move to New York wit' Dom?" Doricka asked.

Tina shook her head no. "I can't do it. I already told him that. That's why I'm still in Virginia, but he likes the NY because business is better for him here. But, this state, this city. Too risky for me, I'm scared I might get myself in trouble again. If you know what

I mean?"

The girls sat in silence for a moment. "So, how does it feel to be twenty-three?" Allexis asked, attempting to change the subject.

Tina shrugged her shoulders, "I really don't feel any different. Just ready for Dom, ass to settle down, I'm not getting any younger. And I feel we have been together long enough to settle down and have one or two babies. I want to have my first and last by the time I'm thirty. But I don't know about the boy. He lives and breathes the streets. So, we will see" she replied, heading back into the dressing room to change her clothes.

Allexis and Passion looked at each other with knowing glances. Doricka peeped this and wondered what they were being so secretive about.

After each girl picked out a dress, along with getting their hair and nails done, they all headed back to the Hyatt to eat, shower, and change for Tina's surprise party.

* * *

"Damn, nigga! You goin' all out for shorty," said Tubby, Dom's younger brother.

"That's my baby," Dom responded smiling. He rented the popular club Platinum, and hired D.J Ill Kidd to host the party. Each table was set up with a bottle of Cristal, along with twelve long white-stemmed roses, held inside a diamond-decorated vase imported from Paris. He also had gift bags for the guest containing his/her Rolexes, along with gift certificates to Bergdorf Goodman and Saks and Fifth Avenue.

Dom decided this year would be the year he retired from the game. Tina had stuck by him for five years, putting up with all the bullshit and drama that comes with a hustler working the streets. He thought about the first time he meant Tina. She was working at a strip club in New York. Dom could tell she was strung out on drugs, but even at her lowest point she was still beautiful. They meant when she tried to cop some product from one of his workers. But, Dom refused to sell her anything. The pain in her eyes wouldn't let him. That day he made her a proposition promising to take care of her if she got herself clean.

She brushed him off at first, cursing him out, but after nearly colliding with death when she nearly overdosed, she quickly changed her mind. It took her awhile to fully trust him, but eventually she did. Dom was glad Doricka was there to support her.

Tina wanted him to get out of the game. She knew what it was like to be strung out, and she couldn't understand how Dom could sell poison to his people. Dom, kept promising her one more year and he was out, but that was three years ago.

He knew he was ready to settle down, and take better care of his future wife, mentally, physically, and emotionally. He already purchased a brand new two-story brick house. The new house contained thirteen rooms, three and a half baths, three-car garage, a basement, and outdoor pool. He also purchased her a 2008 LS Hybrid; he knew she was ready to upgrade her old Lexus. Dom was definitely ready to make Tina his wifey.

He dialed Allexis number, she answered on the third ring. "Yo', when y'all comin'? Some of the guest is already here." He tried to get Tina's mother to come, but she felt Tina wouldn't want anything to do with her.

"Hey, Kane," Allexis said, looking at Tina who was still admiring her figure in the mirror. Allexis rolled her eyes. *This chick need to sit her conceited ass down*, Allexis chuckled to herself. "Ummm, yeah, we are leaving in ten minutes."

"Good. Holla at me before y'all come in so I can get everybody ready. Holla!" he said before hanging up.

"Okay, baby. Love you too," Allexis replied hanging up her phone.

"Is Kane comin'?" Tina asked, slipping into her Manolo heels. She really hoped he wasn't but decided she would ask anyway.

"He said he would try," Allexis lied, adjusting her black Versace jersey halter dress. She and Passion wore the same dress; only Passion's was white. When no one was looking she applied make-up to her body where some of the bruises she had suffered from the severe beating from Kane were still noticeable.

"Y'all ready?" Doricka asked entering the bedroom. Doricka and Tina wore similar dresses, only Doricka's was red and her dress had two straps instead of one.

"Yeah," Passion replied touching up her make-up. She noticed Allexis trying to apply make-up on the sneak. But she already seen the bruises, she made a mental note to talk to Allexis about it later.

"Come on y'all," Tina barked. "I'm ready to roll," she said smacking her ass, about to drop it like it's hot, but opted not to.

"Okay, damn," Doricka said walking with Tina out into the living room. Doricka had to stop and admire her friend for a moment. She remembered the first time she met Tina. She was so pale and thin, strung out. Doricka could see the pain and heartbreak in Tina's eyes. Dom told Doricka about Allexis and Tina's sick father.

She did her best to console Tina during her time of need; sharing and expressing her own troubled past. But, Tina could care less about herself. The only thing she talked about was saving her sister Allexis, whom she had left behind. Tina felt guilty about leaving her behind; she wanted so badly to get herself, so she could save Allexis before it was too late.

Doricka understood how she felt. She never had the chance to save her own sister from her own abusive father; he shot her younger sister in the head with a .38 before turning the gun on himself. Her father wanted to get back at her mother for leaving him, by killing her sister. Doricka remembered crying for days, consumed with guilt. So, when Tina came to her, she was able to sympathize with her, because she understood how she felt. Doricka encouraged Tina to get herself together first, so she could provide a stable environment for Allexis. She remembered the withdrawal process, the cold sweats, and vomiting. Doricka shook her head, thinking about the hell Tina went through, getting herself clean. But, Tina was determined to rid herself of the drugs, and she did.

Twisted Obsession

Doricka sighed as she thought about the day Allexis showed up on Tina's doorstep. It broke Tina's heart, she cried for two days. She felt she had let Allexis down.

"Earth to fuckin' Doricka," Tina said, waving her hand in front of Doricka's face, breaking her train of thought.

Doricka swatted at Tina's hand. "Girl, don't be puttin' your hands in my face! I don't know where them things been."

"In my ass," Tina joked, sticking out her tongue.

"Here," Doricka said, handing Tina a card and a small box.

"What is this?" she asked as she opened the card.

Tina,

Thank you for never judging me and loving me for who I 'am. I just want to say I' am proud of your black ass. Look at how far you have come. Please, don't change for nobody. And hurry yo' ass up and give me some god babies!

L/Y/L/A/S

Doricka

Tina smiled, as she opened the small box. Inside was 14 karat gold friend ship necklace; in diamond letters it read *My Nigga Fo Life*. Tina laughed. "Bitch, you tryin' to make me mess up my make-up," she joked, giving Dorikca a hug. She gave Doricka the other half of her gold heart. "Help me put mines on and I'll put yours on." The girls didn't care if the necklaces clashed with their outfits or not.

Even though she and Doricka had their ups and downs she truly felt Doricka was more of a sister than a friend. Doricka has never once thrown dirt on her about her past life and she was truly grateful for such a friend.

Twisted Obsession

The girls embraced again. "I love you too big head," Doricka said.

"Come on Passion and Allexis," Tina yelled. "I'm ready to get my swerve on. And I want to make it clear now, all my drinks on all of yall tonight," Tina said as she bopped out the door to her own beat with the three ladies in tow shaking their heads.

* * *

Before entering the club Allexis texted Dom, to let him know they were outside. "Aight everybody," Dom broadcasted over the microphone. "Places please, my baby is about to make her grand entrance."

An unsuspecting Tina entered the club first followed by Doricka, Passion and Allexis. "*Surprise!*" everybody yelled as balloons and confetti fell from the ceiling.

Tina looked around in shock at first before covering her face with her hands, in disbelief. Dom scooped her up in his arms, "Happy birthday, baby," he whispered in her ear.

"Dom," she squealed hugging him. "I can't believe you did this!" she said still in shock, planting wet kisses all over his face. He gently put her down and kissed her fore head. Tina turned around to face Doricka. "Why didn't you tell me?"

"Cause ain't nobody tell me," she sulked folding her arms across her chest.

"Good, 'Cause it was a surprise and I wanted to keep it that way," Dom teased. Doricka playfully mushed him in the head.

"I just want to give a shout-out to my girl Tina. Happy Birthday, girl," the D.J blasted over the microphone. "With dat said, we 'bout to turn this muthafucka out." He started out with T.I.'s

Big Shit Poppin'.

"Happy Birthday," Allexis and Passion said in unison giving Tina a hug, handing her an envelope. "You can look at it later," Allexis added.

"Y'all two some sneaky bitches," she said, handing the envelope to Dom. The quartet headed to the bar, ready to get the party started. As soon as they got their drinks, the DJ put on Beyonce's *Get Me Bodied*. Without saying a word each girl left their drink at the bar and made their way out onto the dance floor.

* * *

All eyes were on Joey as he walked into the club decked from head to toe. But he was the least bit concerned about his looks, because he was pissed. He couldn't believe Kane had the balls to stand him up. He pushed his way through the crowded club. Frustrated he headed to the bar, with Frankie and Carlos in tow. He ordered a shot of Bacardi straight. He scanned the room for Kane, and spotted Allexis instead.

"Damn!" he said out loud as he watched Allexis move her body. The dress she wore fit her body just right. His dick immediately began to rise.

"You got an *itch* boss?" Carlos asked. Carlos was Joeys' bodyguard; despite the fact he couldn't stand Joey. Carlos couldn't understand how this mutt had more clout than he did. So what if his grandfather is Joseph D'Amico, Carlos thought to himself. He was a full-blooded Italian and he felt Joey should be working for him.

Joey ignored Carlos's remark. He couldn't stand the limp dick muthafucka anyway. He knew Carlos couldn't stand the fact that he was working for a half-breed. For this reason Joey simply didn't trust him. He made a note to talk to his father about a replacement for Carlos.

Allexis made her way over to the bar after dancing to *Party Like A Rock Star* and the old school jam, *Doin' The Butt*. "Can I have a Rum and Coke?" she asked the exotic looking bartender, as she fanned herself with her hand. She smiled at the good-looking white boy, who sat next to her. She thought he looked familiar, but she couldn't place his face.

Twisted Obsession

He smiled back, "I like the way you move?" he said eyeing her seductively.

The bartender sat Allexis drink down in front of her, nearly spilling it starring at Joey because he looked just that good. Allexis took a sip of her drink. "Thank you," she replied. His deep blue eyes had her hypnotized. "You look familiar. May I ask your name?"

Before Joey could respond, Tina pulled Allexis onto the dance floor so they could do the electric slide. Allexis had a ball watching some of the guest stumbling over some of the dance steps. After the song was over, Dom told everybody that he had an announcement to make.

"Tina, would you mind accompanying me on the stage, please?" Dom asked. Tina made her way to the stage and stood by her man. They both looked nice in their all black attire. Dom took Tina's hand and looked deeply into her eyes. "Baby, I know this has been a long road for the both of us. But, you held a nigga down, even when he was at his lowest. Through all the bullshit, other women, and the late hours I rack in every week. I just want you to know I ain' t been puttin' in work for nothin'," he said getting down on one knee. Tina brought her hands to her face. Dom pulled out a tiny box, opening it, showcasing a seven-karat diamond engagement ring. "Tina, will you marry me?"

"Oh, my God, Dom! Yes, baby. Yes!" she screamed, unable to hold back her tears. The crowd applauded as he scooped her up and spun her around.

"Aight y'all this is for my boy and his wife to be," the DJ announced, as he played Jodeci's *Forever My Lady*.

Allexis smiled as she watched Dom and Tina dance. It was truly a beautiful moment. She immediately thought about Kane, she wished he had come. Just then Allexis felt a pair of strong arms grope her from behind. Mind set on Kane, she leaned back into him, and swayed with the sounds of Jodeci.

"You smell good," Joey whispered in her ear.

Twisted Obsession

Not recognizing the voice Allexis quickly released herself from Joey's embrace once she realized it wasn't Kane she was dancing with. She turned around only to see the fine white boy she met at the bar earlier. Allexis wasn't into white men, but she had to admit the specimen standing before her was fine. Obviously, he didn't want to keep his life to long messing with Kane's property, she thought.

"I'm sorry I thought you were someone else," Allexis said, looking around nervously. She knew Kane had people everywhere.

"I can be," Joey said seductively, eyeing her erect nipples through her dress.

Allexis rolled her eyes. "You obviously don't know who I' am or who I belong to," she said folding her arms across her chest. *Umph, he is too fine!*

Joey cocked his head to the side. "You honestly don't remember me do you?" he asked.

"Should I?" Allexis asked with a cocky grin on her face.

He leaned forward; he was so close his lips touched her ear. "*Paris*. I've seen you naked," he said kissing her ear lobe.

Allexis snapped her neck back; "You work for Kane?" she panicked. *Damn, I knew he looked familiar. Lord, please don't let him run back and tell Kane,* she silently prayed.

Sensing her discomfort he touched her shoulder, "I apologize. Damn, I didn't mean to get you all worked up. *And*, Kane works for me," he said adjusting his collar. *Damn, she damned near pissed herself when I mentioned Kane.* But then he thought back to the incident at the hotel. He wondered if he should say something to Dom. He decided he would when he got Dom alone, Allexis deserved a real man in her life and he felt he was it.

She smiled nervously. "It's okay. It's just that . . . you know how people talk," she said heading to the bar, again. She seriously needed a drink. She noticed Joey following her. *Fuck, why doesn't he go find some other yellow chick to hassle?* She thought rolling her eyes to the ceiling. She sat to the bar and ordered another drink, with Joey standing two steps away.

"Yo' whad up, nigga? I ain't think yo' white ass was gon' make it?" Dom said, giving Joey some dap followed by a hug.

"I had to congratulate Tina on ending your pussy chasin' days," Joey laughed.

"You not supposed to say that in front of her," he said referring to Allexis.

Allexis looked from Dom to Joey, "You know him?" Allexis asked Dom.

"Yeah, this is my cousin."

"But he's . . . white!" Allexis stated a little confused.

Both Dom and Joey laughed. "His mother is my aunt; she was a mutt just like you. Ain't that right Carlos?" Dom asked giving him both his middle fingers. Carlos grabbed his dick in response. Dom couldn't stand that pussy ass muthafucka.

Allexis scrunched up her face, "What?" she asked still confused.

Dom put his head down. "Okay," he said, placing his hands on her shoulders. "His mother," he said pointing to Joey, "and my mother," he said pointing to himself, "were sisters. They had the same mother different fathers. You feel me?"

Allexis nodded letting him know she understood. "And don't be calling me no mutt," she said, playfully punching him in the arm.

"Awe, girl you know I love you."

"So, you gon' hook me up or what?" Joey asked.

"NO," Allexis answered for him. She had to admit he was looking some kind of fine, but she was with Kane and there was no way she could jeopardize her relationship. She knew Kane has his flaws but she honestly believed he only wanted the best for her.

Twisted Obsession

"Come on Allexis. Give him a chance. Besides Kane couldn't even begin to touch Joey's cash flow," Dom said seriously. He would even put her on Tubby if he could get her away from Kane's no good ass.

"Whatever Dom I'm not that shallow. I'm not all about the money. It's more than that with me and Kane."

"Come on baby this is our song," Tina said, running up behind Dom and wrapping her arms around him. *Sexy ass Joey. I ain't see his fine ass in a minute*, Tina thought.

"Please, because he is over here trying to play matchmaker," Allexis smirked, playfully sticking her tongue out at Dom.

"Hey Joey where have you been hiding at?" Tina asked, kissing Dom's ear.

Damn, it's a small world, Allexis thought to herself.

"Business as usual." he replied. "Tell your sister I'm a nice guy. I won't bite, but only if she wants me too. Where's Doricka? She could come over and say hi to me. She knows I don't get to see her that often."

"She's around here somewhere attempting to shake her booty. But I'll tell her to come by and say hi."

"Thanks," he said nodding his head towards Allexis before taking a sip of his drink.

Tina smiled getting the hint, "Go ahead and give him some play," she whispered into Allexis hear, leading Dom to the middle of the dance floor. Before getting lost in the crowd she came back and told Allexis, "Passion is *fucked* up. Just thought I would let you know," she said, as she turned on her heels.

Allexis laughed, "So, where are you from?" she asked hopping onto one of the barstool. Her feet were killing her and she regretted wearing the four inch heels.

Joey watched her breast jiggle. "Miami. You should come and visit me sometime?" He

Twisted Obsession

was doing everything in his power not to reach out and touch her. But he knew pretty soon his urge to touch her was going to take over soon.

"I don't know you like that," she said snapping her neck back and rolling her eyes. "Besides my man wouldn't like that," she added hoping he would get the hint, but he didn't.

Joey positioned himself between her legs, placing the palms of his hands on the bar, turned on by her sassy tone. "I want to get to know you," he whispered.

Allexis felt her thong getting moist. He smelled so good. She swallowed the lump in her throat. "Look, I already have a man, which you already know. How are you going to push up on me when you do business with Kane?," she asked, trying to give off as much attitude as possible.

Without warning he began to nibble on her neck. Allexis accidentally let a moan escape her lips. "You like that?" he whispered, ignoring the bartender, and the few party guest who were staring at them.

"Please, stop," she begged as she gave him a hard push, but he bounced right back. "Oh,!" Allexis moaned while Joey continued to nibble on her neck. *Allexis you can't do this. I know but it feels so good. But what about Kane*, she argued with herself. She pushed his hand away with as much force as possible when he attempted to take things a little further by placing his hand up her dress. He definitely tried to take it a little too far when he did that. She looked around guiltily, hoping no one had witnessed what just happened. She looked down at her drink and immediately set on the bar. She was definitely done for the night.

"Why did you stop me?" he asked, attempting to kiss her neck. "I'm getting to you aren't I?" he asked confidently. He looked at her wanting to take her right then and there. He would have if he could.

Allexis closed her eyes. *Lord, please help me,* she prayed as Joey attempted to place his hand up her dress. She knew she should try to stop him, but for some reason she couldn't bring herself to do it. He smelled so good.

"You know what? You're different from what I remember last time," Joey whispered in her ear.

Allexis cocked her head to the side giving him a sideways glance. She wondered how he could even possibly say something like that. It wasn't like they were around each other for hours at a time, but she was curious as to why he said it. "Different. What do you mean different? You don't even know me."

"I don't know. You seemed so shy and innocent. Maybe I'm bringing out the dirty in you."

Allexis laughed. "Yeah whatever."

"Oooooh, I'm telling," Passion said covering her mouth.

"Hey cuz," Joey said backing away from Allexis a little.

The way Joey said cuz made Passion giggle. She didn't know much about him except that his mother and her mother were half-sisters and that his family had ties to the mafia.

Thank you Jesus! "Girl, shut-up with your *drunken* ass," Allexis replied laughing, trying to fix her dress. Coming to her senses she pushed Joey out of the way. She looked down unable to look Joey in the eye. She felt so guilty if Kane ever found out she knew he would have every right to stomp a hole in her ass. He was always trying to instill in her to act like a lady here she was getting buck wild with his business partner.

"Girl, I ain't drunk," she replied, attempting to sit on the stool next Allexis, but she missed the stool completely and fell flat on her ass, spilling a few drinks in the process.

"*Oh my Gosh*! Are you okay?" Allexis asked, laughing at her friend as she struggled to get up but she couldn't catch a good grip on her heels and fell again.

Twisted Obsession

Joey helped Passion up, while trying to suppress a laugh himself. She snatched her arm away from him, and then proceeded to get up in the face of a girl who happened to be sitting a few seats down from. "Bitch, you knew I was about to sit there!" she said slurring her words.

The girl looked from Passion to Allexis with a confused look on her face. Allexis mouthed *sorry,* to the girl, grabbed Passion and headed to the nearest exit. Patrons looked on as Passion continued to yell obscenities at the girl, while heading out the club. Joey did his best to calm a drunken Passion down while escorting them to their ride. It was quite amusing for Joey to see Passion acting out like this, because from what he remembered of Passion she was the quiet laid back type, but she was definitely showing her ass tonight.

"Can you tell Dom and my sister that I and Passion are going back to the hotel?" Allexis asked. Joey nodded his head. "Thank you," she replied closing the door. He watched the Limo pull off. He was going to get Allexis, besides he knew Dom wouldn't mind passing her number along.

Passion threw up twice in the limo before making it to the hotel, where she threw up again. *Kane is going to piss a bitch. Kane, shit I forgot to call him.* Somehow Allexis and Passion managed to make it to their suite. She awkwardly helped Passion into bed, wiping vomit from her mouth and taking off her shoes. *Ugh, she stank,* Allexis thought turning her nose up in disgust.

It had been a long night. Allexis washed her hands and took a quick shower before hopping into bed herself. After dealing with Passion she was too tired to even think about Kane. She took off her dress, laid down and drifted off into a peaceful sleep, dreaming of Kane. She never even bothered to check her cell phone. If she had she would have noticed the twelve missed calls.

* * *

Twisted Obsession

Chapter 16

"Pick up that damn phone," Doricka yelled. "It's too damn, early in the morning for that shit," she snapped, before diving back under the covers.

Allexis rolled over onto her stomach and groggily reached for her phone, not bothering to look at her caller id, "Hello," she answered, still half asleep. The foul odor of Passions vomit made her curl her nose in disgust.

"You don't know how to answer your damn phone," Kane's voice boomed in her ear. Which instantly had Allexis wide-awake now. She immediately sat up. "Bring yo' muthafuckin' ass downstairs," Kane yelled hanging up the phone without waiting for a reply.

She looked over at Passion who was still asleep. She hopped out of bed and quickly threw on some clothes, and slipped her feet in a pair of sandals. She gathered her shopping bags, swiftly walking toward the door. Her heart was beating a mile a minute.

"Where you goin'?" Doricka asked. Allexis jumped. "Damn, girl. You aight," Doricka asked looking at her strangely. "You all pale and shit."

Allexis smiled nervously, "Yeah, Kane's downstairs waiting for me. Oh, the limo will be here to take you guys back home. Bye," she said quickly darting for the door.

That was strange, Doricka thought. She shrugged her shoulders and rolled back over in order to catch some more z's.

Allexis's heart beat rapidly as she approached Kane, as he impatiently waited in the lobby. "Hey, baby," she said nervously. But the look in his eyes said it all. She ducked a little in anticipation.

Kane roughly grabbed her by the collar. Allexis felt her heart drop to her stomach. "*What*? You can't call a nigga. You don't have fuckin' caller id? I called yo' ass *twelve* times," he spat pointing his finger in her face. "I let you go out and have fun, and this is

Twisted Obsession

how you repay me? Got me out here looking stupid and shit behind yo' ass." Kane had a feelin' he shouldn't have let her come in the first place. He thought the company she kept was too fast and would eventually rub off on her and in his opinion he was right.

"Kane, I'm sorry," she whined, looking around nervously, embarrassed by the attention they were getting. She put her head down too afraid to make eye contact with anybody. She felt Kane had every right to be mad at her, but for him to be acting like this in public, Allexis thought was a bit much. But she wouldn't dare say that to him..

"What the fuck y'all looking at?" Kane barked, glaring at the hotel front desk agents and hotel patrons. They quickly looked the other way or scurried out of Kane's wrath.

He roughly escorted Allexis outside to their awaiting limo. Allexis dropped one of her shopping bags, but she was too scared to even think about stopping to pick it up. The driver held the door open. "Get yo' ass in the car. *Get in the car. Get in the car*," he yelled slapping her upside the head.

Allexis scooted as far away from Kane as possible. Tears began to swell in her eyes. She bit her bottom lip to keep them from falling. All kinds of thoughts clouded her mind, as Kane hopped in the limo. Her first thought was that someone saw her and Joey together last night. She cringed as she thought back to how severely Kane had beaten her for being in Dom's face.

"Ain't no need in cryin' now," he spat.

Damn Allexis this is your entire fault, she secretly chastised herself. *You were all up in some other mans' face now you see why Kane doesn't want to let you go out?*

She looked over at him to scared to say anything but she figured she would at least try. Her bottom lip quivered, "Kane, I'm sorry. I-,"

A quick slap to the face ended Allexis attempt to apologize. He wouldn't allow her to finish her sentence he was so angry. He honestly tried to be nice and she didn't appreciate it. So until she learned her lesson she was getting the cold shoulder.

Twisted Obsession

"Get yo' ass over there and shut the fuck up," he roared.

Allexis grabbed her face, as tears stung her eyes. She looked at him once more before putting her head down. She never uttered another word the entire ride back to Virginia.

As soon as Kane and Allexis entered the house, he grabbed her roughly by the arm causing her to drop her bags. "Get . . . yo' . . . ass . . . upstairs," he said spanking her as if she were five instead of nineteen. Allexis quickly ran up the stairs, tripping and stumbling over a few.

"Go stand in the corner," he barked. "Facin' the wall. You better not move until I tell you too," he said sitting on the bed, and turning on the television.

* * *

"Kane."

Silence.

"Kane."

Slience.

"Kane," Allexis pleaded, bouncing around, "I really have to go to the bathroom." She had been standing there for three hours. He continued to ignore her as she bounced around. He kept his eyes glued to the TV screen, watching reruns of Martin. The pressure on her bladder was unbearable.

"Kane, please," she begged. She took a step towards the bathroom.

"You better not move," he said, his eyes never leaving the TV.

Allexis sniffled and continued to bounce around. The pressure she was feeling in her stomach was unbearable. She looked up at the ceiling as she tried to think of everything except the uncomfortable feeling she was experiencing in the pit of her stomach, but it

Twisted Obsession

didn't help. She fought against her bladder for as long as she could. But, she eventually lost. Allexis held her head down, whimpering softly as warm urine trickled down her legs, making a puddle around her feet.

"Clean that shit up, and take your ass in the bathroom," he snapped at her, with a look of disgust on his face. He snickered a little as he watched her with her head hung low walk into the bathroom. "That's what you get. Have me out here lookin' like some damn fool."

Quietly and quickly Allexis did as she was told. She took a long nice hot shower and washed her hair. She stayed in the bathroom for an hour, praying Kane had calmed down. She dried off and put on one of Kane's old tee shirts, and pulled her wet hair back into a ponytail. Still a little hesitant she eased down onto the bed next to Kane.

He turned and looked at her like she was crazy. "Allexis, what the fuck are you doin'?" Allexis looked at him with a confused look on her face. "You better carry yo' ass downstairs. You weren't a good little girl this weekend so you're on punishment."

"Kane," she said in disbelief. He gave her that look letting her know he was dead serious.

"This is what happens to little girls that don't listen," he said as he looked at her with a strange expression in his eyes.

But I'm not a little girl, she thought to herself as she stared down at him with sad puppy dog eyes, but Kane didn't seem to be the least bit phased by the sad expression that loomed on her face. He already warned her about playing games and the talking obviously wasn't working so he was forced to use tough love.

The way he was staring at her had Allexis on edge. She felt chills run down her spine. "But Kane I'm sorry. I didn't mean it. Things got hectic last night. Passion got drunk and I was being a friend."

"That's exactly why I don't want you hanging with those fast ass girls. But you have

Twisted Obsession

gone see what happens when you get grown and do things your way and not mine. I'm the boss you got that?" Kane watched as Allexis unwillingly nodded her head yes. "You can sleep down stairs in one of the guest rooms, no matter face sleep on the sofa in the living room."

Allexis slowly got up and opened the bedroom door. She looked at him one more time before walking out into the hallway.

"Allexis."

"Yes, Kane," she said popping her head back into the bedroom with a hopeful expression on her face, hoping Kane had forgiven her. But her hopes were quickly shattered.

"Make sure the TV stays off. This is all for your own good Alex. You'll thank me one day. Goodnight."

Allexis felt her heart drop into her stomach. She closed the door, and picked up some blankets from the linen closet. She couldn't hold back the dam of tears that flowed freely down her face, soaking her shirt. She made her way downstairs and sank into the love seat. As she cried herself to sleep, Passion's words stayed on her mind. *"Just remember, it doesn't stop."*

* * *

Twisted Obsession

Chapter 17

Natalia bit her bottom lip and arched her back as a loud moan escaped her lips. She shuttered in agony while her precious gem continued to be penetrated over and over again. To muffle the screams that so desperately seeked an outlet she placed her lips around her hardened nipple. But the pleasure she was receiving down below was too intense too be pacified. "Sssss," she hissed grabbing handfuls of hair that belonged to her perpetrator.

Slowly she grinded her hips in sync with the fat moist tongue that held her pulsating clit hostage. Her eyes rolled in the back of her head once she felt long skinny fingers enter her juicy pussy. Natalia wanted to scream at the top of her lungs as she moved her hips up and down and around and around. The gushy sounds of her dripping wet pussy being invaded only intensified her pleasure.

"Yes," she moaned seductively as she looked down to the culprit that has taken her hostage on this exquisite sexual voyage. She looked deep into their eyes letting them know that she was more than pleased with their work. "Oh, damn, don't stop," she pleaded once her clit began to pulsate growing bigger as she neared a climax. Kane flashed before her eyes and she had to catch herself before she slipped up and said his name.

"Ahh . . . ahh . . . ahh," she panted. "Please lick it slow," she squealed wanting to enjoy and savor this orgasm. She rubbed her nipples while her pussy continued to drip from pleasure, she wound her body seductively as she reached under her pillow and grabbed her 9mm.

"Was it good for you?"

Natalia nodded her head as she looked down between her legs. "Damn that was the best head I've had in a long time," she smiled still in a daze. Natalia jumped once her jewel was licked repeatedly again; temporarily she released her hold on her gun while she was brought to another powerful climax. *Damn I wish I we would have met sooner.*

Twisted Obsession

"That's enough," she said out of breath as she quickly grabbed her gun and cocked it, pointing it between her legs.

"What the fuck?' Marissa said with a bewildered look on her face. But her shock soon turned to fear as she starred down the barrel of the gun.

Natalia continued to stare down at a frightened Marissa with a wicked grin plastered on her face. Unbeknownst to Marissa Natalia had been stalking her the last couple of days just to feel her out, which in the end worked out in her favor. She twitched a little still high from the pleasure she just received.

"Before I take you to meet your maker," Natalia paused momentarily to reposition herself, but kept the gun aimed at Marissa. Sitting up she threw her hair over her shoulder. "Much better, now where were we? Oh yea," she said answering her own question. "I take it you are familiar with a fellow by the name of Dion "Kane" Stanley?"

Marissa didn't answer the question right away. Her mind drifted off to thoughts of escape. Her eyes shifted from side to side seeking any hopes of an outlet. She cursed herself for being so stupid, but her hormones were lonely and in definite need of a tune up. So when the young vivacious Natalia proved to be more than a willing mechanic to fix her engine she couldn't resist the sultry temptress.

"No need," Natalia said reading Marissa's thoughts. "You wouldn't make it out alive anyway. So let's cut the bull shit you know the man in question."

"Maybe," Marissa stated bluntly staring up at the exotic beauty. She couldn't help but to wonder how someone so beautiful could be so devious.

Before responding to Marissa a blank expression crossed Natalia's face, causing Marissa to shift uncomfortably. Without warning Natialia brought the butt of her gun down hard across Marissa's forehead. Blood immediately gushed out of the deep open wound.

"Aaargh," Marissa groaned in unspeakable pain. The sight of so much blood made her slightly nauseous. She tenderly touched her wound, as tears stained her eyes. Her life flashed before her eyes as she came to the realization that she wasn't going to make it out of this alive.

"Please, don't fuck with me," Natalia said calmly. "Now," she said reaching under her pillow grabbing her jack knife and flashing it in front of Marissa. The pure look of fear locked in Marissa's eyes moistened her pussy. She wiggled a little in order to control herself. "Don't let this little sharp thing be the last dick you feel," she smiled.

Sensing that Natalia would make good on her threat she hung her head low before quickly ratting off her cohorts and witnesses involved with the case. Tears flowed freely down her cheeks. She was literally about to die and over nothing, because she really didn't have a case. But she knew it wouldn't do her any good to explain that to Natalia because she held the look of death in her eyes. She actually appeared to be enjoying herself which made her even more sick to her stomach.

"I don't get it Natalia. How could you? You're a-,"

"I know," she laughed mechanically. "That's what makes it so fun." She laughed as she riddled Marissa's body with holes.

* * *

Twisted Obsession

Chapter 18

Defeat could clearly been seen in Dame's face. He pulled hard on his laced blunt while he tenderly stroked the back of his homosexual lover, Lala who laid beside him butterball naked. He so badly wanted to cry as he thought about his current situation, but his pride wouldn't allow him too at that particular moment.

So many hard efforts he put into causing the demise of Kane, and unfortunately each and every last attempt had failed. He was more than shore Kane was onto him by now. He thought about his plan to recruit Neko to kill Kane that failed. The attempt to ruin the shop in Baltimore failed because Kane came up with the idea of running the shop longer hours with twenty four hour security. Then he thought back to how happy he had been once he finally had Kane's drug transaction with his *connect,* so he thought on tape, but to his dismay Kane's face wasn't visible and the tape couldn't be used.

And Detective was a joke, because his case wasn't holding up either. All of a sudden evidence comes up missing, key witness are having severe cases of amnesia. But Dame

Twisted Obsession

felt that was all Detective Coleman's own undoing. If he had just went ahead and prosecuted Kane with the evidence he had then maybe just maybe things would have panned out differently for them both.

But things seemed to be fucked up for him all around, including his personal life. He was disgusted with himself sinking into what seemed like a deep depression. His cash flow was slowly but surely dwindling down to nothing. Kane was now handling all the majority of the cash and often told Dame business was slacking that's why everybody pockets was hurtin'.

"Cash just ain't flowin' like it use to," Kane told him with an arrogant smirk on his face.

But Dame knew that was bullshit because Kane had just purchased a three hundred fifty thousand dollar yacht and another home in Atlanta right beside his. Thoughts of his house in Atlanta made him even more depressed. He hated Kane with a Passion. He despised the ground he walked on and cursed the bitch that birthed him.

"You okay daddy?" his latin lover Lala asked turning his head to look up at Dame. He was still enjoying the aftermath of the rough and kinky sex Dame had just delivered.

"I'm good," Dame said staring off into space at nothing in particular. "Do you believe in God?"

"Huh?" Lala asked, caught off guard by the question.

"Do you believe in God?" Dame repeated, this time looking down at his lovers' petite frame. It was petite and gorgeous. It was very smooth and curveous in all the right places, just like a woman.

"I don't know I guess so," Lala said turning his head in the opposite direction, so that Dame wouldn't see him rolling his eyes.

Here we go again with all these damn questions.

"You don't think there is a higher power out there, judging us for our past sins and all our wrongs?"

"Dame, why do you always do this?" Lala asked becoming a tad bit bored and utterly turned off about the religious talk. He wasn't a very religious person and didn't believe in God. It was hard for him to believe that two people created thousands and thousands of people. He just simply ran with the flow of things and lived life to the fullest. Besides he didn't even come to discuss religion he only came for one thing and that was sex.

"I'm not here to talk Dame," Lala added.

Dame didn't respond he laid there and thought back to how his grandmother would pick him up every Sunday and take him and his little sister to church. But never believed any of the bullshit the preachers talked about. They always preached about a God is always with you during your time of need, and during your struggles, and how all you had to do was call on him and he would answer. Dame shook his head, recounting plenty of times when did call on God, but in his mind he never answered.

A smile crept on Dame's face and caused him to laugh. Lala looked at him, wondering what was so funny. But the look in Dame's eyes put Lala in a frozen state, it was almost scary and Dame really didn't look like himself.

"Are you okay?" Lala asked a frown creasing his forehead.

"Yeah, I'm good," he answered, putting Lala at ease.

"Good, you were looking some kind of creepy for a second there. But, I came here to enjoy myself," he said licking his sensual lips. "I'm ready for round two," Lala stated holding up two fingers.

"Oh, really," Dame replied. "Where are your handcuffs?"

Lala giggled, "Make sure you frisk me," he winked as he reached down on the side of the bed retrieving the handcuffs before handing them to Dame. He then proceeded to

Twisted Obsession

place his hands behind his back. Dame leaned over and placed the handcuffs around Lala's wrist. "*Ouch* there too tight," Lala stated seriously.

But Dame remained silent as he climbed on top of Lala with his dick in hand. He inserted his hard rock shaft in into Lala's ass, before placing his hands around Lala's throat. But it wasn't Lala he was seeing, it was Kane.

"Too tight Dame," Lala panted once he realized Dame's grip around his neck kept getting tighter and tighter.

But Dame was in his own sick twisted fantasy. For once he felt he had power over Kane, he was in charge, and he was in command. He laughed to himself from the pure look of terror that was displayed upon Kane's face. He continued to grip Kane's neck tighter and tighter, while his strokes became longer and deeper. It excited him to see Kane struggling to breathe. A deep moan escaped his lips, once the life was completely drained from Kane's body. He rolled off the lifeless corpse and lay on his back, his eyes staring at the ceiling. He began to laugh, a creepy laugh that the filled the still and silent room while tears continued to steadily flow down his face.

Once again his thoughts traveled back to his grandmother, how she was always talking about God. He wondered where was God when his father up and walked on his family, leaving his mother broken hearted. He wondered where was God when his mother fell into a deep depression after his father left, eventually putting a gun to his head. He wondered where God was when his beloved Grandmother died a slow and painful death from cancer. And he wondered where God was when his sweet innocent sister was beaten, raped and murder by rival gang members.

"Where were you now?" Dame yelled. But he was only greeted by the stillness of the quiet and dark room.

Depression invaded his mind, body and soul, clouding his head with negative thoughts. He got up and walked up to the dresser and stared at his pitiful reflection in the mirror. He really didn't recognize himself, dark circles were formed under his eyes and the color was slowly but surely leaving his face. And his hair no longer had that shine that it once had. His eyes landed on the gun that lay upon the dresser and he picked it up. His heart continued to beat rapidly and his hands shook violently as he raised the gun up, placing it to his temple.

"*Do it! Do it!*" he willed himself as the mixture of sweat and tears stained his face. But he couldn't bring himself to pull the trigger. He released the gun from his hand. It bounced hard off the dresser with a loud thud before landing in the floor.

Dame dropped to the floor and curled his body into the fetal position. Softly he began to sing, "Jesus loves me this I know, for the bible tells me so. Little ones to him they go for they are weak but he is strong."

* * *

Twisted Obsession

Chapter 19

It has been a little over a month since Tina's party and things couldn't be any better, for Tina that is. She and Dom have settled into their new home. Dom promised her this would be his last year living the street life. He said he was ready to settle down with her, and he was ready to start a family. She knew Dom was holding out on her as she looked around at her lavish home, it could easily swallow the previous house Dom had purchased for her. She had a feeling that he was rolling in more dough then led her to believe he had.

She came downstairs wearing a pair of seven jeans, a belly shirt, and a pair of flip-flops. Their cook prepared a dainty lunch for her and her guest. Chicken, ribs, pork chops, neck bone, macaroni and cheese, chitlins, greens, string beans, mashed potatoes, corn, stuffing, and candied yams. For dessert there was lemon cake, chocolate cake, vanilla cake, apple pie, and cherry pie. Her stomach began to growl thinking about the meal. She looked at the clock in heavy anticipation ready to chow down. Looking at the food she wasn't about to wait. She went to the cupboard and grabbed a plate.

I hope Allexis makes it this time, she thought. Tina has noticed a drastic change in her sister over the last few weeks. All of a sudden Allexis stopped returning her phone calls, and constantly canceling their lunch dates. In her heart she knew something was wrong, but Allexis never talked to her anymore. If Allexis didn't show tonight she was going to make an unannounced trip over to Kane's house, making sure that her Beretta traveled along.

Bang! Bang!

"Damn, that ain't nobody but Doricka ghetto ass," she said as she took her time opening the door.

"Damn, bitch. I know ya high yellow ass heard me knockin'," Doricka snapped walking past her and into the living room. "Shit, it's *hot* as hell outside." Which it truly was, lines of sweat flowed freely down the sides of her face. "You cheatin," Doricka stated taking note of the plate in Tina's hand. "Your not supposed to start without us."

Tina rolled her eyes and ignored Doricka's comment about the food but decided to jump on the current matter at hand. "*Girl, I bet ya muthafuckin' ass is hot, with that fuckin' hot ass leather on,*" Tina replied, rolling her eyes. She could only laugh and shake her head because she knew no matter what people said about her, Doricka would remain the same.

"Shut-up," Doricka replied, taking the sticky leather off. "I couldn't help it. This leather looks too good to be sittin' up in my closet. Dame paid a grip for this shit too," she said as she walked out into the kitchen and grabbed some paper towel. "Whew," she said out loud as she walked back out into the living room with Tina.

"Whateva," Tina replied. Tina had to admit the leather was nice, but the orange and yellow shirt with the lime green skirt Doricka wore was a tad bit loud. "You still fuck wit' that lame ass nigga?" Tina had met Dame a couple of times and she didn't like him. She couldn't place it but he just gave off a really bad vibe.

" First of all he is not wack and yeah I fucks him on occasions when MiMi get's lonely," she answered.

"Whoa!" Tina said throwing her hands up in the air. "Too much information. Plus I don't think Mariah Carey wants to be named after that stank ass pussy." Doricka flipped her middle finger. "When are Passion and Allexis gon' be here?"

"Hmmm . . . I don't know. I called Passion before I left Dame's house. I think Passion was on her way to go get Allexis then."

"I hope Allexis doesn't cancel this again. I miss my baby sis," Tina pouted flopping down on the couch beside Doricka, who was rolling a blunt.

"Yeah, she has been distant lately. The day after the party, she left the hotel all in a hurry and shit. And I miss that ole white heifer too, wit' her non cussin' ass," and which she truly did. Allexis and Passion kind of reminded her of her and Tina's relationship. Which Doricka was truly grateful because Passion never really had a lot of friends, but

Twisted Obsession

she saw a true friend in Allexis just like she has a true friend in Tina.

Tina mushed Doricka in the head, "Don't be talkin' bout my sister." Doricka handed Tina the blunt. Tina was hesitant at first considering her situation, but she decided to hit it one last time for the heck of it.

* * *

Allexis grabbed her purse and hopped downstairs. She was going to see her sister today. All of the bruises that she had sustained at the hands of Kane had healed, so hopefully she could make it out of the house today without getting any fresh ones.

Before leaving her bedroom Allexis looked at herself long and hard in the mirror. The once happy and bright girl she once knew, no longer existed. She thought back to happier times in her life and honestly wished she could go back. She had been so caught up and infatuated with Kane that she overlooked a lot of his flaws. She realizes now that in a lot of ways she was shallow because of Kane's power, and his money.

She closed her eyes as tears fell down her face, because despite how badly he treated her she still loved him. And why she didn't know. Whenever he hurt her he would always apologize and tell her that he only wants the best for her. Allexis sympathized with him once he finally broke down and told her the story about his sister's child hood and her untimely death. And that she was the reason he was so protective of her. He didn't want to fail her the way he did Alex.

Taking a deep breath she walked into the bathroom and threw some water on her face. She wiped her tears and walked downstairs.

Kane greeted her at the bottom of the stairs. "Where are you goin'?"He asked.

Shit, I forgot to tell him Tina wanted me to come over for lunch. Allexis you think you would learn by now. "Ummm, Tina invited me over for lunch," she said nervously. "I-,"

Twisted Obsession

He cut her off mushing her in the head, causing her to stumble backwards onto the steps. She felt a sharp pain in her hip. "Next time make sure you ask ahead of time. You never know what I might have planned for you." Allexis nodded her head. He helped her up. "Make sure you are home by eight. You know I don't like you hanging out with a whole bunch of unclassy females."

"Okay." she said, forcing herself to kiss him on the cheek. She heard Passion honk the horn. "That's Passion," she said walking towards the door, while checking her purse to make sure she had her cell phone.

He grabbed her arm, squeezing until it hurt. It excited him to see Allexis wince from the pain. That's what he wanted so that she would no he meant business. "Eight, Allexis. Not a minute later," he warned.

"Okay, I will Kane, I promise" she said nervously kissing him again. She rubbed her shoulder where a little before walking out the door.

Kane watched her leave with an unsettling feeling brewing in his stomach those were the last words his sister said to him before she died.

"*Hey girl*," Passion said, as Allexis hopped into the car.

"What's up?" Allexis asked leaning over to give Passion a hug.

"You chick! What's been up with you? Can't anybody ever get a hold of you anymore? You don't answer your phone or check your emails. What is really good?" Passion asked looking at Allexis suspiciously. She noticed the glow Allexis once had was now gone. To her Allexis looked tired and defeated.

Allexis sighed, she knew this was coming. "I know it's just that Kane doesn't like me to be out and about too much, considering his type of employment."

Passion looked at her. "*Bullshit* and you know it. I peeped that bruise on your neck the other day." Passion had noticed the bruise on Allexis neck when she unexpectedly

Twisted Obsession

stopped by Kane's house to check on Allexis. She was worried about her girl and wanted to know why she hadn't been out lately.

"Um . . . it . . . uh," Allexis' words were caught up in her throat. She couldn't think of anything to say. The lies that had so easily flowed before could no longer escape her lips. She continued to stutter at a loss for words. Tears welled in her eyes as she looked at Passion.

Passion touched her shoulder. "Don't. I already know. But, Allexis why?" Passion asked trying to hold back her own tears. She knew first-hand what Allexis was going through, because for the longest time she couldn't leave her own abusive boyfriend until Dom came to her rescue. He opened her eyes to a lot of things.

"He loves me Passion. I know he does. And a lot of the things he says and does, he really doesn't mean. He just wants the best for me," Allexis said still trying to convince herself that her staying with Kane when deep down inside she knew it wasn't. She was honestly tired, her spirit broken.

"Allexis. That's not love sweet heart. How can you honestly think that any man that puts his hands on you loves you? Trust me. I had to learn the hard way. The first time he put his hands on you should have been a sign."

"It's just that . . . it's just that . . .," Allexis couldn't hold back her tears any longer. She couldn't breathe, and started gasping for air.

Passion quickly pulled over, "Allexis are you okay?" she asked with a look of concern on her face.

Allexis turned to face her friend. "Passion, I'm so scared. I can't . . . I don't know how much longer I can do this. It's like walking on pins and needles," she sniffled. "It's like I'm constantly in fear of doing something wrong. He's been on this kick about people in his life doing him wrong and leaving him, so he's been extra possessive not allowing me to leave the house alone, or one of his boys tagging along. It's a miracle that he even let me come out today. The thing of it is Passion, I don't know what's wrong with me because despite all the horrible shit he has put me through . . . I still love him," Allexis said with tears in her eyes.

Passion embraced her friends with open arms. "Okay, Allexis, you are not going to like what I have to say. But, somebody had to tell me the same thing."

"What?" Allexis asked wiping her eyes.

"Girl, you are crazy as *hell!* How can you honestly say you love this nigga when he *boxin'* wit' yo' ass. You know I love you so don't get me wrong. Allexis, Kane is the only man you have ever known, so I understand why you feel connected to him. But, he's not showing you love," Passion replied. "Allexis, leave his ass before it's too late."

Allexis took a deep breath and looked out the window. "Who else would want me?" I'm not as pretty as some of these other girls out here," Allexis said letting her insecurities show. " And Passion I'm scared. The two men I actually trusted in my life both did me wrong. Am I a magnet for bad men or what?" she asked honestly.

Passion rolled her eyes. *Why is it, all the pretty bitches think they ugly, and the ugly bitches think they pretty?*

"If you don't shut-up! You are a very beautiful woman. No Allexis I just truly think you were infatuated with Kane. And I just happen to know that most good girls like bad guys. There are plenty of good men still out there. Matter of fact I know one. *Joey*," Passion teased.

Allexis looked at her, "No he doesn't." Allexis tried to tell herself. Joey had actually called her a couple of times after Tina's birthday party. She so badly wanted to talk to him, but her fear of Kane wouldn't let her.

"Yes, he does," Passion, taunted sticking out her tongue. "He asks about you all the time."

Twisted Obsession

"I can't do that. He and Kane do business together."

Passion snapped her neck back, "*And*?"

"Passion, I can't do that to him. Despite all the bull shit Kane has done to me, I think that would be a little extreme. Don't you think?"

"I'm going to tell you what I think. I think you are stupid, if you're going to let a good man like Joey pass you by."

"*Passion!*"

"What you want me to say. For one my cuz is fine as hell, and sittin' on top of more money than Bill Gates himself. *Shiet,* you stupid!" Passion said giving Allexis a sideways glance.

Allexis laughed, "I think I need to take a break from men."

"Yeah, you do that. Just give yourself sometime to relax and do you. I'm sure he will wait for you. He seems like that type of guy. And also take things slower this time. You just up and rushed things with Kane. Before I could blink you were playing mommy and y'all were up and living together."

"Your right," Allexis said nodding her head in agreement. "Thanks girl you are the best. I love you."

"Anytime, I know you would do the same for me. And I love you too chick."

* * *

"It's about damn, time," Doricka yelled swinging the door open. She immediately changed her tone when she saw Allexis's red and puffy face. "What's wrong Allexis?" Doricka asked concerned.

"Move so we can get in here first," Passion snapped brushing past Doricka.

Twisted Obsession

"Asshole," Doricka mumbled stepping aside.

"Yo'mama," Passion yelled back.

"Well, my mama is yo' aunt."

"Well-,"

"Okay," Tina yelled, rolling her eyes. She walked over to Allexis and gave her a hug. "What's wrong with baby girl?" she asked.

Allexis walked Tina to the kitchen, with Doricka following close behind. Allexis broke down and told Tina everything, starting with Paris. Tina listened silently. She felt so guilty all the signs were telling her something was wrong but she never pressed the issue; she grabbed her sister and hugged her. Once again, she felt she had let her baby sister down.

"It's alright baby girl, you don't have to worry about that no good nigga ever again," she cooed softly stroking Allexis hair.

"Doricka, you alright?" Passion asked.

"Yeah," Doricka replied. She was feeling guilty about the ill feelings she had for Allexis when she first found out she and Kane were messing around. In some sorted twisted way God must have spared her life. She went over and hugged Allexis too.

After watching how pitiful the three of them looked hugging and crying on each other. Passion wiped the lone tear that escaped her eye and joined the trio and cried too.

* * *

Twisted Obsession

Chapter 20

Joey had just gotten off the phone with Carlos. Joey ordered Carlos to take care of Dame while he found someone to take care of Detective Coleman. He initially wanted Frankie to handle the job, but his father needed him for an out of town job. For some reason he just didn't trust Carlos. He wanted the job handled quickly and quietly. He didn't want a whole lot of attention. But with Carlos handling things he wasn't so sure. He knew Carlos would go out of his way to piss him off.

"What's on your mind?" Dom asked, taking note of the troubled look on Joey's face.

"Nothin'. Just don't want Carlos to fuck this up," he answered honestly. "He always on some double o' seven bullshit."

"I feel you. I can't stand that bitch man. Where did Uncle Johnny pick him up from anyway?"

"His father use to work for my Grandfather. A good hard working man who earned his place as a made man, whereas Carlos feels someone should hand it to him."

Dom pulled into his driveway parking behind Doricika's car. "Yo' who truck is that? It looks familiar," Dom asked Joey.

Joey shrugged his shoulder's opening the door, "I don't know. You ready for tonight. Yeah, ready as I'll ever be." They were supposed to be meeting up with some Mexican cats.

Dom and Joey weren't sure what to make of the situation when they walked into the house. "

"What's wrong wit' y'all?" Dom asked.

"Baby . . . can . . . I . . . talk . . . to . . . you upstairs?" Tina asked in between sniffles.

Look at all that snot, Dom thought. "Sure, baby," he said wrapping his arm around her

Twisted Obsession

neck, escorting her up the stairs.

"Is everything okay?" Joey asked. The three ladies nodded their heads in unison. His eyes connected with Allexis. Even with puffy eyes and a snotty nose, she was still beautiful. He wanted to wrap his arms around her, to console her to let her know that whatever she was going through would be okay.

"That nigga is dead," Dom yelled, hopping down the steps two at a time, his eyes loaded with rage. He was two seconds away from punching a hole in the wall.

"Dom, don't do anything stupid," Tina pleaded. She knew how Dom got when he was angry and would react first and think later. She wanted to see Kane six feet under, but she knew Kane had deep pockets and she didn't want to see anything bad happen to her man.

"What! Fuck dat," Dom spat. He was doing everything in his power to keep from crying. His mother died at the hands of an abusive man and he vowed never to put his hands on a woman and thought that any man that did was a pussy and a coward.

"What's up?" Joey asked.

"*Kane*! That's what's up. He been puttin' his hands on her yo'," Dom replied, pacing back and forth. He was so mad, he knew if Kane so much as stepped foot in the house he would crack his back in two.

Joey put his head down and shook his head. He felt partially responsible because he never told Dom about the incident in Paris. In his mind he hoped that was just a one-time incident, besides Allexis didn't seem like the type that would give Kane much reason to go around like a ticking time bomb. But then again he should have realized that it didn't take much for abusive men to be set off. Once again he shook his head as he made brief eye contact with Allexis.

"Dom, please! I know your upset right now, but he's just not worth it," Allexis pleaded. Allexis remembered what Passion said about Dom taking care of her abusive boyfriend. Despite what Kane has put her through, she still loved him, and she didn't want him to get hurt.

"Damn, Allexis. Don't you know your worth more than that pussy ass nigga?" Allexis put her head down. "You stayin' here tonight," Dom said giving her a hug. It hurt that she was going through all of this right under his nose. He loved Allexis just like a sister and would do anything in the world for her. He just couldn't understand how some women put themselves through such abuse. His mother did the same before falling to a terrible demise. The remembrance of the situation was painful.

"Are you sure that's a good idea?" Joey asked, rocking Dom out of his thoughts, remembering that they had a very important meeting tonight.

"Whatchu mean?" Dom asked, still reeling inside from the disturbing news he just received. He just couldn't believe why Allexis of all people would even subject herself to abuse.

"Well, we have some necessary business to take care of tonight. So, I was thinkin' maybe we should send the girls on a mini vacation. You guys look like you could use one right about now. Uh, I didn't mean it like that," Joey replied, when he noticed the girls were giving him funny look. "I have a condo down in Mexico I sure you girls would enjoy." Passions eyes lit up, she only had a couple more weeks until school started.

"Okay," Dom said nodding his head. "Tina get my American Express Card and book y'all flights. Call me and let me know when y'all flight leave. Come on Joey," he said giving each girl a hug and Tina a quick peck on the cheek. "The only reason that nigga still breathing is 'cause you saved his life sis. You still got that toy I gave you?" he asked Tina, referring to the .22 he gave her. Tina nodded her head. Dom didn't feel comfortable about leaving the girls. He pulled out his cell phone searching for Frankie's number. He hoped the drug deal goes smoothly and quickly.

Joey gave Allexis his number. "Use it if you need it," he said sincerely. He kissed her on the forehead and left.

Twisted Obsession

The girls were excited about going to Mexico, despite the circumstances. Tina was able to get their plane tickets. They had to be at the airport by 10:30.

"Damn, Tina cuz got it like dat?" Doricka asked.

Tina looked at her watch ignoring Doricka's question. "Okay," she said clapping her hands, "the time is now 5:00 p.m. If you have some shit to do, do it now, but make sure you are back in this house by 9:15. Comprende?"

"Comprende," Allexis, Doricka and Passion answered in unison.

Make it last forever, Allexis cell phone sang. She instantly knew it was Kane.

"Don't answer it Allexis," Tina said.

Allexis held the phone in her hand, until it stopped singing. She then turned her cell phone off. It took every in her being not to turn on her phone and dial Kane's number. That's how on edge he had her. She was at his every beck and call, she was ashamed to say that if he did ask her to jump she would say how hi.

Once Allexis phone was safely tucked away the girls decided to go ahead and discuss their plans for the evening. Tina and Allexis decided to go to the Hill Top Shopping Center, while Doricka and Passion opted to go to the mall.

"Passion can you follow me? I have to take Dame back his truck first."

"Yeah," Passion replied fishing around in her purse for the keys. "Are you sure that's a good idea, being that he works with Kane."

"It's cool we will be in and out. Besides I don't think he and Kane are Seeing Eye to eye anymore. A lot of shit has gone down between the two of them. He says Kane has been on some foul shit lately. And to top it off somebody's raided two of his homes, one in Rhode Island and one in Atlanta."

Passion could only frown she knew Dame was a bitch ass nigga. From the way Doricka was ranting on she knew he gossiped just like a woman. "Okay, that's enough," she said holding up her hand. She knew if she didn't stop Doricka they would be there all day. "Let's go so we can hurry up and get back."

Doricka hopped in the Denali, while Passion hopped in the Tahoe. "What the fuck is she doing?" Doricka said, when she noticed Passion hadn't moved yet. She put the window down, "What's wrong?"

"It won't start."

That's when Doricka noticed the lights were on. "That's because ya dumb ass got the lights on," she said sucking her teeth. *Why the fuck she have the lights on, in broad day light anyway? Only Passions ass.*

Passion giggled, "Oops, my bad. You have any jumper cables?"

"Nah, just get yo' ass in the truck. I'll just call Tina and tell her we gon' me her at the shopping center."

"Are you sure that is a good idea?"

"What other choice do we have?" Doricka asked with an irritated look on her face.

Passion shrugged her shoulders and shook her head, as Doricka put the truck in drive.

* * *

Twisted Obsession

Chapter 21

Looking at the picture of his baby sis, Kane couldn't help but to smile yet feel sad at the same time. How he wished she was here. Allexis reminded him so much of his sister that is why he wanted to protect and mold her. But she couldn't do that if she continued to hang around her sister and them other two bitches, to him they were a bad influence. He decided today would be the last time Allexis would interact with her sister other than a phone call on occasions. He kissed the picture of his sister before putting it back in the jewelry box. A tear escaped his eye and he shook his head hating how he let his memories get the best of him.

Things were looking on the up and up with Kane, despite his place in Baltimore being vandalized with the help of Dimples he was able to remodel and beef up the place. Now it was up and running and he was in the process of expanding to New York. And with his business doing well he wanted to settle down and finally make Allexis his wife. He wanted to have another little shorty and had plans of taking full custody of little Sean despite any objections Rasheeda may have.

Looking at the clock he realized that it was almost time for Allexis to come home, so he decided to give her a call just to remind her. *Your call has been forwarded to an automated voice messaging system.* After several attempts Kane slammed his phone down in frustration. "Fuck," I don't need this shit today," he said out loud. "Damn, every time I try to give her ass some space," she goes and fucks it up.

Glancing at the clock he paced the kitchen floor smoking a blunt, his whole persona doing a complete 360. His nerves were on edge now as he thought about Dame's sheisty ass along with some other individuals that had did him wrong in the past. Of all the people that crossed him, Dame's betrayal was the one that was eating at him the most. But hed decided not to kill Dame right away. He slowly but surely has made Dame's life a living hell, by punking him front of the workers and actually making him work out on the street corners. A smile crossed his face as he thought how he raided both of Dame's homes in Rhode Island and Atlanta. But instead of stealing the money he had it shredded

Twisted Obsession

and spread all over the front lawns. A deep laugh escaped his lips at the thought. But the time was now for the games with Dame to end. Against his will Joey took it upon himself to handle it, or so he thinks.

Kane had to admit that his business slacking off was somewhat his own fault, since he has been with Allexis he has been slippin. It seemed like she consumed his every being. She was constantly on his mind. He really wanted to be around her twenty-four seven. He just didn't want to make sure the same mistake with her that he did with his *sister*. He wanted to make sure Allexis turned out to be a nice wholesome girl, nothing like his *mother*.

"K I'm hungry. We never have anything to eat."

"*I know Alex. But one-day things will get better, I promise. You just wait and see. I'm a give you the world and so much more.*"

How he wished he could have made things different. He shook his head to relieve himself of painful memories and dialed Allexis number; again it went straight to her voice mail. He angrily threw his phone, mad because he told Allexis to have her phone on at all times. But somehow he knew this was going to happen. She only acted like this when she got around her sister, and those other two bitches, but all this hanging out and partying was about to come to a drastic end.

"I try to be nice," he said nodding his head vigorously. "And look how she repays me." He grabbed the keys to the Lexus and headed out the door. His mind kept drifting as he thought about how everybody in his life was tryin to up and disappear. But if this was the case with Allexis she had another thing coming if she thought he was going to let her go that easily. In his mind Allexis was already his, he wasn't about to let her leave, not without a fight anyway.

* * *

Twisted Obsession

Chapter 22

A funny feeling overtook Detective Coleman as he looked outside. He scanned his quiet neighborhood and nothing seemed to be out of place. But that still didn't ease his tension. Just a few months ago he was happy, because he thought he was about to break the biggest case of his life. He had Dion Kane Stanley caught on tape transporting drugs across state lines, along with Frank Buttafucco, who he found out works for Joseph D'Amico. One of the largest mafia families know nationwide.

He had all kinds of evidence stacked against the two, but without warning things went from bad to worse. First Dame loses the video tape he has of Kane buying drugs but from what he heard it wouldn't them any good for Kane, but he could have gotten Frankie Kane's connect. Which he could have eventually persuaded Joey to testify against Kane, but that dream was long gone. Then the evidence he himself collected against Kane alone goes missing. Then every single one of the officers who were working with him on the case backed out. Once Detective Curise found out that DA Marissa Coleman was missing he resigned, and moved him and his family out of state. Then there was officer Dionte who happened to have a hard on for his mistress Shelly who was actually born as Sheldon. She wrote him a nice note telling him to fuck off and that she and Dionte planned to get married down in the Bahamas. And that Dionte promise to pay for the surgery that *he* had been promising her for years.

Grabbing the bottle of Jack Daniels off of his table, he didn't even bother to get a cup he simply took off the top and took a long swig. The brown liquid glided down his throat, it burned but to him if felt good. He rubbed his unshaven chin, before ruffling what little hair he had left on his head.

He looked at the clock; it was almost 7:00.

"Where the fuck is she?" he mumbled scared his only lead may have slipped out of his hands. He was determined to bring down Kane even if it cost him his life.

A knock on the door startled him out of his thoughts. "It's about damn time," he said as he hurriedly set the bottle back down on the wobbly table. He maneuvered through his bachelor pad side stepping articles of clothing, and other foreign objects.

He opened the door and she walked in nervously. She was well disguised, dressed in all black, a black wig that was styled into a bob that sat right above her neck, black long sleeve shirt, leggings, black sun glasses, and black *gloves*. She swung her purse over her shoulder. Despite the simple appearance she looked exotic and the deep red lipstick looked perfect against her dark skin.

"Well it's about time. I didn't think you were going to show."

"I didn't either," she said nervously, wringing her hands together. "I don't know about this. If Kane find's out I'm here he will wring my neck."

"Look you will have more than Kane to worry about. Did you get it or not?" Detective Coleman asked agitated. He wanted Kane so badly he could taste it. He was willing to go the distance even risk a couple of lives if it landed him a conviction.

"How the hell are we supposed to do this with everybody backing out? I mean is there anyone left who is willing to go all the way with us on this one?"

"No, it's just me and you. And honestly I've been after that black bastard for a long time. He's killed and hurt a lot of people, not caring who or what was in his path. My brother worked for that son of a bitch," Detective Coleman sat on his lounge chair as he reminisced about the first man he ever truly loved. "We weren't blood brothers but we grew up together. Same upbringing, matter of fact he was treated slightly better but somehow he still ended up on the wrong side of the tracks."

"What happened?" she asked sympathizing with the pain in his eyes.

"He called it the black man's struggle," he laughed. "I told him he was too old to get caught up in the game as they call it. But he had just gotten fired from his job and he had a wife and family to feed."

Twisted Obsession

"What did he do for Kane?"

"The usual, transporting drugs across state lines and he eventually got caught. Kane found out that the feds offered him a deal to testify against him. His wife hasn't heard from him since," a glossy look came over his eyes. He shook his head as he thought about his brother. He was a hardworking man who happened to get caught up in the struggles of life. "So you see, this is why this monster has to be stopped. He has to be taken off of the streets so that no one will have to go through what I've went through and what that poor man's wife went through. So can you help me put this monster behind bars before he kills someone else?" he asked pleading with his eyes.

"I'll do my best."

Detective Coleman clapped her hands causing her to jump. "I will see what I can do about protective custody. No one knows you are here right?" he asked. She nodded her head yes. "Good girl. Now what do you have for me?" he smiled glad to be back in business. He has become completely obsessed with Kane. He vowed he wouldn't sleep until that son of a bitch was six feet under. He thought about his brother and how proud he was to finally have some sort of justice. He often wondered how his brother would have reacted if he revealed his true feelings about him, but that he will never know.

While Detective Coleman wasn't looking she reached in her purse. "Here it is," she smiled.

Detective Coleman returned his attention to her, but his smile quickly evaporated once his eyes made direct contact with the barrel of the .38 she was holding. He was at a loss for words as he watched as a devislish smile appeared on her face.

"Natalia," Detective Coleman whispered shaking his head. He would have never thought. She graduated at the top of her class, a model fellow officer who was well on her way to the top. He only agreed to allow her to work with him because of her smarts and quick wit and when she told him that she had arrested one of Kane's street corner flunkies and possibly had evidence against Kane he was all for it.

Twisted Obsession

"Cat got your tongue," she smiled loving the look of pure shock on his face. It was always the same look whenever she killed one of her own.

"Do you know what you've done? Why Natalia. Why?"

"Sorry about your brother, I'm truly am, but unfortunately this story doesn't have a happy ending. And besides being bad is fun," she laughed mechanically. And with the pull of the trigger she silently sent Detective Coleman to his death.

Officer Natalia Roberts an officer who fucked both sides of the law.

* * *

Dame was walking outside when Doricka pulled up. He walked out the door, ready to hop in his Armada, one of his last prized possessions. The house he currently lived in and two other trucks were items that he still currently owned.

"Hey baby," Doricka cooed. "You on your way somewhere?" she asked.

"Yeah," Dame replied, stealing glances at Passion. Passion simply rolled her eyes. "Where y'all on y'all way to?" he asked starring at Doricka's ensemble. He couldn't believe she actually left the house looking like that. It was a good thing her looks backed her up.

"We need a ride to Hill Top Shopping center," she walked over and hooked her arm in heads planting a juicy kiss on his cheek. He lost some weight but he still looked good.

Dame scratched his head. He was about to go hit some Spanish dude he just met. He checked his watch, he had time, "Yeah, come on," he said walking toward his truck.

"Can I have the keys to the house? I gotta piss," she said giving him a sideways glance, doing her best to appear as innocent as possible.

He looked at her for a moment. He really didn't trust her slick ass. "It should be one on those set of keys," he said pointing to the keys she held in her hand.

Twisted Obsession

"Oh, I see it.," she said turning on her heels. Passion started to follow her. Doricka stopped her once Dame was out of ear shot. "What are you doin'?" she asked.

"I gotta piss too. Besides, I don't want to be alone with him," Passion said an with and attitude as she folded her arms across her chest.

"Come on. Please. I'm 'bout to raid this nigga house. I know where he keeps his shit. He be leavin' it out, all obvious and shit."

Passion sucked her teeth. "Alright, bitch. But, I want half, if I have to put up with that stank ass breath. I don't see how you do it." Passion climbed into the back of the truck. It's a shame, all this money and he can't do anything about his halitosis.

"What's up ma?" Dame asked, eyeing her seductively as Passion climbed into the back seat. He couldn't keep his eyes off her thighs and the way they filled her dress out just right. "Why you sittin' in the back you can sit up here." He stated knowing damn well Passion wouldn't give him the time of day but he couldn't help but to fuck with her

"Nothin and I'm good," Passion said rolling her eyes. She wasn't really trying to stay cooped up in the truck with him giving off that foul odor. But the truck was nice.

"Why you act like that?"

"Like what?" she asked with an attitude.

"Like you don't like me," he said licking his lips. "You be acting all stuck up and shit, when you know you want a nigga."

"You wit' my cousin, and I don't get down like that," Passion said with an attitude to keep from laughing. She could only roll his eyes and it was obvious he was conceited. But Passion was the least bit interested. He looked different to her, he looked tired and a little on the sickly side. She just hoped he didn't have that bug, for Doricka's sake.

Dame turned around in his seat. "What the fuck is taking Doricka ass so long? It's all

Twisted Obsession

hot out here and shit," he said opening the door ready to get out.

"No," Passion said quickly, Dame gave her a strange look. "Why don't you turn the air on?" she said seductively, pulling down the front of her dress just a little, advertising a little cleavage. Dame smiled putting the key in the ignition, but never taking his eyes off her thighs. He was about to put the windows up when he noticed the startled look on Passion's face. "What's wrong shorty?"

"Kane sends his regards," Dame turned his head and immediately recognized him as one the guys Kane met with during the drug deal in New York.

But Dame just sat there and didn't flinch. In his mind all along he knew this day was going to come. The first thing that came to his mind was the poem *Foot Prints in the Sand*. A tear escaped his eye as he laid his head against the head rest and mouthed *Lord forgives me of my sins*.

Carlos smiled reading Dames lips, "I hope he does," he said before pulling the trigger splattering Dame's brain matter inside the truck, some of which landed on Passion.

To in shock to move Passion trembled violently in the back seat. "Please don't kill me," she whispered as tears strolled down her face. Fear was evident in her eyes as she stared down the face of death.

"Passion, is it?" She nodded her head yes. "Well, you don't have anything to do with this. And you're such a beautiful lady. But unfortunately you're a nigger. I mean don't get me wrong you're a beautiful nigger and any other day I would take you and fuck your brains out. But you happen to be related to another nigger, well two niggers that I don't like. So this is nothing against you."

A blood curtling scream escaped Passions lips once the bullet pierced through her chest. She looked down as the burning sensation that consumed her entire body. Tenderly she covered her wound with her hand, which immediately became soaked with her blood. Carlos watched with a sinister smile on his face as he backed away from the truck.

Twisted Obsession

Passion somehow managed to open the door, blood staining her Christian Dior dress. She was so weak she immediately fell to the ground as she slowly crawled towards the house.

"Help . . . me," she wanted to yell but her voice was barely above a whisper.

"Here I come," Doricka yelled a little upset because she didn't score not one red dime of Dames money. He uses to leave it hanging around but today she couldn't find anything. And without the slightest warning the blast sent her flying backwards. A little shaken up, she slowly sat up, dazed and confused. She shook her head a little and coughed. Smoke was everywhere but that didn't prevent her from seeing her cousin lying in the middle of the yard. She quickly got up and crawled over to where she was lying. "Passion, no. . . No . . .No . . ." she cried cradling Passions near lifeless body in her arms. Immediately she became covered in Passions blood and she panicked. She managed to find her cell phone and quickly dialed 911.

"Doricka . . ." were Passion's final words before the life completely left her body.

911 emergency what is your emergency?

But Doricka couldn't hear anything. Tears spilled from her eyes. "PASSION," she screamed at the top of her lungs shaking Passions body violently. "PASSION!" But no matter how hard she shook nothing could wake Passion from this everlasting sleep.

Ma'am what is your emergency?

Doricka couldn't speak her cousin was dead; all she could do was hold her in her arms. This was her fault. She should have been in that truck, not Passion.

* * *

Chapter 23

"Where the fuck is these bitches at?" Tina said out loud dialing Doricka's number. Her phone went straight to voice mail. "I tell you color people."

"Something's wrong! I can feel it," Allexis said, pacing back and forth. Ever since they left earlier that afternoon she has had a funny feeling in the pit of her stomach. She knew something was wrong. This was like Doricka, but she knew Passion would have at least called by now. Something told her they should have all stayed together.

Your love is a one in a million, Tina's cell phone sang. "Hello," she answered. "Whoa, Doricka calm down. What!" Tina screamed into the phone. Tina dropped to her knees. Allexis immediately went to her sister's side. "No! No! No!" she screamed. Tina dropped her phone and placed her face into her hands. "This can't be happening," she mumbled rocking back and forth.

"What is it?" Allexis asked, taking note of the distress in her sister's eyes.

Tina took a deep breath, not wanting to believe the words that were ready to come out of her mouth. With tear stained eyes she looked at Allexis and whimpered, "Passion . . . she's dead."

Without much effort Allexis slumped to the floor beside Tina. "How?" she cried. Allexis shook her head. "Tina please tell me this isn't true. I mean she was just here," she said with pleading eyes. Knots formed in her stomach once she realized what Tina said was true.

Tina wrapped her arms around her baby sister and did her best to tell her what Doricka told her over the phone. The two girls cried and did the best to console one another. Allexis was a complete wreck. She couldn't believe Passion was dead. She didn't want to believe it, they were just together.

"Tina where are they?" she asked still hoping that this was some sort of mistake, or misunderstanding.

"I don't know Allexis, just calm down," she said even though she was a complete wreck herself. "We have to call Dom to see if he knows anything."

Allexis wrung her hands nervously. A loud banging on the door made them stop dead in their tracks. Without going to the door they both knew who it already was.

"Allexis," Kane roared. "I know ya yellow ass is in there."

"Could things get any worse," Tina mumbled.

"Oh my God!" Allexis immediately began to panic. She had a feeling he was going to come after her; she just wished they could have left sooner. "Tina," she said grabbing onto her sister for dear life. The way Kane was pounding on the door made Allexis entire body tremble in fear. Flashes of the previous abuse ran through Allexis mind, and she knew if he made it through that door it would be ten times worse.

Tina quickly dialed Dom's number, her hands shaking the entire time. Any other time she probably would have opened the door to give him the business. But the way he was pounding and the authority in his voice let her know to do otherwise. Besides she was in no way emotionally prepared to go against Kane.

"Hello," Dom said into the phone.

"Dom, Kane's here. He's going crazy please get over her now."

"Okay baby calm down," Dom said taking note of the panic in Tina's voice. "Of all days this has to be a fucked up day. Aight stay right there were on our way. You got your heat right?"

"Open the GOT DAMN DOOR!" he yelled. Kane kicked and pounded on the door until his fist were bloody. "Bitch, you better open this muthafuckin' door," he yelled, spit oozing out the corners of his mouth.

Tina jumped every time he pounded on the door. She was more than sure Dom heard

Twisted Obsession

him in the background. "Yes," she whispered.

"Good, y'all go upstairs were on our way."

"Kane, just go away, please," Allexis begged.

"Alex! Alex!" he yelled "Open the door," he said going wild as he got lost in own twisted obsession.

Kane continued to bang on the door like a mad man. Tina got her purse and pulled out her .22. She didn't want to use it, but if she had to protect her sister. A neighbor overheard the commotion and quickly dialed 911. Both girls jumped when they heard the hinges on the door break. Tina aimed her gun, ready to fire.

Before Kane could enter the house, he felt the barrel of a gun, pressed firmly against the back of his head. "Yo' what the fuck?" Dom said, punching Kane in the back of the head.

Dom was glad they were able to cancel their meeting. Something in his gut told him something was about to happen. Kane turned around to see Joey standing behind him, pointing a gun at him.

"Look I just came to get my girl," Kane said, his chest pounding up and down. He was so full of rage ready to stomp a hole in Allexis ass.

"She's no longer yours," Joey said, hitting Kane in the back of the head, with the butt of his gun, dropping Kane to his knees.

Kane screamed in agony. "Fuck, what you mean she's no longer mine," he roared. "She's my property. She belongs to me." Kane was pissed he left his gun at home.

"Are you serious nigga? You think we gon' let her leave wit' yo' sorry ass. When you been putting yo' hands on her and shit yo'." Dom began to pound Kane, stomping him with his Timberland boots. "Fuck you, punk ass nigga. You ain't never gon' put yo' hands on her again." Dom was in a zone, he was no longer thinking of Allexis but of his

Twisted Obsession

beautiful mother, that was sent to her grave early because of punk ass men like Kane.

Joey stopped Dom, when he heard sirens. Cops soon flooded the entire driveway. "Put down your weapon. And place your hands behind your head?" Joey threw down his gun and placed his hands behind his back. Dom was glad they decided to switch cars, and he quickly followed suite throwing down his gun and putting his hands behind his back.

Kane managed to stand up despite the severe beating he just received. He spit out the blood that had settled in his mouth. He smiled wickedly at both Dom and Joey, before placing his hands behind his head. A sinister grin spread across his face, "Until next time . . . ."

* * *

Twisted Obsession

Epilogue

Hello all it's me Allexis, Purity decided to let me go ahead and bring you up to speed on a few things. Well let me say first off thanks to all who have journeyed with us thus far. Now with that said lets' get down to business.

Passions untimely and unnecessary death has left us all feeling a little guilty. We all feel in some sort of way responsible for what happened to her. And poor Doricka she hasn't spoken a word since it's happened. A witness saw the entire incident take place now both Dom and Joey know that this was no accidental shooting. Now both Dom and Joey are beside themselves with rage because they appointed Carlos to take out Dom, but Carlos allowed his hatred for the two to get the best of him and of course Passion was caught in the crossfire. Needless to say there is definitely a hit out on Carlos.

As far as Dom and Tina are still together but with Carlos out on the prowl and revenge fresh on Dom's mind he doesn't plan on retiring from the game anytime soon. Joey's life is pretty much the same as usual with an addition of me in it. Fearing for my safety Tina convinced me to move to Miami with Joey which I'm enjoying so far. I know what you're thinking, but no Joey and I are not in a relationship, but feel that this move is for my own protection. That's what got me in trouble last time, jumping in head over heels. The next time I fall in love I'm making sure that I do it right.

As I look back over the last few months of my life, I can't help but to think that so much has taken place in such a short time. If I had just listened to my sister about Kane, that fateful night where I lay on my bed with Kane's number in hand. I go back to that night all the time and wish that I never made that phone call. Sometimes I still can't help but to sit and wonder if I never made the call would things in my life have gone differently. I can't help but to wonder, would Passion still be alive?

Lord, knows I am going to miss that girl. She was a true friend, matter of fact one of a kind. And she truly didn't deserve to go out the way that she did. May she rest in peace.

But I do hope that this story reaches someone out there. That someone in my situation

Twisted Obsession

may learn from my mistake and think twice. It's never okay to let a nigga put pounds on you. You better up and do what Tina did to Ike in the limo, seriously. But for right now I'm okay, just hope I can break this cycle, so that this tradition of somehow being attracted to abusive men doesn't carry on to my daughter.

And as for Kane id did find that his twisted obsession stems from his little sister Alex, who happened to look just like me. I do know that she died a long time ago, but I'm not too sure about the details. I do hope that he eventually gets help for his aggressive behavior, because that is one deranged mutha- well you know the rest.

But since him being incarcerated for breaking into Dom and Tina's home, much hasn't been heard or said about him. But for some reason I do have this strange feeling that I haven't quite seen the last of him, just yet.

Twisted Obsession

Twisted Obsession

Please stay tuned because this isn't over. Sneak peek of what to expect in the page turning sequel.

Twisted Obsession

Twisted Obsession Pt 2

Prologue

Kane sat in the police car with his hands secured tightly around his back. His head screamed in agony from the severe beating he was just delivered. But he could care less about that. He could only watch with envious eyes as Joey wrapped his arms around Allexis, stroking her hair as if he were taunting him. He couldn't stand the sight of another man touching Allexis, especially a white man.

"You mutha fucka," Kane yelled as spit flew out of his mouth and onto the window. He began to yell and kick the police car uncontrollably. "YOUR MINE ALLEXIS! MINE!"

Allexis looked on in horror as Kane continued to rant and rave. Seeing her sister's discomfort Tina grabbed her and escorted her inside the house. Tina could only shake her head. This was supposed to be a happy day for them. They were all supposed to go down to Mexico and kick back on the beach and have some fun. But instead one of their friends is dead and it looked like Allexis has found herself caught up in a real live fatal attraction.

A police officer banged on the window, "Hey keep it down in there." *Niggers*, he thought to himself.

Both Dom and Joey stood outside staring at Kane with smug grins plastered across their faces. In their minds they had won but the battle was just beginning.

Kane continued to look on with glossy eyes. "I just want to protect you Alex," he mumbled to himself. "Can't you see that?" He closed his eyes and laid his head back on the seat, and allowed painful memories of his childhood to invade his mind. A tear strolled down his cheek, as he remembered like it was yesterday.

"Sss . . . shit! Yes, you white bastard, you betta pound this black pussy." Kane could hear his mother through the thin walls of their tiny apartment as she turned one of her daily tricks. Ever since Kane could remember his mother has always been a fiend. They

often had to sit in the dark or scramble for food because his mother spent her government check to support her habit.

Kane did his best to block out the noise his mother was making by singing to his five year old sister Alex. Her beautiful brown eyes lit up as he sang, "The itsy bitsy spider went up the water spout. Down came the rain and washed the spider out. Out came the sun and dried up all the rain, and the itsy bitsy spider went up the spout again."

Alex clapped her hands, "Again, again," she begged.

Kane started to sing again, but stopped when he heard his mother yelling. "She's only five."

"I like kiddie pussy," a gruff voice responded. "Besides, I'll give you two vials of crack and a thousand dollars if you let me break that little bitch in."

A worn out looking Vera Stanley hung her head low. This wasn't the first time she had been offered this proposition but she had fallen on hard times. A once beautiful woman, with a banging body to match looked well beyond her years and her once curveacous body hung like a bag of bones. Without saying another word she put on her robe and headed down the hall towards the bedroom where Alex and Kane were playing.

A nine year old Kane listened intently, when didn't hear his mother object any longer to what the man was implying, but instead heard heavy footsteps coming down the hall he sprang into action. He locked his bedroom door and grabbed Alex and hid in his small cramped closet.

Alex giggled, "Are we playing hide and seek?" she asked oblivious to the danger she was in.

Kane didn't answer he just held onto his sister for dear life. She was his everything, until she came into the world he felt he had nothing to live for. He looked down at her oval shaped face and huge brown eyes, a tear fell down his cheek. Alex easily sensed Kane's discomfort and buried her face in his arm.

Twisted Obsession

"Kane! Open this got damn door!" he heard his mother yell.

"I'm scared K," Alex began to whimper.

"It's okay," he said stroking her long black hair, doing his best to be brave when he was scared himself. But despite how scared he was, he knew he had to do everything in his power to protect his baby sister.

Vera quickly ran to retrieve her spare key. Her hands shook violently as she struggled to put the key in the lock "Lord, please forgive me," she said as she unlocked the door to her daughter's innocence.

Kane hung on to his sister for with all his might; as the closet door swung open nearly knocking them both out of the closet. Kane did his best to block Alex, but in one swift motion the big burly white man grabbed Kane by the arm and literally threw him across the room, momentarily knocking the wind out of him.

"Come here you yellow bitch," the white man hungrily growled as he snatched Alex up as if she were a rag doll.

"K," she screamed. She tried to fight, her small fist beating the man in the back of the head. But this only excited the white man more, causing him to laugh.

By the time Kane was able to catch his breath he heard the click of the door lock. He kicked and banged on the door until his poor fist were bloody. "Alex," he cried softly as he sank to the floor, a steady stream of tears flowing down his face.

Alex frantic screams could be heard through the thin walls. Kane covered his ears. He cried as he sang. "The itsy bitsy spider went up the water spout."

A lone tear landed on Kane's arm bringing him back to the present. He shook his head a few times before wiping his face on his shoulder. "I just want to protect you Alex," he mumbled again starring out the window at nothing in particular.

Twisted Obsession

"What you say?" one of the police officers in the front asked turning to face Kane. He thought Kane was holding up pretty good, considering one of his eyes was nearly swollen shut and he had a deep gash in his head that was now bleeding. He continued to look but shook his head once he saw that Kane wasn't paying him any mind.

But Kane was oblivious to the pain of his wounds, but was succumbing to the pain of his broken heart. He ignored the police officer focusing his attention on Dom and Joey. He stared on with daggers in his eyes; he smiled as the police car rode past. He hoped they could read lips if they did it was obvious Kane wasn't about to back down and leave his woman that easily. "This ain't over, let the games began."

Made in the USA
Charleston, SC
25 October 2012